Die Illustration stammt aus *Recueil des Pieces qui ont remporté le Prix de l'Académie Royale des Sciences*, Band VIII, 1751–61 (1769), Nr. 3 und begleitet die *Recherches sur la manière la plus avantageuse de suppléer à l'action du vent sur les grands vaisseaux* von Daniel Bernoulli. Die Abbildung verdeutlicht die Experimente, die Bernoulli durch Gabriel Cramer auf dem Genfer See ausführen liess, um seine Theorie über die Kräfte der Ruderer zu beweisen. Die Arbeit wurde mit dem Preis der Pariser Akademie ausgezeichnet.

Die gesammelten Werke der Mathematiker und Physiker der Familie Bernoulli

Herausgegeben von der
Naturforschenden Gesellschaft
in Basel

Birkhäuser Verlag
Basel · Boston · Berlin

Die Werke von Daniel Bernoulli

Im Auftrag
der Naturforschenden Gesellschaft in Basel
und der Otto Spiess-Stiftung

Ediert von
Patricia Radelet-de Grave

Birkhäuser Verlag
Basel · Boston · Berlin

Die Werke von Daniel Bernoulli

Band 8
Technologie II

Bearbeitet und kommentiert
von Frans A. Cerulus

2004
Birkhäuser Verlag
Basel · Boston · Berlin

Library of Congress Cataloging-in-Publication Data
A CIP catalogue record for this book is available from the Library of Congress, Washington, D.C., USA.

Bibliografische Information der Deutschen Bibliothek
Die Deutsche Bibliothek verzeichnet diese Publikation in der Deutschen Nationalbibliografie; detaillierte bibliografische Daten sind im Internet über http://dnb.ddb.de abrufbar.

Publiziert mit Unterstützung des Schweizerischen Nationalfonds
zur Förderung der wissenschaftlichen Forschung.

Gedruckt auf säurefreiem Papier, hergestellt aus chlorfrei gebleichtem Zellstoff. TCF ∞
Buchgestaltung: Heinz-Dieter Ecker, Bonn
Printed in Switzerland
ISBN 3-7643-7007-6

9 8 7 6 5 4 3 2 1

INHALTSVERZEICHNIS

Vorwort der Generaleditorin

Der vorliegende Band, der achte über die *Werke von Daniel Bernoulli*, vervollständigt und vereinigt gemeinsam mit dem siebten Band die technologischen Schriften. Vorwiegend enthält er Antworten auf Anfragen der *Académie Royale des Sciences de Paris*, die bestrebt war, Wissenschaften zum Schiffswesen zu entwickeln. Im Gegensatz zum siebten Band, der sich mit der Zeitmessung auf See auseinandersetzt, deckt der vorliegende Band ein breiter gefächertes Spektrum an technologischen Problemen ab. Der wichtigste Teil dieses Bandes ist den drei Auszeichnungen von Daniel Bernoulli gewidmet.

Die erste Auszeichnung erhält er für seine Forschungen zur Formgebung von Ankern. Zwar schlägt er keine neue Ankerform vor, doch weist er auf die Qualitäten der traditionellen Anker hin. Die hohe Wahrscheinlichkeit, dass ein Anker in eine Position fällt, die guten Griff zur Verankerung gewährt, ist wegen dessen spezifischer Form gegeben. Daniel Bernoulli beweist ein weiteres Mal seine Orginalität in Bezug auf Anwendungen von Wahrscheinlichkeiten. Er verweist auf die wichtige Rolle des Gewichts und auf die Länge der Ankerkette.

Seinen zweiten Preis erhält Daniel Bernoulli für Untersuchungen zu den Einflüssen auf die Bewegung von Schiffen. Welchen Beitrag leistet das Segel, wieviel tragen die Ruderer bei? Er beurteilt, wieviel Arbeit ein Mann während einer Stunde bis zu einem Tag unter normaler Anstrengung leisten kann. Demgegenüber schätzt er die Leistung bei extremer Anstrengung während kurzer Zeitabschnitte ein. Er erinnert an den Vorschlag aus seiner *Hydrodynamica*, um die Reaktion auf einen Wasserstrahl aus einem Boot zu nutzen, um dieses zu bewegen.

Die dritte Auszeichnung gilt dem Rollen und Stampfen eines Schiffes. Diese Arbeit betrifft einige Forschungsschwerpunkte die Daniel Bernoulli am Herzen liegen: das Gleichgewicht schwimmender Körper, die Bewegung fester Körper und natürlich deren Schwingungen.

Der letzte Text dieses Bandes ist keine Antwort an die *Académie*, sondern eine Publikation in den *Novi Commentarii* von Sankt Petersburg. Daniel Bernoulli analysiert darin die Leistung einer von Andreas Wirz in Zürich konstruierten Spiralpumpe.

Diesem Band ist auch ein Text beigefügt, der sich nicht mit Technologie befasst. Er gehört in den ersten Band zu den astronomischen Werken. Es handelt sich um eine Antwort auf eine Rede von de L'Isle an die Akademie von Sankt Petersburg. Die Quelle wurde leider erst nach Erscheinen des ersten Bandes von Martin Mattmüller entdeckt, weshalb wir sie hier außerhalb der Systematik veröffentlichen. De L'Isle stellt sich die Frage *si l'on peut démontrer par les seuls faits Astronomiques, quel est le vrai Système du monde? Et si la Terre tourne, ou non?*

Der Kommentar zu den Texten zur Navigation, die in diesem Band zusammengetragen sind, wurden ursprünglich Herrn André Lederer, Professor für Schiffskonstruktion an der *Université catholique de Louvain*, anvertraut. Von ihm stammen die mit [A. L.] gekennzeichneten Anmerkungen. Nach dessen Tode 1997, hat David Speiser die Aufgabe Herrn Frans Cerulus, Professor für Theoretische Physik und Wissenschaftsgeschichte an der *Katholieke Universiteit Leuven* anvertraut. Wir danken ihm herzlich für seinen Beitrag.

Dank

Zuerst möchte ich Herrn Dr. J.-L. von Planta, dem Präsidenten unserer neugegründeten Editionskommission und des Kuratoriums der Otto Spiess-Stiftung für seine ständige Disponibilität herzlich danken. Ebenfalls sei den Herren Prof. H.-J. Güntherodt, permanenter Delegierter der Naturforschenden Gesellschaft Basel bei der Bernoulli-Edition, Prof. H. Striebel, Präsident des Fördervereins, Prof. A. Tammann, Präsident des wissenschaftlichen Beirats, sowie Frau K. Leonhard, Schatzmeisterin des Fördervereins für ihre aktive Mitarbeit zugunsten der Bernoulli-Edition gedankt.

Dem Schweizerischen Nationalfonds zur Förderung der wissenschaftlichen Forschung danken wir für die finanzielle Unterstützung und insbesondere Herrn Prof. M. A. Knus für die Betreuung dieses Bandes.

Frau Dr. Gaino-Marcq hat die Produktion des vorliegenden Bandes von der Satzvorbereitung bis zur Erstellung der Register betreut. Ihr sei hierfür mein bester Dank ausgesprochen. Ebenso möchte ich den Herren H.-P. Thür und E. Mazenauer sowie allen Mitarbeitern des Birkhäuser-Verlages, vor allem Herrn E. Klementz und Frau V. Anklin, für ihre ausserordentlich fruchtbare und vertrauensvolle Zusammenarbeit danken.

Schlussendlich möchte ich allen Editoren, vor allem Herrn N. Guicciardini aus Siena, der kürzlich das Mandat angenommen hat, die Streitschriften von Johann Bernoulli mit den Engländern zu kommentieren, für ihre wohlwollende und ungebrochene Unterstützung unserer Edition meine profunde Dankbarkeit zum Ausdruck bringen.

Louvain-la-Neuve, im Mai 2004 Patricia Radelet-de Grave

Die Herausgabe des vorliegenden Bandes wurde durch Beiträge der folgenden Institutionen ermöglicht, denen wir herzlich danken:

Schweizerischer Nationalfonds zur Förderung
der wissenschaftlichen Forschung

Freiwillige Akademische Gesellschaft, Basel

Lotteriefonds des Kantons Basel-Landschaft

Lotteriefonds des Kantons Basel-Stadt

Max Geldner-Stiftung, Basel

Bank Julius Bär & Co. AG
im Auftrag eines nicht genannt sein wollenden Spenders

Zur Wiedergabe der Texte

Das Hauptziel unserer Edition ist es, die wissenschaftlichen Schriften der Mathematiker und Physiker der Famille Bernoulli dem modernen Leser zugänglich zu machen. Im Folgenden seien die Kriterien vorgestellt, denen wir bei der Wiedergabe der hier vorgelegten Texte gefolgt sind.

Die Quellen

Für die gedruckten Arbeiten Daniel Bernoullis wurde jeweils der Text der ersten veröffentlichten Fassung verwendet.

Die Textfassung

Der Text der zugrunde liegenden Quellen wurde in der vorliegenden Ausgabe orginalgetreu wiedergegeben. Einige wenige Ausnahmen haben zu Veränderungen der Inhalte aus den Originalen geführt. Diese Ausnahmen werden im Folgenden genauer erläutert. Insbesondere haben wir Orthographie und Formelsprache der Zeit um 1750 streng beibehalten, um die zahlreichen Fehlerquellen auszuschliessen, die sich aus einer modernen «Umschreibung» von Wörtern und Formeln ergeben können. Um dem heutigen Leser die Lektüre zu erleichtern, wurden jedoch folgende Veränderungen vorgenommen:

- Das in Texten des 16. bis 18. Jahrhunderts verwendete Zeichnen «ſ» wurde durch ein modernes «s» ersetzt; «u» und «v» wurden entsprechend dem modernen Gebrauch unterschieden; griechische Buchstaben wurden mit modernen Typen wiedergegeben.
- Akzente wurden im französischen Text originalgetreu wiedergegeben; fehlende Akzente wurden nur da ergänzt, wo es nötig war, um Fehllesungen vorzubeugen («ou» für «où» etc.). Im lateinischen Text wurden die Akzente, mit Ausnahme der Dierese, weggelassen.
- Gross- und Kleinschreibung folgen dem Original. Stets gross geschrieben wurden Disziplinen, Titel, Verwandtschaftsbezeichnungen, Funktionen von Personen, Werk- und Werkteil-Bezeichnungen. Jeder Satz beginnt mit einer Majuskel.
- Die vielfach benutzten Ligaturen und Abbreviaturen wurden aufgelöst.
- Offensichtliche Druckfehler wurden stillschweigend behoben; sie werden in eckigen Klammern [ ] nur dann erwähnt, wenn sie für das Verständnis des Textes von Wichtigkeit sind.

- Zwei Texte, nämlich W: *Discours lû dans l'assemblée publique de l'Academie des Sciences, Le 2. Mars 1728* und St. 48: *Mémoire sur la maniere de diminuer le Roulis et le Tangage d'un navire*, sind voller Druckfehler. Daniel Bernoulli selbst äussert sich darüber in einem Brief an Johann-Albrecht Euler, *«...j'en ai trouvé toute l'impression extremement remplie de fautes; les formules irreconnaissables, les nombres alterés; la seconde table y manque; Mr de Lalande me marque qu'elle a été detournée &c. et tout cela apres un delay de 18 ans»*. Wir haben die offensichtlichen Druckfehler in den Formeln und im Text (z. B. eine Nichtübereinstimmung der Formel mit der Figur) ohne weitere Angaben korrigiert. Weniger triviale Fehler sind in einer Fussnote angegeben.

Die Formeln

Die unsystematischen mathematischen Notationen Daniel Bernoullis wurden genau so wiedergegeben, wie sie in der Textvorlage erscheinen.

In manchen Fällen werden Formeln, die im Original im fortlaufenden Text standen, im Interesse der Lesbarkeit herausgestellt und zentriert. Auf folgende Besonderheiten der Formelschreibweise um die Mitte des 18. Jahrhunderts sei hier hingewiesen:

- Mehrere Multiplikationszeichen wie $\times$, $\cdot$ und die Schreibweise yy für y^2.
- Die Schreibweise für Proportionen $n\,.\,m :: a\,.\,b$ und $n : m :: a : b$, sowie $n : m = a : b$.
- Das «6» wurde durch «β» und «ς» durch «σ» ersetzt.
- Die Form der Wurzelzeichen mit oder ohne *vinculum*, wurde respektiert. Nur in einem Fall wurde dem Wurzelzeichen im Interesse der Verständlichkeit ein *vinculum* angefügt.
- Werte der trigonometrischen Funktionen werden, wie im 18. Jahrhundert üblich, in Relation zum Radius angegeben, der dann als «*sinus totus*» bezeichnet wird; manche unserer Autoren setzen diesen *sinus totus* ausdrücklich gleich 1, wie es sich später durchgesetzt hat.
- Die Dezimalstelle war uneinheitlich mit Punkt oder Komma angegeben. Wir verwenden einheitlich Punkte. So ist zum Beispiel aus 2,92 eine 2.92 geworden.

Die Abbildungen

Die Figuren stammen aus der Orginalfassung der Texte. Ursprünglich waren die Abbildungen auf separaten Tafeln gedruckt. Zur Verbesserung der Lesbarkeit sind die Figuren in den vorliegenden Text eingebunden. Abbildungen, auf welche der Autor an mehreren weit voneinander entfernten Stellen Bezug nimmt, haben wir aus demselben Grund wiederholt.

Abkürzungsverzeichnis

In der Einführung, den Fussnoten und den Registern dieses Bandes werden folgende Abkürzungen benutzt:

AAN	Archive of the St. Petersburg Branch of the Russian Academy of Science
Académie	Académie Royale des Sciences de Paris
Academy	Imperial Academy of Sciences in St. Petersburg
A. L.	André Lederer
Fig.	Figuren aus Originaltexten Daniel Bernoullis
Fig.	Neugezeichnete Figuren
h. v.	*hujus voluminis* (im vorliegenden Band)
IAHS	International Academy of the History of Science
Ms	Manuskript
r	*recto* (Vorderseite)
UB Basel	Öffentliche Bibliothek der Universität Basel
v	*verso* (Rückseite)

Werknumerierungen

St. 1–75, R–Z, Nr. 76–82	Numerierung der Werke von Daniel Bernoulli nach dem ergänzten Verzeichnis von H. Straub (vgl. pp. 351–358 h. v.)
Jac. B. Op. I–CII	Numerierung der Werke von Jacob Bernoulli nach dessen *Opera* (Genevae 1744)
Med. I–CCLXXXVI	Numerierung der *Meditationes* (Jacob Bernoullis wissenschaftliches Tagebuch: Ms UB Basel L I a 3)
Joh. I B. Op. I–CLXXXIX	Numerierung der Werke von Johann I Bernoulli nach dessen *Opera* (Lausannae & Genevae 1742)
E 1–866	Numerierung der Werke von Leonhard Euler nach G. Eneström, *Verzeichnis der Schriften Leonhard Eulers*, Leipzig 1910

Werke

Werke 1–8	*Die Werke von Daniel Bernoulli*, Basel: Birkhäuser, 1982–
Jac. B. *Briefe*	*Der Briefwechsel von Jacob Bernoulli*, Basel: Birkhäuser, 1993
Jac. B. *Opera*	*Jacobi Bernoulli Basileensis Opera*, Genevae: Cramer & Philibert, 1744
Jac. B. *Werke* 1–5	*Die Werke von Jacob Bernoulli*, Basel: Birkhäuser, 1969–
Joh. I B. *Briefe* 1–3	*Der Briefwechsel von Johann I Bernoulli*, Basel: Birkhäuser, 1955–
Joh. I B. *Opera* I–IV	*Johannis Bernoulli ... Opera Omnia*, Lausannae & Genevae: Bousquet, 1742
Descartes, *Œuvres* (A. T.)	*Œuvres de Descartes* publiées par Ch. Adam et P. Tannery, Paris: Vrin, 1897–1913
Euler, *Op. Omnia*	*Leonhardi Euleri Opera Omnia*, Leipzig: Teubner, Zürich: Orell Füssli, Basel: Birkhäuser 1911– (Series: römisch / Band: arabisch numeriert)

Fuss, *Correspondance* I–II	P.-H. Fuss (ed.), *Correspondance mathématique et physique de quelques célèbres géomètres du XVIIIème siècle*, St.-Pétersbourg 1843
Galilei, *Opere* (Ed.Naz.)	*Le Opere di Galileo Galilei*, Edizione nazionale pubblicata da A. Favaro, Firenze 1890–1909
Huygens, *Œuvres*	*Œuvres complètes de Christiaan Huygens*, publiées par la Société Hollandaise des Sciences, La Haye 1888–1950
Mariotte, *Œuvres*	*Œuvres de Mr. Mariotte*, revues et corrigées de nouveau, Leide 1717
Newton, *Correspondence* I–VII	*The Correspondence of Isaac Newton*, Cambridge 1959–1977
Newton, *Principia* (1687)	Isaac Newton, *Philosophiae naturalis Principia mathematica*, Londini 1687
Newton, *Principia* (1713)	Isaac Newton, *Philosophiae naturalis Principia mathematica*. Editio secunda auctior et emendatior, Cantabrigiae 1713
Newton, *Principia* (1726)	Isaac Newton, *Philosophiae naturalis Principia mathematica*. The third edition (1726) with variant readings, Edited by A. Koyré and I. B. Cohen, Cambridge 1972
Newton, Principia	*The Mathematical principles of natural philosophy*, by sir Isaac Newton (translated into english by Andrew Motte) (B. Motte), London 1729
Torricelli, *Opere*	*Opere di Evangelista Torricelli*. Edite in occasione del III centenario della nascita col concorso del Comune di Faenza da Gino Loria e Giuseppe Vassura, Faenza 1919

Zeitschriften

AE	*Acta Eruditorum*, Lipsiae 1682–1731
AP	*Acta Academiae Scientiarum Imperialis Petropolitanae*, Petropoli 1778–1786
CP I–XIV	*Commentarii Academiae Scientiarum Imperialis Petropolitanae*, Petropoli, I 1726 (1728) – XIV 1744/6 (1751)
Hist. Paris	*Histoire de l'Académie Royale des Sciences*, Paris 1666 (1703) – 1789 (1793)
Mém. Berlin I–XXV	*Histoire de l'Académie Royale des Sciences et des belles lettres de Berlin ... avec les Mémoires tirez des Registres de cette Académie*, Berlin, I 1745 (1746) – XXV 1769 (1771)
Mém. Paris	*Mémoires de l'Académie Royale des Sciences*, Paris 1666 (1703) – 1790 (1797)
NCP I–XX	*Novi Commentarii Academiae Scientiarum Imperialis Petropolitanae*, Petropoli, I 1747/8 (1750) – XX 1775 (1776)
Nouv. Mém. Berlin	*Nouveaux Mémoires de l'Académie Royale des Sciences et des belles lettres de Berlin*, Berlin 1770 (1772) –
Phil. Trans.	*Philosophical Transactions*, published by the London Royal Society, London 1665–

Prix Paris	*Recueil des Pieces qui ont remporté le Prix de l'Académie Royale des Sciences*, Paris 1720 (1721) – 1772 (1777)
P. V. Acad. sci. Paris	*Procès-Verbaux de l'Académie royale des sciences de Paris*, Paris 1667–1788

Die von Daniel Bernoulli verwendeten Masseinheiten

In den Quellentexten, die in diesem Band abgedruckt oder in den Kommentarteilen zitiert sind, treten öfters nicht-metrische Masseinheiten des 17. und 18. Jahrhunderts auf. Im letzten Abschnitt ist die Umrechnung einiger obsoleten mechanischen Einheiten des 19. und 20. Jahrhunderts beschrieben.

1. **Längen-, Flächen- und Volumeneinheiten**

Alle in diesem Band besprochenen Quellentexte verwenden als grundlegende Längeneinheit den **Fuss**; dessen Länge unterliegt jedoch erheblichen regionalen Unterschieden. Im Einzelnen kommen vor:

– die französische Längeneinheit:

1 pied (*du roi*)		= 32.484 cm
1 pouce (*Pariser Zoll*)	= 1/12 pied	= 2.707 cm
1 ligne (*Linie*)	= 1/12 pouce	= 2.256 mm
1 toise	= 6 pieds	= 1.949 m
1 perche de Paris (*Ruthe*)	= 18 pieds	= 5.847 m
1 brasse (*französischer Faden*)	= 5 pieds	= 1.624 m
1 mille (*französische Meile*)	= 1000 toises	= 1.949 km
1 lieue commune		= 4.445 km

– die englische Längeneinheit:

1 foot (*ped Lond.*) (*Fuss*)		= 30.480 cm
1 inch (*poll. Angl.*) (*Zoll*)	= 1/12 foot	= 2.540 cm
1 line (*Linie*)	= 1/12 inch	= 2.117 mm

– der Basler Fuss:

1 Fuss		= 29.82 cm

Von den Längenmassen wurden folgende Flächen- und Volumeneinheiten abgeleitet:

– vom französischen Fuss:

1 pied carré	= 10.552 dm^2
1 pied cube	= 34.28 dm^3

– vom englischen Fuss:

1 cubic foot	= 28.32 dm^3
1 imperial gallon	= 4.5435 liter (dm^3)

2. **Gewichtseinheiten**

Die Quellentexte verwenden als grundlegende Gewichts- bzw. Masseneinheit das **Pfund**; auch hier gibt es beträchtliche regionale Unterschiede. Es treten auf :

– das englische Pfund:

1 pound (avoirdupois)	= 453.59 g

– das französische Pfund

1 livre		= 489.51 g
1 once	= 1/16 livre	= 30.59 g
1 tonneau de mer	= 2000 livres	= 979 kg

3. **Zeiteinheiten**

Die von Bernoulli verwendeten Zeiteinheiten (Stunde, Minute, Sekunde) entsprechen im Rahmen der Messgenauigkeit den heutigen.

4. **Einheiten für Kraft, Druck, Arbeit und Leistung**

In bezug auf Begriff, Definition und Messung dieser Grössen bestanden im 17. und 18. Jahrhundert grosse Unklarheiten, die in den Kontroversen um das richtige Kraftmass eine beträchtliche Rolle spielten.

Eine Kraft wird meist durch die Angabe von Masse oder Volumen eines Körpers spezifiziert, der die entsprechende Gewichtskraft ausübt. Ebenso wird der Druck («*elater*») durch die Höhe der entsprechenden Wasser- oder Quecksilbersäule angegeben. Für die Umrechnung in moderne Einheiten gilt:

1 Fuss (*du roi*) Wassersäule	= 3185.2 Pa	= 0.0315 Atm.
1 Zoll Quecksilbersäule	= 3612.4 Pa	= 0.0356 Atm.

Arbeit («*puissance absolue*») wird als Produkt von Kraft und Länge (aber oft als Produkt von Kraft, Zeit und Geschwindigkeit) definiert und meist in Einheiten von (Gewicht × Höhe) gemessen:

1 französisches Pfund × 1 französischer Fuss = 1.56 J.

Als Mass für die Leistung ergibt sich damit (Gewicht × Höhe / Zeit). Die in der St. 47, *Recherches sur la maniere la plus avantageuse de suppléer à l'action du Vent* (p. 187 h. v.) angegebene Leistung eines Arbeiters, der pro Sekunde 20 Pfund um 3 Fuss hebt, beträgt ca. 93.6 W = 0.127 PS.

5. **Zyklometrie und Trigonometrie**

Bernoulli verwendet die gewöhnliche Kreisteilung ($1° = 60' = 3600''$). Für «1°» finden sich auch die Schreibweisen «1 d.» und «1^{d}».

Werte der trigonometrischen Funktionen werden, wie im 18. Jahrhundert üblich, in Relation zum Radius (dem «sinus total») angegeben; oft verwenden die trigonometrischen Tafeln den Wert 10^7 für diesen Radius. Wo Bernoulli die moderne Konvention verwendet, gibt er das ausdrücklich an: «le sinus total = 1».

General Introduction

(Frans A. Cerulus)

General Introduction

1 The contents

This is the second volume on technology. It is mainly devoted to three prize-winning memoirs, answering questions related to shipbuilding which were proposed by the French *Académie Royale des Sciences* in the years 1737, 1755 and 1757. The volume hosts two further items: a popularizing text on astronomy, written with Delisle, from 1723 – the early St. Petersburg period of our author – and the theory of the spiral pump, a technical item from 1772, one of the later works. We have chosen to present all of them in chronological order, but this also happens to group the nautical works together.

2 Nautical prizes

Why should a Swiss scientist be interested in problems of ship building? His only acquaintance with naval matters was as a resident of St. Petersburg, an important harbour then being built. There was also his voyage from Kronstadt to Danzig, an eventful fortnight in July 1733, which he described in his letter to Delisle of July 17th, 1733.

The author of the *Hydrodynamica* realized fully that insights in mathematics and mechanics could buttress and correct the empiricism of the naval architects; a hope shared by the *Académie* when it put up the questions and confirmed when it awarded the prizes. Those prizes of the *Académie* provided of course a mighty incentive. Bernoulli participated thirteen times in the Paris competition and competed once for a prize of the Berlin Academy; he won ten times in Paris and the one time in Berlin.

The history of the prizes has been well described by Ernest Maindron[1]. We recall the salient facts. In 1614 Rouillé de Meslay bequeathed to the *Académie* the sum of 124 000 *livres* on the charge of putting up for competition two questions every year; one on astronomy and the principles of light and movement, the other on longitude and all discoveries helpful to navigation and great voyages. The winner of the first prize should obtain 2000 *livres*, the other 500. Shortly after occurred the bankruptcy of Law and the ensuing financial disaster forced the *Académie* to adapt her rules. From

1 E. Maindron, *Les Fondations de Prix de l'Académie des Sciences, Les Lauréats de l'Académie 1714–1880*, Paris: Gauthier-Villars, 1881.

1623 onwards there was only one prize per year, alternately on astronomy (of 2500 *livres*), and on navigation (of 2000 *livres*). Sometimes a prize was split among two contenders. Worthwhile papers who did not make it to first place were awarded an *accessit*, which was quite honourable but afforded no financial reward. The winning memoirs and the *accessits*, were published in a special series.

Bernoulli's others memoirs on nautical matters[2] appeared in *Werke* 3 and in *Werke* 7.

Some years ago professor André Lederer, professor of naval engineering at the Université Catholique de Louvain, undertook a study of the shipbuilding works, checking especially the results of Daniel Bernoulli with the methods now in common use by naval engineers. He had made numerous notes and comments, but regretfully passed away before that work was finished. We were fortunate to be able to consult those notes. We shall incorporate some of them in our comments, where they will provide the reader with the critical view of a man aware of the state of the art in the 20th century. The notes of professor Lederer are reproduced in the original French and marked with his initials (A. L.).

The ideas proposed by Daniel Bernoulli are part of the very early history of theoretical naval architecture. Some of them belong now to the daily practice of naval engineers, who – as a rule – are unaware of their origin. Others, based on approximations not really applicable or forgetful of some of the constraints inherent in the trade, would be considered wrong by modern practitioners, as pointed out in the notes of A. Lederer.

3 Acknowledgements

We are indebted, for their informative remarks, to F. Nagel, D. Speiser, J. Dhombres, P. Radelet-de Grave and L. Ferreiro. In this last volume of the *Werke of Daniel Bernoulli* it is fitting that we should thank our colleague and friend, David Speiser, for the many discussions we had, which led ultimately to our participation in the grand enterprise of the Bernoulli Edition of which he was the driving force for so many years.

2 D. Bernoulli St. 33, *Sur le flux et le reflux de la mer*: Prix Paris IV (pièce VII) 1740 (1741), pp. 55–191 – *Werke* 3, pp. 327–438; St. 8, *Discours sur l'égalité du mouvement des clepsidres*: Prix Paris I (pièce IV) 1725 (1725), pp. 3–21 – *Werke* 7, pp. 221–239; St. 39, *Memoire sur les boussoles d'inclinaison*: Prix Paris V (pièce VIII) 1743 (1748), pp. 3–61 – *Werke* 7, pp. 67–111; St. 42 (a, b), *Recherches mécaniques et astronomiques*: Prix Paris VI (pièce I) 1745–47 (1750), pp. 3–78/79–110 – *Werke* 7, pp. 241–293/295–319; St. 49, *Sur les nouvelles aiguilles d'inclinaison*: Acta Helvetica, vol. III (1758), pp. 233–249 – *Werke* 7, pp. 140–151.

Introductions

Discours lû dans l'Assemblée publique de l'Académie des Sciences, Le 2. Mars 1728
Par Mr. DE L'ISLE avec la réponse de Mr. BERNOULLI *(W)*

1 Historical Note

Peter I, the Great, instituted the Academy of St. Petersburg by a decree of January 22nd[1], 1724. He wanted to gather in this Academy some of the best minds of (western) Europe in order to create a kernel around which science in Russia could develop. He initiated an active search for scientists of the first rank which could be lured to St. Petersburg by the attractive conditions. An advertisement appeared in the *Neue Zeitung*[2] and several well known scientists were invited to nominate candidates.

N. Delisle was one of the first to be invited, but he managed to delay his departure, at the same time bargaining for a very favourable contract. He arrived on January 23rd, 1726. In the meantime Peter the Great had died on January 28th, 1725. He was succeeded, as reigning monarch, by his wife Catherine I who carried out faithfully the plans for the Academy laid out by her late husband. The brothers Nicolaus and Daniel Bernoulli had both been recruited as members of the Academy and arrived on October 27th, 1725. The empress died on May 17th, 1727, and the crown befell Peter II, the son of Alexis the eldest son of Peter I by his first wife. The young emperor was only twelve and a half years old at the time, the state was ruled by the Supreme Privy Council but actual power was held by Prince A. D. Menshikov, the sovereign's guardian.

The Academy decided to organize a festive public session in honour of the new monarch and invitations were published in the press. The coronation took place in Moscow on February 25th, 1728 and festivities were organized in St. Petersburg on March[3] 4th. The Academy held its public session on March[4] 2nd in the way which had been resolved at its ordinary session of February[5] 20th:

1 *Матеріалы для исторіи императорской Академіи наукъ* (1725–1743), vol. 6, St. Petersburg 1890 (*Beiträge zur Geschichte der kaiserlichen Akademie der Wissenschaften* (1725–1743), Bd. 6, St. Petersburg 1890), p. 14.

2 Neue Zeitung von gelehrten Sachen (1715–1784), Leipzig N° XXXI, 1724, p. 311.

3 At first the plans were for an earlier date, both for the festivities and the session of the Academy.

4 That is according to the Julian calendar, observed by the orthodox church in Russia; it was a Saturday, corresponding to March 13th in the Gregorian calendar.

1. Discourse by N. De l'Isle[6], in French.
2. Reply, in the name of the Academy, by Daniel Bernoulli, in French.
3. Panegyric, in Latin verses, by Gottlieb Bayer.
4. Words of thanks to the audience by Jacob Hermann.

The text was published in a separate booklet. It is obvious that the occasion dictated the style of both Delisle and Bernoulli; no formulae are used and theoretical developments are eschewed. As Delisle explicitly said:

> «If I say that I proved this proposition, I think I proved it only in the manner that is suited to an audience such as this. I used a simple and familiar comparison instead of the exact demonstration I should have made; but its thread could have seemed too long and too tedious to most of those who honour me by listening.»[7]

The eulogies on Peter I and Peter II («which our zeal required») that both orators inserted in their text belong also to the imposed style.

Müller, in his *Матеріалы* of the Academy[8] which sound as his personal reminiscences, reports that Delisle, as a devout catholic, asserted that astronomy could not be at variance with the doctrine of the Church; Bernoulli had no such qualms – always according to Müller – and stated straightforwardly the opinion of the Academy that only the Copernican system is tenable. The careful reader of the texts will find that both authors present well founded and thoughtful arguments which complement each other. Müller's impression might have arisen because Delisle never mentions Galilei whereas Bernoulli refers explicitly to him and his persecution right in his first paragraph[9].

2 The subject (short survey)

> «Can one determine with absolute certainty whether the earth is moving or at rest?»[10]

5 *Procès-Verbaux des Séances de l'Académie Impériale des Sciences. Depuis sa fondation jusqu'à 1803*, T. I (1725–1742), St. Petersburg 1897, p. 15.

6 This is the spelling mostly used in the proceedings of the Academy.

7 Cf. p. 141 h. v.

8 Cf. *op. cit. supra* p. 7, note 1.

9 Cf. p. 146 h. v.

10 Cf. p. 135 h. v.

This is the only question being debated, which is undoubtedly inspired by the religious and philosophical discussions of the day, as Delisle mentions in his introduction. On the other hand the claims of Flamsteed and others for the observation of a parallax made it a matter of contemporary scientific controversy, as we show in §4 *Scientific background of Delisle's and Bernoulli's presentations*, p. 12 h. v.

2.1 *The* discours *by Delisle*

Delisle defends the thesis that there cannot be an absolute certainty, as long as we use only the observed positions of celestial objects and their changes in time. His argument is essentially that a *perceived* change of position can always be interpreted either as due to a motion of the object as well as to a motion due to the observer; of course the words «relative» or «observer» do not belong to the vocabulary of the authors, but the idea is clearly present. At the same time they confidently speak of the motion of the earth, the sun, the stars, as if these notions had a meaning of their own, relative e.g. to an absolute space, but this notion is not explicitly stated either.

So, argues Delisle, by observation only we shall never know. No doubt the system of Copernicus can do with a lesser number of basic movements than the system of Ptolemy, but an obdurate partisan of a fixed earth could always invoke yet another movement – if necessary among the fixed stars – to explain any present or future observation.

Among those future observations Delisle expects a parallax of fixed stars will be found, and he reviews some of the claims that had been made up to that time; but all of them have been criticized and he considers the existence of a star parallax an open question.

2.2 *The* réponse *by Bernoulli*

Bernoulli concurs with Delisle and proposes to their audience still more convincing analogies. He elaborates on the star parallax: none has been observed; as the telescopes of that period – he claims – could only distinguish angles in excess of 20 arc seconds, it follows that the fixed stars must be farther away than 40 000 times the distance from the earth to the sun[11].

Such huge distances should not impress us, because «small» and «large» have no meaning by themselves: «it could be that in a grain of sand, in an

11 True indeed; the closest star, Proxima Centauri, is 4.3 light years away, i.e. 272 000 times that distance.

atom, there would be a world similar to our own and that the world as we see it is but a particle of air or of ether in another world»[12]. An idea which impressed already Pascal and Fontenelle and which may have inspired Bohr.

But, even if there were a parallax, stubborn Ptolemeians could always invent – and they proved very inventive in the past – an epicyclical movement of the fixed stars responsible for a yearly varying parallax. All this, as was said, if we insist on a purely geometrical interpretation of the astronomical data. But the picture changes completely, says Bernoulli, if we consider metaphysics and physics as well.

His metaphysics takes the form of a principle of simplicity. In support of this he adduces a twofold argument: nature always acts in the shortest way (he does not specify what phenomena he has in mind) and could not do otherwise, as it is directed by a Being supremely wise. Such an idea was already proposed in 1657 by Marin Cureau de La Chambre in his treatise on light[13]. It set Fermat thinking and by using a metric – as we now say – which depended on the medium being traversed he derived the law of refraction; to be true without referring to metaphysics[14].

It is interesting to note that Bernoulli, 70 years afterwards, still believes in this almost metaphysical principle. It is the sort of idea that Maupertuis took up again, more than twelve years later[15].

12 Cf. p. 149 h. v.

13 M. Cureau de La Chambre, *La lumière…*, Paris: P. Rocolet, 1657; 2nd ed. Paris: J. d'Allin, 1662, Ch. VI p. 311, cited from the 1662 edition. *«En effet, se peut-il rien dire de plus conforme à la raison que lors qu'ils asseurent que l'égalité des angles dans la reflexion se fait selon les lois que la Nature garde en tous ses mouvements. Car comme celle-cy emploie en toutes ses actions les moyens les plus courts, elle fait mouvoir les choses par l'espace le plus court:…»*

14 Consult the biography of Fermat: M. S. Mahoney, *The mathematical career of Pierre de Fermat (1601–1665)*, Princeton (N.Y.): Princeton University press, 1973, pp. 382–390.

15 A first publication – on statics – appeared in 1740 (P. L. M. de Maupertuis, *Loi du repos des Corps*: Mém. Paris 1740 (1742), pp. 170–176); his principle of least action was published in 1744 (P. L. M. de Maupertuis, *Accord de différentes loix de la nature qui avoient jusqu'ici paru incompatibles*: Mém. Paris 1744 (1748), pp. 417–426.) and his metaphysical and theological considerations in 1746 (P. L. M. de Maupertuis, *Les loix du mouvement et du repos, déduites d'un principe métaphysique*: Mém. Berlin 1746 (1748), pp. 267–294). Those lead to fierce controversies; consult e.g. P. Brunet, *Etude historique sur le principe de la moindre action*, Paris 1938 = Actualités Scientifiques N° 693. J. O. Fleckenstein, *The introduction to Euler's Opera Omnia* (Orell Füssli), II/5, Lausanne 1957, pp. XV–XLVI. Marco Panza, *De la nature épargnante aux forces généreuses: le principe de la moindre action entre mathématiques et métaphysique. Maupertuis et Euler, 1740–1751*: Revue d'histoire des sciences 48 (1995), N° 4 pp. 435–520.

From the side of physics comes the argument that the earth is in no essential way different from the other planets. To start with, they are all opaque and their brightness is due to the reflected sunlight, but the decisive argument is that they all – including the earth – obey the same law of motion. More specifically, Kepler discovered a general law relating the radius of a planetary orbit to the time of revolution of the planet: all planets, including the earth, obey it. Much more important, however, is that Kepler's laws were shown by Newton[16] «that illustrious Englishman, that all nations call the ornament of our century»[17], to be simply consequences of one single law – universal gravitation – and the universal laws of mechanics that Newton discovered[18]. So, the earth is a planet and consequently circles around the sun as do all other planets, a fact no more open to doubt than the obvious truth that man has a heart, lungs, a liver, etc.

Bernoulli, having said this, comes back to the religious motives one might have to believe the Ptolemaic world system. If all men would have the knowledge required to understand the Newtonian theories, they would be filled with admiration for the works of the Creator of the Universe. Compared with the beautiful simplicity of the Newtonian-Copernican world system, Ptolemy's system is obscure and confused, almost as much as the movement of the dust particles in a ray of sunlight and is more suggestive of an Epicurean world view, where everything is a mere fortuitous assemblage of atoms. How unfair then to tax the Copernicans of feelings opposed to the Sound Religion[19] while quite the reverse is true. Indeed, a case against Ptolemy might be made: if one would have read the Holy Scriptures without preconceptions, one could have found they contradict Ptolemy; this is exactly what Cosmas of Egypt[20] did in his *Topographia Christiana*, written around 548. But those were the times when it was heretical to claim that the earth is round, that there are antipodes and that the earth revolves; the leaders or the originators of such false [religious] zeal deserve less than praise.

16 Newton, *Principia* (1726).

17 Cf. p. 151 h. v.

18 D. Speiser, in a private communication, pointed out that Tycho Brahé's system is in accord with the laws of Kepler, but not with the mechanics of Newton. Bernoulli is therefore right in emphasizing the latter.

19 This is the accurate translation of the words *Saine Religion* i.e. a religion free of unwholesome traits; an idea that would be in line with the philosophy of the enlightenment. However, a simple printer's error could have corrupted an intended *Sainte Religion*, i.e. Holy Religion.

20 Commonly known as Cosmas Indicopleustès; see *Topographie Chrétienne* (critical text, notes and French translation by W. Wolska-Conus), Paris: Ed. du Cerf: Livres I–IV 1968, Livre V 1970, Livre VI–XII 1973.

3 The concluding *Avertissement* by Delisle

After the *Discours* was delivered (and before it went to the printer) a new Danish treatise[21] became known that purported, thanks to a new kind of observations, to determine the distance to the stars. Delisle does not take sides in this discussion, but confirms his thesis that no purely geometrical arguments could settle the question of the moving earth. But he adds that it would be utterly silly to refuse the proofs based on mechanics and physics, which are as certain and convincing as those from pure geometry. Because, states Delisle, pure geometry gets its first principles from mechanics and metaphysics[22].

4 Scientific background of Delisle's and Bernoulli's presentations

4.1 *Parallax*

If we consider (see Fig.1) the triangle defined by the earth E, the sun S, and a celestial object O (a star in the present context) the angle $\alpha = \widehat{EOS}$ is called the parallax of the object. If a star were seen at slightly different angles from two opposed positions of the earth on its orbit around the sun, this difference would give 2α. Knowing ES, we could then find the distance OE of the star from the earth.

There is a somewhat less direct method to put a star parallax in evidence. Suppose a star O would be close to the earth, and a star O' would be so far away its parallax would be unobservable. As a rule O will be a bright star and O' a faint one. To an observer on the earth, O would seem to change its position slightly over a time of half a year, while O' would stay fixed. It would then only require to measure regularly the (small) angle between O and O' and observe a yearly periodicity in order to find evidence for a star parallax.

21 This refers almost certainly to P. N. Horrebow, *Copernicus triumphans sive De parallaxi orbis annui Tractatus epistolaris...* Hafniae (i.e. Copenhagen) 1727.

22 Compare with Newton's words in the preface to the *Principia* (Translation by A. Motte, 1729) (1934), p. XVII:

> «Therefore geometry is founded in mechanical practice and is nothing but that part of universal mechanics which accurately proposes and demonstrates the art of measuring.»

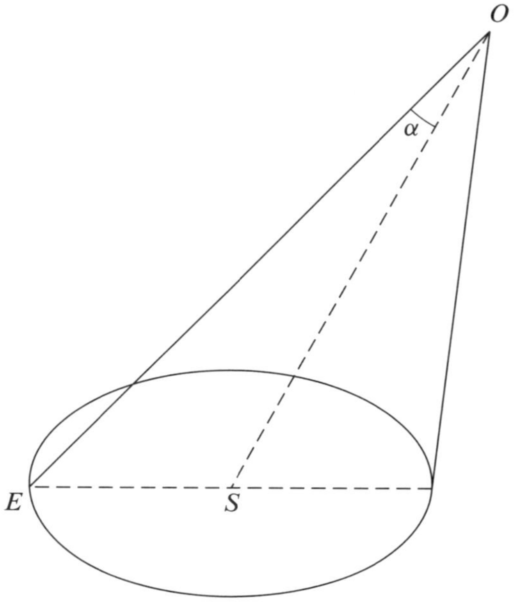

Figure 1: The parallax α of a star O as seen from the earth E in its – almost circular – orbit around the sun S.

4.2 *Aberration*

There is another effect which changes the position of a star in the sky: the velocity of the earth in its orbit, when this makes an angle with the line of sight of the star, combines with the velocity of light from the star to give an apparent direction to the starlight, different from the position we would observe from e.g. the sun. This effect is independent from the distance of the star to the earth. As a result of this phenomenon all stars are seen to describe little ellipses in the course of the year. The changes which Flamsteed and others observed were mostly due to this[23]. The first to give the correct interpretation was James Bradley in a letter to Halley, published in 1729[24], less than a year after Delisle held his *Discours*. It was confirmed the

23 Cf. P. Acloque, *L'aberration stellaire*: Cahiers d'histoire et de philosophie des sciences N° 36, 1991.

24 J. Bradley, ...*an Account of a new discovered Motion of the Fix'd Stars*: Philosophical Transactions of the Royal Society 35 N° 406 December 1728 (1729), pp. 637–661.

next year by Manfredi, who coined the name «aberration»[25]. It is important to realize that aberration gives changes in angles of up to 20 seconds of arc, which was the limit of the accuracy at the time. Parallaxes are mostly fractions of a second and were only detected in the 19th century.

4.3 *The experimental evidence used by Delisle*

Delisle has given an extensive list of the reports on changes of positions up to 1727. A thorough review of the history of the subject appeared[26] in 1853. We rely on it for the following short note on the astronomers mentioned by Delisle:

1. Riccioli[27] reported that he did not observe any variation of the position of Sirius.
2. Flamsteed observed changes in the zenithal distance of the polar star, from 1689 to 1697. They were published as a letter to Wallis[28] and in his *Historia Cœlestis Brittanica*[29]. Recalculation by Peters showed they were in fact accurate observations of the aberration effect.
3. Gian Domenico Cassini published also on the changing position of the polar star[30]. According to Peters those measurements are unreliable.
4. Jacques Cassini measured the parallax of Sirius[31]. Calandrelli showed in 1806 that those results are inconsistent[32].

25 Euler pointed out that the phenomenon by itself is inconsistent with the concept of an earth at rest and hence supports the Copernican system; see: L. Euler E 127, *Explicatio phænomenorum quae a motu locis succesivo oriuntur*: CP XI, 1739 (1750), pp. 150–193 – Euler, *Op. Omnia* III/5, p. 80.

26 C. A. F. Peters, *Précis historique des travaux sur la parallaxe des étoiles fixes jusqu'en 1842* in *Recherches sur la Parallaxe des étoiles fixes*: Mémoires de l'Acad. Imp. de St. Petersbourg, 6e série, sec. Math. et Phys., V (1853), pp. 1–65.

27 G. B. Riccioli, *Almagestum Novum astronomiam veterem novamque complectens observationibus aliorum, et propriis novisque theorematibus, problematibus, ac tabulis promotam, in tres tomos distributam*, t. I, Bologna: ex typ. hrdi V. Benatii, 1651, pars II, p. 452.

28 J. Flamsteed, in *Opera of Wallis*, vol. III, Oxford 1699, p. 701.

29 J. Flamsteed, *Historia Cœlestis Brittanica volumen primum-[tertium]*, vol. III, Londini: typis H. Meere, 1725, p. 108.

30 G. D. Cassini, *S'il est arrivé du changement dans la hauteur du Pole, ou dans le cours du Soleil?*: Mém. Paris 1693 (1730) pp. 360–375 first published in: Mém. de Math. et de Phys., tirez des Registres de l'Acad. Roy. des Sci. 1693 (31 juillet), pp. 113–127.

31 J. Cassini, *De la grandeur des Etoiles fixes, & de leur distance à la Terre*: Mém. Paris 1717 (1719), pp. 256–268.

32 G. Calandrelli & A. Conti, *Opuscoli astronomici e fisici*, Romae: il Salomoni, 1803–1824.

5. Horrebow used in fact observations, made from 1701 to 1704, by Ole Rømer on the meridian heights and the passage times of α Lyrae and Sirius. Their relative angular distances were computed and these displayed seasonal variations. Peters found out this was due to length differences of the uncompensated pendulum in the clock used by Rømer; those were caused by temperature differences, one set of observations being done in the early night, another towards the morning.

The first observations of a genuine star parallax were announced in 1838 and 1839 independently by F. W. Bessel, F. G. W. Struve and T. Henderson. Bessel observed 61 Cygni (0.314″), Struve observed Vega (0.125″) and Henderson α Centauri (1.16″); the modern values are resp. 0.292″, 0.121″ and 0.767″.

4.4 *The distance to Sirius and its estimate by Huygens*

Delisle is probably mixing up two results by Huygens. The illustrious Dutchman discovered a great many formulae in geometrical optics, a. o. those related to the magnifying power of telescopes[33]. On the other hand he did estimate the distance of Sirius, but not by comparing size but by comparing brightness. He did this in two ways. First by covering a long tube with a cover in which a tiniest hole was made; in this hole a small spherical glass bead was inserted, which further diffused the light; observing the sun through this tube it would appear to be no more bright than Sirius. The laws of optics allow to relate the observed light to the luminosity of the sun[34]. In a second method he assumed the magnitude of Sirius and Jupiter to be the same; if you make a reasonable guess how much of the sunlight is reflected by Jupiter and knowing its angular diameter, one can again find the ratio of the brightness of Sirius and of the sun. Assuming Sirius to be a star like the sun its distance can be calculated. By these estimates Huygens arrives – in modern units – at distances of 1.6 light years[35]. (Actually, it is 8.7 light years away.)

33 Cf. Ch. Huygens, *De Telescopiis et Microscopiis 1685–1692*, *Œuvres*, t. XIII, part II. Part of the Huygens manuscript was published posthumously (and imperfectly) by De Volder and Fullenius at Leiden in 1703.

34 Cf. Ch. Huygens, *ΚΟΣΜΟΘΕΩΡΟΣ*, La Haye 1698 – Huygens, *Œuvres*, t. XXI, p. 815.

35 Cf. Ch. Huygens, *ΚΟΣΜΟΘΕΩΡΟΣ*, La Haye 1698 – Huygens, *Œuvres*, t. XXI; Avertissement p. 672 et Appendice VII au *ΚΟΣΜΟΘΕΩΡΟΣ*.

Huygens would probably not have used the method suggested by Delisle; he knew quite well the luminous disc of a star as viewed in a telescope, of which he wrote: «*...ex aliqua visus fallacia nasci arbitror*» (...arises, I think, from some fallacy of the view)[36].

36 Cf. Ch. Huygens, *Systema Saturnium*, La Haye 1659 – Huygens, *Œuvres*, t. XV, p. 237.

Réfléxions sur la meilleure figure à donner aux Ancres, et la meilleure maniére de les essayer *(St. 28)*

1 **The prizes of 1735 and 1737**

1.1 *History of the prizes*

For 1735 the *Académie* had put the following question:

> «What should be the best construction of the anchors, taking into account their figure as well as their forging, and what is the best way to test them?»[1]

It was found that the best memoirs received had well answered part of the question but failed on other parts; the *Académie* decided that no piece could obtain a prize unless all parts of a compound question were satisfactorily solved. It was then resolved to put up the question again for 1737, but to split it up into three distinct questions, each endowed with its separate prize of 1900 *livres*. The memoirs handed in for 1735 would automatically compete for the prize of 1737, in the category they had best covered, but the authors were free to amend their text. The three questions were[2]:

1. What is the most adequate form one could give to an anchor?
2. What is the best way to forge the anchors?
3. What is the best way to test the anchors?

The laureates are shown in Table 1. In 1782 Johann II Bernoulli wrote to Lalande[3], giving details of Daniel's life and he recalled:

> «...this time we competed the one and the other, my brother and I, but separately, and I was fortunate to win the first of the three prizes...»[4]

Both brothers were on good terms and arranged to transfer the two amounts won in one operation through a letter of change for 2850 *livres*[5].

1 *«Quelle doit être la meilleure construction des ancres, tant par rapport à leur figure qu'à la manière de les forger, et quelle est la meilleure manière de les éprouver?»*
2 P. V. Acad. Sci. Paris of April 20th, 1735.
3 Autograph in the archives of the *Académie*. (*Dossier* Johann II Bernoulli)
4 Letter of Johann II Bernoulli to Lalande of 1782. «...*Pour cette-fois-ci nous concourumes l'un et l'autre, mon frère et moi séparément et j'eus le bonheur de remporter le premier des 3 prix...*»
5 Letter of D. Bernoulli to Mairan of June 4th, 1737.

Bernoulli (Johann II)	Prize 1st question	1900
Trésaguet	Prize 2nd question	1900
Bernoulli (Daniel)	$\frac{1}{2}$Prize 3rd question	950
Poleni (Marquis)	$\frac{1}{2}$Prize 3rd question	950
Créqui (Comte de)	Accessit	
Anonyme (Mémoire nr 5)	Accessit	

Table 1: The laureates of the prizes for 1737 and the amounts won, in *livres*.

In his memoir Johann II wrote:

> «... after having told my idea to one of my friends, an excellent expert in mechanics, he approved very much of my theory; perhaps he will make use of it, supposed he fancies to work also on the subject in question.»[6]

The «friend» was probably Daniel.

The judges of the contest were: Réaumur, Pitot, Nicole, Cassini and Mairan[7]; they examined 13 memoirs newly submitted in addition to the five for 1735. The decisions of the *Académie*[8] were neatly summarized by Maindron[9]:

> «The third question was, in the wording of the Académie, «not sufficiently answered yet the essays sent in contained interesting and useful [*curieuses et utiles*] investigations on other aspects which should not be lost for the public». Consequently the third price was divided equally between Daniel Bernoulli and Marquess Poleni of Padua. Those four essays were published the same year.»[10]

6 Joh. B.: Prix Paris III (pièce IV) 1737 (1752), p. 17 §21. «*...après avoir communiqué mon idée à un de mes amis, bon connoisseur en fait de méchanique, il approuva fort ma théorie, dont il se servira peut-être lui-même, supposé que l'envie lui prenne de travailler aussi sur le sujet en question.»*

7 P. V. Acad. Sci. Paris of September 5th, 1736, and of September 4th, 1734.

8 P. V. Acad. Sci. Paris of May 4th, 1737.

9 E. Maindron, *Les fondations de prix de l'Académie des Sciences, Les lauréats de l'Académie 1714–1880*, Paris: Gauthier-Villars, 1881, p. 18. Maindron states prize money of 4000 *livres* but both from the procès-verbaux and the letter of Bernoulli (just quoted) it appears that it was 5700, i.e. 1900 for each.

10 Prix Paris III (pièces IV, V, VI, VII) 1737 (1752).

Tresaguet, an engineer responsible for manufacturing anchors, described mainly the metallurgy of that process, to which he contributed. Poleni had sent in answers to all three the questions.

1.2 *The contribution of Johann II*

The memoir of Johann II is – for his time – an original and sophisticated application of the calculus. In designing the arms of an anchor we can choose the curve of an arm, its thickness in the plane of the arm and its thickness perpendicularly. Johann II investigates how this freedom can be used to design the best anchor. In the first part the perpendicular thickness is taken constant over the whole span of the arm and Johann II derives the ideal curve and thickness in the plane by imposing two conditions:

- **Condition one.** The strength of the arm should be uniform: at no point should the arm be weaker than at any other point under the combined effort of the pulling cable and the resistance of the seabed on the fluke; this is important in the first moments, when the anchor starts gripping the soil. For a straight beam the problem had been solved by Galilei[11] which is explicitly cited; in that case the profile of the thickness is a parabola.
- **Condition two.** When the arm is fully dug in, its concave surface is pressed in every point by the mud which resists the anchor being pulled by the horizontal cable. Johann II compares this to the problem of a ribbon or a sail fixed at the edges and subject to the pressure of the wind or a flowing liquid: the normal component of the force, should be proportional to the curvature of the fabric; this produces the surfaces known as *lintearia* and *velaria*. For the essentially one dimensional problem of an anchor with constant perpendicular thickness it leads for the curve of an arm to a catenary (french: *chaînette*). An arm having this form will keep it under the impact of the mud and will not bend.

In the second part the perpendicular thickness is allowed to vary; taking now an arbitrary curve for the arm Johann II derives, from his two conditions, a general differential relation between the curve and the two thicknesses. This relation is in general not integrable. But in three cases he is able to solve it: for an arm in the form of an arc of a circle or of a parabola or a catenary. In the third part the third question is addressed: how to test

11 G. Galilei, *Discorsi e dimostrazioni matematiche intorno a due nuove scienze attenenti alla mecanica e i movimenti locali*, Leiden: Elsevier, 1638, *seconda giornata* – Galilei, *Opere* (Ed. Naz.), vol. VIII, pp. 43–313.

anchors. The mathematical results of the second part allow to derive scaling laws and consequently it would be enough to perform the test on a small model of an anchor: the formulae allow then to predict the behaviour of a real anchor.

1.3 *The contribution of Daniel*

Daniel entered a piece where he examines two questions: why and how does the conventional shape (with flukes at the arms and a wooden crossbar) work and how much of the force exerted by the vessel is transmitted to the anchor (which leads to the condition for safe anchoring). Both questions keep remarkably clear of the problem which Johannn II is addressing. Daniel exploits mainly the geometry of the anchor and the cable, while his brother displays his virtuosity in the differential calculus.

2 **Summary**

The first part (§3 to §13, pp. 155–162 h. v.) is an elementary discussion, couched in geometrical terms, based on the potential energy of the anchor in its two characteristic positions; in the way he uses those results Bernoulli has, however, oversimplified the problem in making several tacit assumptions. He concludes that the traditional form works well, except that the wooden crossbar (the stock) should be replaced by an iron one, and the arms should be rounded, forming a segment of a circle. It is quite remarkable that a hundred years later all anchors did show those features[12]. In the second part (§14 to §17, pp. 163–171 h. v.) the theory of the catenary is adduced to demonstrate how the weight and the length of the cable are essential to safe anchoring, because the cable reacts to the vertical forces exerted by the ship and transmits almost none to the anchor.

In the third part (§18 to §23, pp. 171–181 h. v.) it is examined how variations in the size and the proportions of anchor and cable influence their behaviour. The conclusion (§24, p. 181 h. v.) is that the preceding theory gives a framework to devise tests and interpret them; a possible procedure is described in a bare two pages. It was indeed an accurate assessment by the *Académie* that this was a scant answer to the third question but that the essay nevertheless brought essential ideas for the working of an anchor.

12 G. Cotsell, *A treatise on ship's anchors*, London: J. Weale, 1856.

3 **Review**

3.1 *The anchor, parts and shape*

The list of parts is displayed in Table 2 and Fig. 1, taken from the *Encyclopédie*[13]:

English	French	length from end to end
stock	jas ou jouet ou essieu	$4a$
shank	vergue ou verge	$4a$
arms	croisée (bras ou branche)	$2a$
fluke	patte	
ring	arganeau (organeau)	
rode or cable	cable	

Table 2: The parts of an anchor; the length a scales with the size of the ship, normally it is 1/10 of the width of the ship for the main anchor. Modern French usage in brackets.

It seems that all anchors, whether large or small, are similar and have the same proportion between their parts, as displayed in the table. They are forged from wrought iron, except the stock which consists of a heavy wooden beam, at right angles to the arms; the centre of mass of the anchor lies approximately in the middle of the shank.

3.2 *The stable positions on a horizontal surface*

3.2.1 The anchor on dry land

Let us start by giving the argument according to modern mechanics. There are two stable positions, as Bernoulli explicitly mentions (see Fig. 2): either an arm lies flat and the stock rests with a tip on the surface, pointing upwards (position A); or else the stock and the ring lie flat and the arms rise

13 D. Diderot, D'Alembert, *L'Encyclopédie: recueil de planches sur les sciences, les arts libéraux et les arts méchaniques, avec leur explication. Marine*, Paris 1757, Planche II (partim).

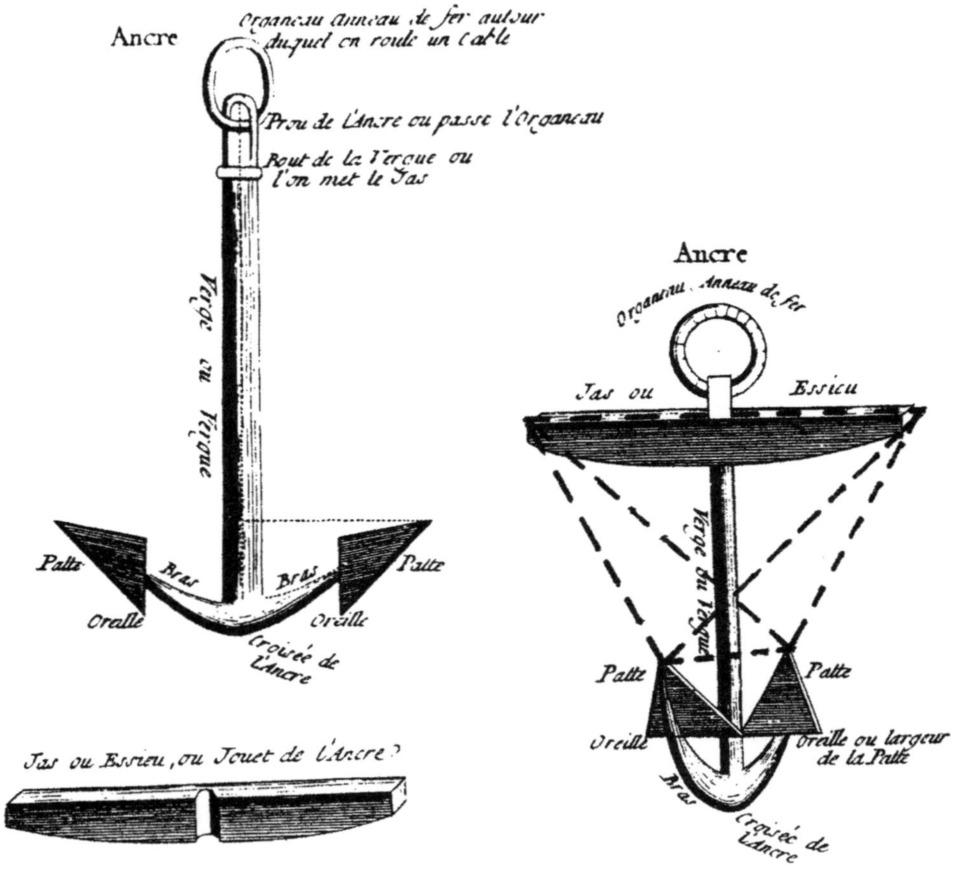

Figure 1: The parts of an anchor according to the *Encyclopédie* 1757. The geometrical «skeleton» used by Bernoulli (his *Fig. 1*) defines a tetrahedron which is drawn in as dashed lines.

vertically above the surface (position B). Only the last position will allow the fluke to dig into the soil and secure a firm mooring. A rotation around a line joining the tip of an arm to the tip of the stock (e.g. EG) will move the anchor from one position to the other. In so doing the centre of mass moves up and then comes down again, i.e. the potential energy is raised from a local minimum over a «hill» to another local minimum (see Fig. 4). The centre of mass of the anchor is higher in position A than in position B. We imagine an anchor thrown with force on the surface of a flat beach, and always in contact with the soil. It will tumble along, rotating haphazard

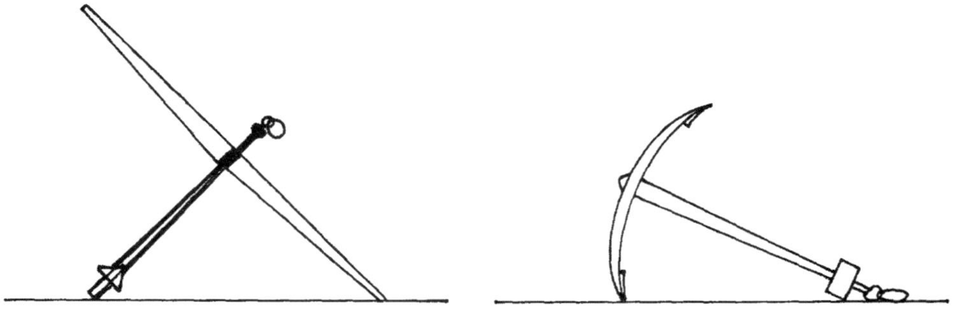

A *B*

Figure 2: The two stable positions of an anchor on a horizontal surface.

around one of those possible axes, then along another. As long as the kinetic energy is large enough the tumbling will go on, moving from one position to the next. By friction, however, kinetic energy is lost to the surroundings up to the point where it is barely sufficient to tilt the anchor from the higher energy state *A* over the potential hill into the lower state *B*; in the course of this motion still more kinetic energy is lost and the anchor is trapped in the lower valley. This lower state is precisely the one where an arm grips into the soil and the anchor fixes itself.

In Figs. 1 and 3 this is made clear: the stock on the one hand and the line joining the two tips of the arms on the other can be considered as two opposite (and unequal) edges of a tetrahedron. An anchor tumbling over a horizontal plane is undergoing successive rotations around the edges of the tetrahedron. Going from the base *DEG* (position *A*) to the base *EGF* (position *B*) means rotation around *EG*. In Fig. 3 the tetrahedron is shown with its side *DEG* as base and the other sides folded down in the same plane. The point *L* is the vertical projection of the centre of mass on the side *DEG* and the point λ the projection on *EGF*. Upon rotation the centre of mass *H* describes an arc with radius h in the vertical plane through *L* and λ, going from a height h_1 to a height h_2, as shown in the lower part of the figure. The heights h_1 and h_2 are a measure of the potential energy in the positions *A* and *B*.

By simple geometry and repeated applications of Pythagoras' theorem, one arrives at the following formulae, where $2a$ is the span of the arms and 2α the width of the stock[14]. The centre of mass *H* lies supposedly

14 We adopt here Bernoulli's notations.

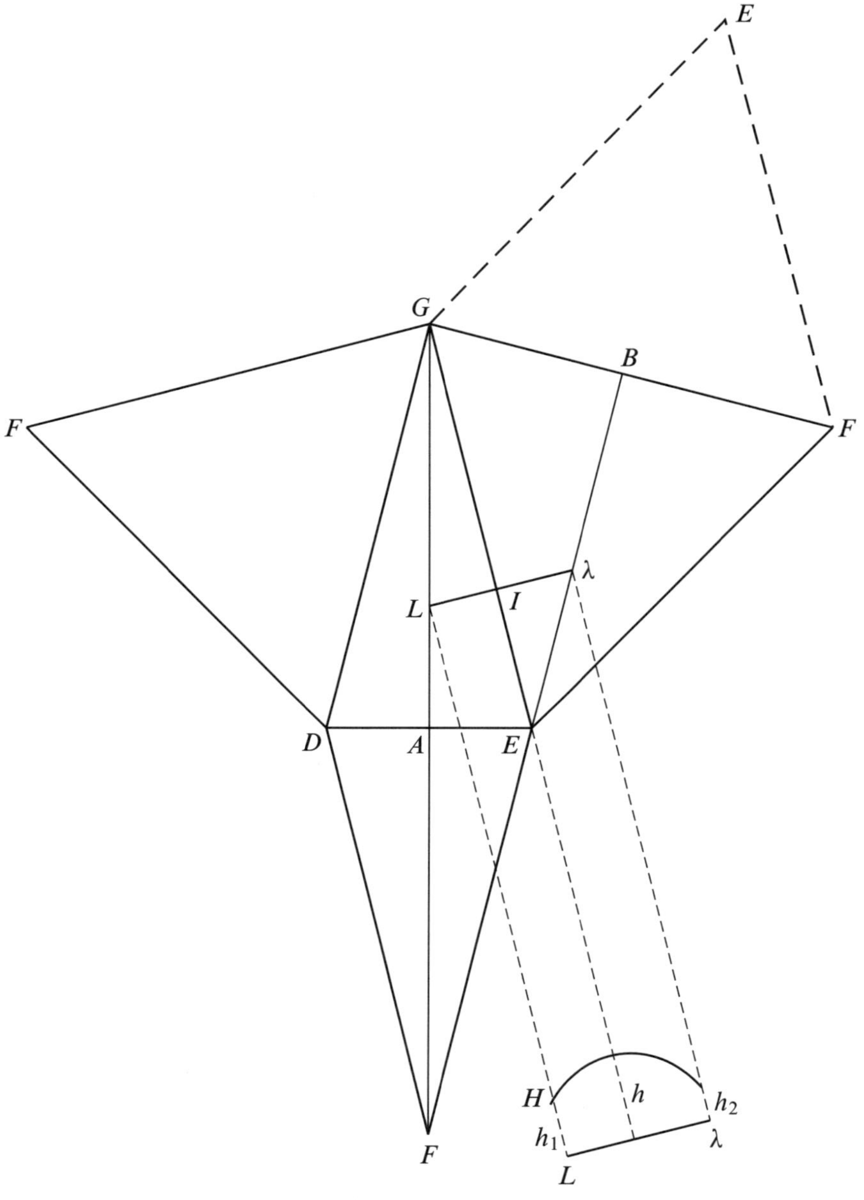

Figure 3: The faces of the tetrahedron which circumscribes the anchor. The edge GF represents the stock, DE the arms. The symbols used in the text are: $a = AE$, $\alpha = GB$, $c = GE$.

h	$1.069a$
h_1	$0.8944a$
h_2	$0.4851a$

Table 3: Values of the height of the centre of mass in various positions (see Fig. 3), in units of a.

a distance b from the arms, and a distance β from the stock (the length of the shank is therefore $v = b + \beta$). We call the length of the edge $c = \sqrt{a^2 + \alpha^2 + v^2}$.

$$h = \frac{1}{2c}\left[2c^2(\alpha^2 + \beta^2 + a^2 + b^2) + 2(a^2 + b^2)(\alpha^2 + \beta^2) - c^4 - (a^2 + b^2)^2 - (\alpha^2 + \beta^2)^2\right]^{\frac{1}{2}} , \tag{1}$$

$$h_1 = b\frac{\alpha}{\sqrt{a^2 + (b + \beta)^2}} , \tag{2}$$

$$h_2 = \beta\frac{a}{\sqrt{a^2 + (b + \beta)^2}} . \tag{3}$$

If we apply those general formulae to an anchor with the standard proportions, as given in Table 2, we obtain Table 3.

By reasoning on the skeleton of an anchor (Bernoulli's *Fig. 1*) our author arrives at the same formulae and the same values.

3.2.2 Bernoulli's stability argument

Bernoulli, in his argument, concentrates on potential and on kinetic energies, or – as he says – on heights and living forces.

To him the important numbers are $h - h_1 = 0.1747a$ and $h - h_2 = 0.5839a$ which represent the differences of the potential energy between the two stable minima and the maximum in between. He views them as obstacles to be overcome by *forces*, i.e. «living forces», proportional to our kinetic energies. He is interested in their ratio, which in the example is equal to 2/7 (within a rounding error of 5%) and in the subsequent argument he simply takes them as 2 and 7. We illustrate the state of affairs in Fig. 4.

His argument runs as follows (see his §8, p. 159 h. v.): the anchor cast on a flat surface will have a large living force which will decrease as it is

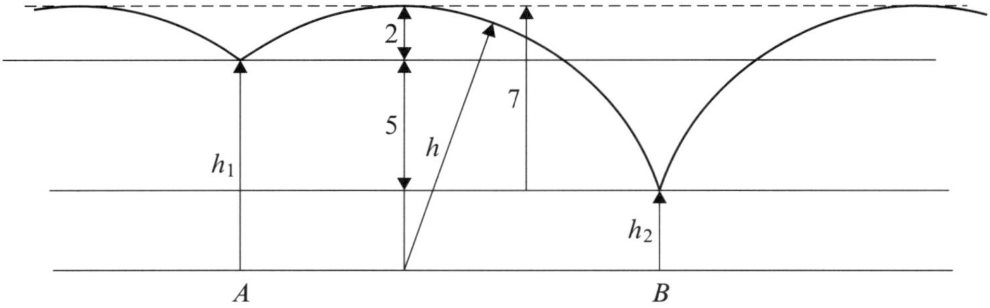

Figure 4: The height of the centre of mass of the anchor as it tumbles from one position to the next. The depths of the potential wells, i.e. 2 and 7, are the approximate values used by Bernoulli.

rolling over the surface until it is smaller than 7 but larger than 2; if it finds itself in the «bad» position A, that energy is enough to topple it and it lands in the «good» position B where – he claims – it is going to stay. The reverse is unlikely to happen: Bernoulli takes the example of an anchor with a force somewhat in excess of 7; if it finds itself in the «bad» state B it is bound to get out of that again; otherwise one should assume that «in this one tumbling it lost more than 5/7 of its force», i.e. it lost so much kinetic energy (actually more than 5 parts out of 7, leaving less than 2) that it can no longer get out from the local minimum A. Bernoulli's conclusion is: «it is morally impossible that the arms of the anchor will stay flat.»[15]

This argument is incomplete: it supposes the kinetic energy of the anchor to be a fixed initial quantity, independent of the orientation of the anchor when reaching the ground. But e.g. a freely falling anchor – dropped in air from a given height – will have a larger kinetic energy in position B because its centre of mass will have fallen over a larger distance. Conversely, the anchor rolling from B to A loses kinetic energy because its centre of mass is rising.

If, however, we think of an anchor as falling through water, i.e. through a resisting medium, it reaches rapidly its terminal velocity. When hitting the bottom it will have a corresponding kinetic energy, the same either in the «good» position B or the «bad» position A. If we imagine this is smallish («between 2 and 7»), it will make the anchor rock a little; this rocking is

15 Cf. p. 159 h. v.

enough to topple it from the «bad» to the «good» position; in this last one it will stay because water resistance and the friction with the bottom absorb hopefully enough energy to confine it to the potential well. If it had hit the bottom in the «good» position to start with it could never, with its low kinetic energy, got out of it.

The argument of Bernoulli has then to do with anchors coming down at a given kinetic energy and is only valid under a tacit assumption on the rate of loss of the energy, which should not be too large («not 2/7 in one tumble») but not too small either.

3.2.3 The anchor submerged

When the anchor is not rolling on a flat surface in air, but on the seabed, immersed in water, the effective weights are lower but this works differently for the wooden stock as for the iron arms and shank. The stock will become almost weightless, the iron loses one part in 7.85[16]. As a consequence the centre of weight will move closer to the arms. In the example mentioned earlier this means that in formula (1) we put $b = \frac{4}{3}a$ and $\beta = \frac{8}{3}a$ resulting in

$$h - h_1 = 0.3352a$$
$$h - h_2 = 0.2848a\ .$$

These two numbers are close one to the other but above all the potential energy in position B is now the higher one. Therefore the chances for the «good» result are now less than for the «bad» result. A firm recommendation by Bernoulli is therefore to increase the weight of the stock either by cladding the wooden beam with iron plates or simply by making the stock from iron, of the same weight as the wooden stock. That way the submerged anchor would keep its centre of weight in its centre of mass and Table 3 remains valid, also under water.

3.2.4 The anchor drawn by the cable

Leaving now the freely tumbling anchor and turning to the anchor dragged by the cable over the floor, one could develop an argument as follows: as the seafloor is admittedly rough, the traction on the cable and the reaction

16 We assume, with Bernoulli, that the density of wrought iron is 7.85 times the density of seawater.

of the soil, on whatever part of the anchor touches it, will generate torques inducing such rotations as we considered above. One assumes that at each moment these torques balance the reaction of the anchor, i.e. the torques induced by gravity; those last torques are proportional to the lever arms IL and $I\lambda$ in Fig. 3, i.e. they are proportional to $\sqrt{h^2 - h_1^2}$ and $\sqrt{h^2 - h_2^2}$. A smaller torque will be needed to topple the anchor out of the «bad» position than out of the «good» position.

Bernoulli does not mention such arguments using torques. He has had a small anchor made and experimented with it and observed how it behaved when dragged over a rough soil by an almost horizontal cable. In the «bad» position the tip of the stock will be arrested at the bottom (Bernoulli remarks that this is easily done because the stock makes a sharp angle with the soil); the traction of the cable will make the arms jump up, thereby easily tumbling in the «good» position. The anchor's arms come then down with a fluke towards the seabed, this will be hit with force under an angle which favours furrowing. In a subsequent paragraph Bernoulli will come back to the angle which is best to drag the fluke deeply into the sea floor.

3.3 *The Cable*

3.3.1 The shape of the cable

Here again we begin by stating the problem and its solution in a modern way. Let a flexible cable, of linear density ρ be suspended at its endpoints in P and in A, such that its tangent in A is horizontal, as in Fig. 5. The weight w of the cable is proportional to its length

$$w = g \cdot \rho \cdot S \ ,$$

where S is the length of the cable and g the acceleration due to gravity. At P a force $\vec{F}$, along the tangent is required to keep the cable fixed; in A similarly a force $\vec{H}$. The tangent at P makes an angle ψ with the horizontal axis. Let x be the horizontal coordinate and $y(x)$ be the curve described by the cable. The equilibrium of the horizontal and the vertical forces requires that:

$$H = F \cos \psi \tag{4}$$

$$g \cdot \rho \cdot S = F \sin \psi \ . \tag{5}$$

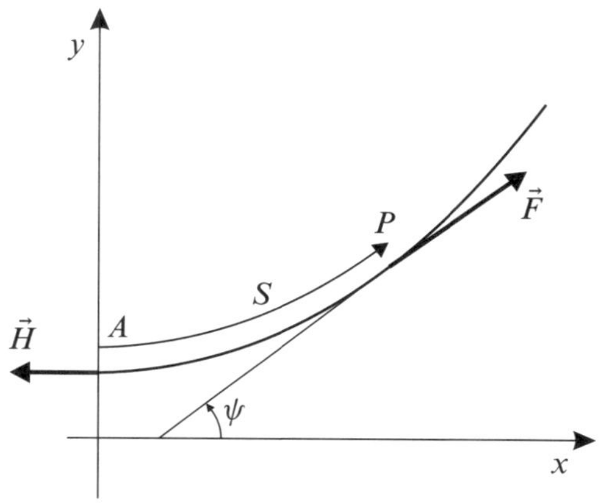

Figure 5: The statics of a hanging cable.

For an elegant demonstration see e.g. the note by G. Cramer in the Works of Jacob Bernoulli[17]. Eliminating F we obtain:

$$g \cdot \rho \cdot S = H \tan \psi \ ,$$

and calling $c = \dfrac{H}{g.\rho}$

$$\tan \psi = \frac{dy}{dx} = \frac{1}{c} S \ , \tag{6}$$

i.e. the slope is proportional to the distance from the apex along the curve. Expressing the curve's length S as an integral of the elementary arc

$$S = \int_0^x \sqrt{1 + \left(\frac{dy}{dx}\right)^2} dx \ .$$

Introducing this into eq. (6) and differentiating, we obtain the following differential equation:

$$\frac{d^2 y}{dx^2} = \frac{1}{c} \sqrt{1 + \left(\frac{dy}{dx}\right)^2} \ . \tag{7}$$

A solution is:

$$y = c \cosh\left(\frac{x}{c}\right) \ , \tag{8}$$

17 Jac. B. *Opera*, pp. 424–426 and Tab. XVIIIb, p. 506.

i.e. the equation for a catenary (*chaînette*); it follows then at once that

$$S = c \sinh\left(\frac{x}{c}\right) , \tag{9}$$

$$\tan \psi = \sinh\left(\frac{x}{c}\right) , \tag{10}$$

showing that this is the solution with the apex of the catenary at $x = 0$.

3.3.2 The catenary: a family curve

The years around 1690 saw a lively battle of wits between the best mathematicians, exploring the realm opened up by the new calculus introduced by Leibniz. In May 1690 Jacob Bernoulli proposed the following conundrum: «to find which curve describes a loose string, freely suspended between two fixed points.»[18] Leibniz announced he had solved it, but would defer publication so that others could join the competition. In June 1691 three articles appeared, by Leibniz[19], Huygens[20] and Johann Bernoulli[21]; all had found the solution by various means. The three contenders concentrated on solving the differential equation but did not elaborate on the argument leading from the physics of the problem to the equation. When the stress of the competition had subsided Huygens did[22] and Johann Bernoulli wrote an extensive treatment of the catenary in his lectures for l'Hospital[23]. When the present essay was written Daniel Bernoulli could safely assume that his readers would know the two laws (4) and (5), to be used below.

18 Jac. B. Op. XXXIX, *Analysis problematis antehac propositi, de inventione lineae descensus a corpore gravi percurrendae uniformiter, (...)*: AE May 1690, pp. 219–223 – Jac. B. *Opera*, p. 426. *«Invenire, quam curvam referat funis laxus & inter duo puncta fixa libere suspensus».*

19 G. Leibniz, *De linea in quam flexile se pondere proprio curvat, ejusque usu insigni adinveniendas quotcunque medias proportionales & Logarithmos*: AE June 1691, pp. 277–281.

20 Ch. Huygens, *Solutio ejusdem Problematis*: AE June 1691, pp. 281–282 – Huygens, *Œuvres*, t. XIX.

21 Joh. I B., *Solutio problematis funicularii*: AE June 1691, pp. 274–276; see also the *Additamentum* by Joh. I B., *ibidem*, p. 288.

22 Ch. Huygens, *letter from Ch. Huygens to H. Basnage de Beauval*: Histoire des Ouvrages des Sçavants, Janvier & Février 1693, pp. 244–257 and Huygens, *Œuvres*, t. X, specifically pp. 413–416; a more explicit manuscript is published in Huygens, *Œuvres*, t. X, pp. 216–218.

23 Joh. I B., *Lectiones mathematicae, de methodo integralium, aliisque, conscriptae in usum Ill. Marchionis Hospitalii, cum auctor Parisiis ageret, Annis 1691 & 1692*, Joh. I B. *Opera III*, especially from the 36th to the 45th lecture.

3.3.3 Two theorems by Bernoulli

For the purpose of the present discussion there is no need to resort to the full solution (8), everything can be derived from the laws of statics, viz. (4) and (5). Bernoulli bases his arguments on his *Fig. 3*, which is reproduced here as Fig. 6 for convenience.

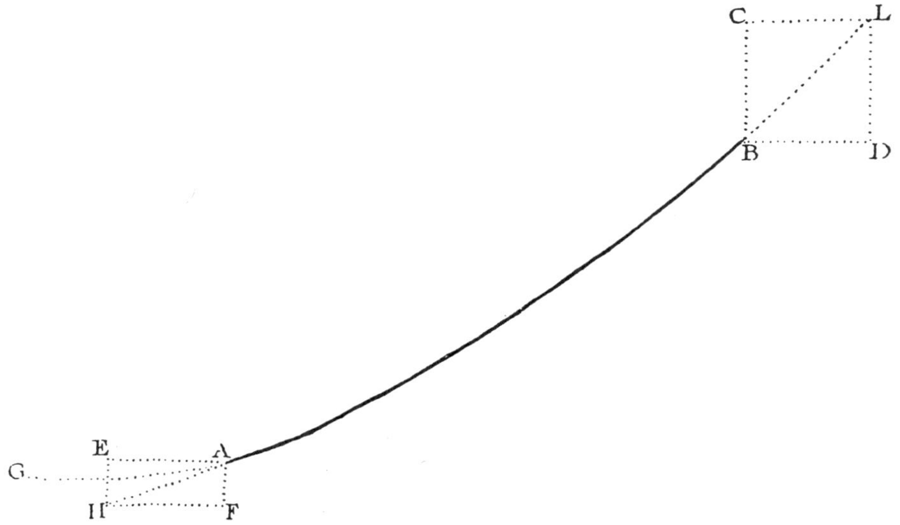

Figure 6: The form of the anchor cable and the forces on it, reproduced from the plate published by Bernoulli.

The lower point of the cable, at the anchor, is usually above the apex of the catenary. Three forces are in equilibrium: the forces at the end points and the weight of the cable; as the weight is a vertical force the horizontal components EA and BD must be equal and opposite: this is

Theorem 1.

$$AE = BD \; . \tag{11}$$

If the cable would extend to the apex G, there would be no vertical force there; consequently the vertical force BC is exactly equal to the weight of the cable GB (as expressed by formula (5)). Exactly the same argument applies to a short stretch of cable GA, but at A the vertical force due to the left part of the cable compensates the vertical force due to the right part. The ratio of the vertical forces in A and P is therefore as the weights of GB to GA, i.e. as the lengths of the curves GB and GA:

Theorem 2.

$$\frac{BC}{AF} = \frac{GB}{GA} . \tag{12}$$

Provided the mariners use a long cable, such that it is almost horizontal near the anchor, the vertical force on the anchor (which tends to raise it from the seafloor) is only a small fraction of the vertical force at the ship which in turn is almost exclusively the weight of the cable in the water. We can formulate this in an alternative way: the weight of AG (i.e. the force AF) is equal to the force BC minus the effective weight of the cable; and the force BC is the total force at B times the tangent of ψ. This «total force» is the force caused by the wind or by currents on the ship; this is horizontal and – in the drawing – proportional to BD. Adopting the notation of Bernoulli: $P =$ total force in B, $s = \sin\psi$, $c = \cos\psi$, $p =$ effective weight of the cable, we obtain his formula:

$$AF = \frac{s}{c}P - p.$$

3.3.4 The lift on the anchor

If

$$\frac{s}{c}\frac{P}{p} = 1 \tag{13}$$

the lifting force on the anchor is zero; if $\frac{s}{c}\frac{P}{p} > 1$ the slope in A is positive and there is an upwards force on the anchor. If $\frac{s}{c}\frac{P}{p} < 1$ the lower end of the cable trails on the seabed.

How much the force AF should be in order to lift the stock can be computed simply for our standard anchor. We recall that in air the arms are as heavy as the stock and the shank weights as much as those two together. One takes into account that:

- the wooden stock is weightless in water,
- the iron's weight is reduced by 6.85/7.85 (due to the buoyancy),
- we denote by π the total weight of the anchor in air[24].

A simple calculation in statics reveals that a vertical force on the ring of $0.218\,\pi$ (which Bernoulli rounds off to $\frac{1}{5}\pi$) will lift up the stock. Hence the force on the moored ship should not exceed

$$P < \frac{c}{s}\left(\frac{\pi}{5} + p\right) .$$

The practical conclusions drawn from these formulae, especially from (13), by Bernoulli are:

1. the angle ψ of the cable near the ship with the sea's surface should be kept small;
2. the weight of the cable should be increased.

3.4 *How to test anchors and sundry items*

Based on the arguments and the formulae given hitherto, Bernoulli investigates what effects a change of parameters would have on the stability of an anchor position, on the best angle by which the fluke penetrates the sea bed, on the ease of lifting the anchor. He estimates what should be a typical value of the force P for a frigate, withstanding drift in a current of 6 feet/second.

He concludes that the traditional form is on the whole a good compromise; two improvements suggest themselves:

- the stock should be made of iron instead of wood, without altering the length and the weight (the thickness, of course, will be smaller);
- the arms should have the form of an arc of a circle, so that the arm should readily follow in the hole dug by the fluke[25].

3.4.1 Alternative anchors

Could one improve on the traditional concept of an anchor? Bernoulli answers wholeheartedly yes, and presents such a novel project in his *Fig. 4*. But he has no illusions on the acceptance of such inventions:

> «There would be much to say on the nature of such an anchor and on the proportions its parts should have; but things too novel in which the

24 This is the notation of Bernoulli, at the time π was not yet a symbol reserved for the ratio of circumference to diameter in a circle.

25 The modern reader will note the similarity with a surgeon's curved needle.

> *Sçavants* had any part are seldom heeded when it comes to execution; it would be a lost comment.»[26]

3.4.2 Testing anchors

The theory just given will allow to carry out the experiments on land and draw firm conclusions for the behaviour of the anchors on the seabed. One should arrange for a suitable ground similar to the seafloor. The cable should be hauled by means of a capstan placed a distance away and higher up, such that the angle of the cable with the ground is between 0 and 20 degrees. The buoyancy experienced by the stock could be simulated by applying a vertical upward force on the stock (equal to 7/8 of its weight) or one could replace the original stock by a lighter one, or pull on the anchor at a larger angle. One could then experiment with various forms or proportions and try out different procedures for casting or raising the anchor.

> «To carry out these experiments and trials with a better profit one will be able to consult the arguments and the calculations which I made in the body of this discourse. I trust they will thereby be confirmed.»[27]

26 Cf. §23, p. 181 h. v.
27 Cf. concluding phrase, p. 183 h. v.

Recherches sur la maniere la plus avantageuse de suppléer à l'action du Vent sur les grands Vaisseaux, soit en y appliquant les Rames, soit en y employant quelqu'autre moyen que ce puisse être. *Fondées sur une nouvelle Théorie de l'économie des forces et de leurs effets.* *(St. 47)*

Introduction

We have split the comments in four parts. In the first we give a general review of the work, accompanied by brief remarks. In the next three parts we go deeper into points which are specially relevant in the memoir:

- The concept of power. It was introduced in the *Hydrodynamica*[1], but here it is central to the discussion, together with the notion of the efficient use of power.
- The new forms of oars or mechanical fins invented by Bernoulli and his evaluation of their efficiency.
- The use of waterjets to propel a ship by reaction. This was proposed in Section XIII of the *Hydrodynamica* and very briefly taken up here. We point out an error of principle in the reasoning by Bernoulli.

1 **Part 1: review and introductory remarks**

1.1 *Historical remarks*

The question which the *Académie* asked for the year 1753 was:

> «The most advantageous ways to substitute for the action of the wind on naval vessels, either by applying oars or using whatever other means.»[2]

The judges (*commissaires*) were: Camus, Bouguer[3], Clairaut, Le Monnier and d'Alembert; nine memoirs were submitted[4]. The prize of 2000 *livres*

1 D. Bernoulli St. 31, *Hydrodynamica*, Strasbourg 1738 – *Werke* 5, pp. 91–424.

2 *«La manière la plus avantageuse de suppléer à l'action du vent sur les vaisseaux soit en y appliquant les rames, soit en employant quelque autre moyen que ce puisse être.»*

3 Bouguer was clearly the genuine expert; he had worked on the problem; cf. p. 48 note 26.

4 P. V. Acad. Sci. Paris of September 6th, 1752.

went to Daniel Bernoulli[5]. In the same meeting the question for 1755 was announced (see next item in this volume). Three accessits were awarded: to Mathon de la Cour, Pereyre and Leonhard Euler[6]. The text of Bernoulli was published in the *Prix de l'Académie de Paris pour 1753*[7] and mentioned in the *Histoire de l'Académie Royale des Sciences, année 1753*[8], printed the same year, where it is stated that Euler's contribution was the next best. The *Académie* noted in its minutes words of warning to the effect that there was still much incomplete, mainly as practice is concerned[9].

When Bernoulli wrote the memoir he was well settled as a professor in his home town of Basel, perfectly inserted in the life of the city which he did not leave any more. As Condorcet said in his *Eloge*[10] of 1783:

> *«Depuis ce moment*[11] *l'histoire de sa vie n'est plus que celle de ses travaux.»*

He lived not in scientific isolation, however, as testified by his ample correspondence.

1.1.1 The Euler memoir

It is interesting to compare the prize-winning contribution of Bernoulli with the accessit sent in by his friend Euler. Euler has divided his memoir in three parts:

1. The speed of movement at which a man is strongest; oars and other devices should be conceived in order that men manipulating them could do this at maximum strength.
2. Euler proposes some alternative schemes to the conventional oars:

5 E. Maindron, *Les fondations de prix de l'Académie des Sciences, Les lauréats de l'Académie 1714–1880*, Paris: Gauthier-Villars, 1881, p. 20.

6 L. Euler E 413, *De promotione navium sine vi venti*: Prix Paris VIII (pièce II) 1753 (1771), pp. 1–77 – Euler, *Op. Omnia* II/20, pp. 190–228; the editor of the volume attributes wrongly a first prize to this memoir.

7 Prix Paris VII (pièce III) 1753 (1769), pp. 3–99.

8 Hist. Paris 1753 (1757), p. 303.

9 *«Mais l'Académie, en rendant au travail de ces Auteurs la justice qu'il mérite, croit devoir avertir qu'il n'y a aucune des pièces qui lui ont été envoyées, où elle n'ait trouvé beaucoup à désirer, surtout pour ce qui concerne la pratique et elle renouvelle à cette occasion qu'elle a fait plusieurs fois, qu'elle ne prétend point adopter ce qui est contenu dans les ouvrages qu'elle couronne.»* P. V. Acad. Sci. Paris of September 6th, 1752.

10 *Eloge de M. Bernoulli par Condorcet*: Hist. Paris 1783 (1786), p. 84.

11 The return to Basel in 1733.

- vertical oars alongside the ship, with blades as Venitian blinds;
- rotating oars with four blades;
- the first instance of a ship's propeller.

The forces exerted by all of these are calculated assuming the same law (2) as Bernoulli.

3. A thorough discussion of propulsion by water jets, based on Section XIII of the *Hydrodynamica*, with either a system of reservoirs, filled by pumps or two coupled pumps feeding directly two jets.

Euler's thinking concentrates on maximizing the forces. Bernoulli's main concern is the most efficient use of the power available. Johann Albrecht Euler made a French translation of the Latin memoir by his father[12].

1.2 *The mechanics*

1.2.1 The living forces

Bernoulli distinguishes a living force (*vis viva*) from a dead force (*vis mortua*), a common distinction at the time. At the origin dead forces were those which were continuously active, such as gravitation: a weight on scales, a falling pebble are under the influence of a dead force. Living forces have to do with percussion or collision: the force is active for an instant only, without duration. Euler showed in 1747 that there is no fundamental difference between both[13]. In the present essay the «living force» is used in a precise technical sense: the living force of a mass m moving with speed v is the quantity mv^2, in fact double the kinetic energy, as we now say. If that mass acquired its speed by falling from a height s, we have that $s = \frac{1}{2g}v^2$ where g is the acceleration of gravity. The conservation of the living forces means that the sum of the living forces of a system of masses in a constant gravitational field increases by exactly $2Mgh$ where M is the sum of the masses and h the distance the centre of mass has come down[14]. Because of this we

12 J. A. Euler A 15, *Mémoire sur la manière la plus avantageuse de suppléer à l'action du vent sur les grands vaisseaux*: Mém. Berlin XX 1764 (1766), pp. 240–294 – *Op. Omnia* II/20, pp. 229–275.

13 L. Euler E 82, *De la force de percussion et de sa veritable mesure*: Mém. Berlin I 1745 (1746), pp. 21–53 – Euler, *Op. Omnia* II/8, pp. 27–53. For a historical discussion see e.g. G. Maltese, *La storia di « F = ma». La seconda legge del moto nel XVIII secolo*, Firenze: Olschki, 1992, pp. 93–95.

14 D. Bernoulli St. 43, *Sur la conservation des forces vives*: Mém. Berlin IV, 1748 (1750), pp. 356–354 – *Werke* 3, pp. 197–206. This theorem is extended to n masses exerting Newtonian gravitational pulls on each other. See the Introduction in the *Werke* 3, pp. 95–101 for an extensive discussion on the conservation of living forces.

can express the living force of a mass equivalently as a height. The height from which a falling mass will acquire the velocity v but also exactly the same height the mass would reach if it were thrown up with the given speed. But not only the gravitational pull produces motion: the muscular force of a man, gunpowder in a cannon, a spring under tension, compressed gases in a cylinder, they all will set masses in motion and produce living force. Bernoulli assumes furthermore that each of those sources of motion produces a characteristic amount of living force until exhausted; all of them can therefore be compared in this respect and the comparison is made by stating that a given source could raise so many pounds by so many feet.

1.2.2 Work and Power

As just said, a weight (i.e. a force mg due to gravitation) acting (in a free fall) over a distance h produces a living force:

$$mv^2 = 2mgh \; , \tag{1}$$

i.e. the living force is equal to the weight times twice the height needed to acquire the velocity v in a free fall, a rule often referred to as Galilei's law. This formulation shows that the notion of conservation of energy (i.e. kinetic $\frac{1}{2}mv^2$ plus potential mgh) was not yet clear; the central notion was the living force with the laws by which it increased or decreased. One can consider the gravitational force, mg, either as having raised the weight mg over a distance $2h$ or as having given the mass m a quantity of motion measured by its *vis viva* mv^2; both terms are equivalent and Bernoulli will usually express a living force by its equivalent height, according to formula (1).

At this point Bernoulli introduces the concept of «work». A mass has either acquired a *vis viva* or has been raised to a height $2h$ due to work being done on it against the gravitational force. Similarly, work being done by any «push»[15] (*pression*) is measured by the intensity of that force (mg for the weight) and twice the distance over which the force displaced the object being moved (supposedly the force and the direction of motion are parallel). Remark: that gives double the value compared to the modern definition.

If the displacement of the object happens at constant speed – as is usual with ships – we can equal, disregarding the factor 2,

$$\text{work done} = \text{force} \times \text{velocity} \times \text{time}.$$

15 In D. Bernoulli St. 31, *Hydrodynamica*, Strasbourg 1738 – *Werke* 5, pp. 91–94, this «push» is called *potentia movens*, i.e. moving force.

Because of the equivalence just discussed work done can always be expressed in units of weight lifted (pounds) times units of length (feet). Central in the memoir is the notion of power which was introduced in the *Hydrodynamica*[16]. It was not called by that name, but presented as the amount of work (i.e. generation of living force) a source can produce in a given time. An able-bodied man e.g. can raise twenty pounds over three feet every second. The measure of power is therefore force (measured in pounds) times distance over time or, equivalently, force times velocity[17].

Every source of power is seen as a reservoir of work, i.e. ultimately a reservoir of living force. This reservoir has a fixed capacity; within reasonable bounds it can be tapped at various rates, produce large or small forces, but in the end it will have produced its innate, unvarying amount of work.

This work capacity can be determined experimentally; e.g. a galley-slave can produce his 60 pound feet/second during eight hours a day, but he could also have short bouts of 120 pound feet/second, with rests in between; the total amount of work per day will invariably stay the same: the equivalent of raising 1 728 000 pounds by one foot. Another example is gunpowder: a cubic foot of it is able to raise – according to Bernoulli – 200 million pounds by one foot.

1.2.3 Dissipation

We have to apply those notions to movements in a resisting medium: ships moving through the water. The living force generated by any source (wind power or muscular power or even steam) will not accumulate in the ship. This will reach a terminal velocity v and all the living force communicated to the ship will thereafter be transformed into living force of the water masses, set in motion by the sailing ship. To a given sailing speed will correspond a given consumption of living force per second, i.e. a given motive power or, as Bernoulli states it, a number of pounds (= force) times velocity v; in a modern formula (W = power, $\vec{F}$ = force exerted by the cause of propulsion):

$$W = (\vec{F}, \vec{v}) ,$$

the modern notation with the scalar product denoting the fact that it is only the projection of the force on the direction of motion that produces work.

16 D. Bernoulli St. 31, *Hydrodynamica*, Strasbourg 1738 – *Werke* 5, Section 9, pp. 274–312.

17 In the *Hydrodynamica* Bernoulli used the term *potentia absoluta* for power. Here he prefers to go back to the definition each time he has to mention the concept.

1.3 *The resistance of a ship*

1.3.1 The law of the resistance in fluids

The motive force required to move a plane surface perpendicularly through the water is proportional to the area, say A, and to the velocity squared. It is equal to the force exerted on a stationary area A by a steady flow of velocity v. This law was known in Bernoulli's time and can e.g. be found in the *Principia*[18]. The question is how to derive the numerical constant. This depends on fairly detailed assumptions how the oncoming fluid loses momentum when meeting the surface. This was discussed in *Dissertatio de actione fluidorum*[19] and the result of the argument, confirmed by experiment, was that:

$$F = \frac{1}{2} A \rho v^2 \qquad (2)$$

(ρ = the density of the liquid)[20].

1.3.2 The equivalent surface

A ship resists being towed through the water. This resistance can be measured, e.g. by evaluating the tension in the towing cable; this can be done as a function of the speed and it turns out that also this resistance varies as the square of the speed. Therefore the law (2) can be applied where A stands for an area equivalent to the hull; as one would expect this area is not too different from the area under the water-line of the largest section of the ship (perpendicular to its keel). Bernoulli insists this «equivalent area» is a quantity that has to be determined experimentally; careful computations of the area of the section are not sufficient, because the form of the hull influences the resistance[21].

18 Newton, *Principia* (1726), p. 239.

19 D. Bernoulli St. 14a, *Dissertatio de actione fluidorum*: CP II, 1727 (1729), pp. 304–342 – *Werke* 4.

20 In modern texts one finds $F = C_D \frac{1}{2} A \rho v^2$, where C_D is called the drag coefficient; it depends a.o. on the shape of the object. It is in fact not constant but a function of the Reynolds number. See Part 2, p. 47 h. v.

21 See §2.2, p. 48 h. v. for further comments.

1.4 *Questions of power*

1.4.1 The power required

Suppose we would know the equivalent area A and we find that a ship, in a calm sea, develops a steady speed of v feet/second. As the speed is constant, the applied force must be exactly equal, and opposite to, the resistance F; multiplying this force with the velocity we obtain the power needed to maintain the steady course. Hence we can state that a ship of equivalent area A, proceeding at a speed of v needs a motive power of

$$W = \frac{1}{2} A \rho v^3 . \tag{3}$$

Sometimes this formula is used to find the magnitude of A, when the power can be estimated (e.g. the number of oarsmen) and the speed of the ship is known.

1.4.2 Power used and power wasted

The formula (3) states the strict minimal power required from the propelling system. The laws of nature will not allow to attain the velocity v on that particular ship with anything less, whatever the means of propulsion used.

This does not mean that if a machine would deliver this power it could impart this speed to the ship. If a small boat were moved forward by pushing with poles on the bottom of a canal, the power of the boatmen would indeed be used entirely in maintaining the boat's speed, because they are pushing against an unyielding, fixed and infinite mass. If the same boat is rowed along, the oars are pushing against the water and some of this water is thereby set in motion. One communicates living force to the waters not only by means of the boat moving through the waters but also by means of the oars. So the power is now divided between a «useful» part, imposed by the movement of the boat, and a «wasteful» part, absorbed by additional, so to say parasitical, movements of the water.

In the sequel a number of devices will be considered: rows of rowers (as in a trireme), gangs of men moving vertically suspended oars through pulleys and ropes up to and including a steam engine driving a reaction motor. Any such device trying to «supplement the action of the wind» will be judged by its ability to keep this wasteful expense of power to a minimum. Bernoulli goes very far in this: he neglects almost entirely other considerations of convenience or of economical advantage.

This is, in short, the *Nouvelle théorie de l'économie des forces et des effets* that is referred to in the title[22].

1.5 *Devices*

1.5.1 Oarsmanship

Bernoulli takes a critical look at the form and dimensions of the oars used in the man-powered warships, triremes and the like. He computes the power «wasted» by the water sent whirling by the oars and makes suggestions to increase the area of the blades and decrease the weight of the oars by taking into account elements of the strength of materials. After each stroke the oars have to be pushed forward by the oarsmen (the feather); as these are fairly heavy objects they gain an amount of kinetic energy which is lost, because the movement has to be stopped before the oars dip again into the water. Bernoulli neglects the effort required by this stopping and concentrates on the power required for setting the oars in motion. He considers they move in a horizontal plane and applies the mechanics of pivoting solid objects, such as they are used for pendulums. He makes some educated guesses as to weight and size of an oar and computes that in a galley the power is expended as follows:

1. the hull setting the water in motion: 45%,
2. the oar blades setting the water in motion (stroke): 30%,
3. kinetic energy of oars when pushed back (feather): 17%,
4. sundry wasted power, e.g. friction at the tholes: 8%,

The ratio of 1. to 2. is trustworthy, because it rests only on (3) and the geometry of the oars (interpreted rather loosely by our author). 3. is good geometry combined with inspired guesses. The absolute values of 1. and 2. rest on the hypothesis that a man (any man) delivers 60 pound feet/sec.

22 Shortly after the article became available in print, Bernoulli wrote to Johann Albrecht Euler:

> *«Je vous avoue que j'ai toujours eu quelque predilection pour ce mémoire; il y a des idées toutes neuves sur l'œconomie des forces animales, sur l'action la plus avantageuse des rames...»*

letter of D. Bernoulli to J.A. Euler of February 16th, 1771.

1.5.2 Paddle wheels

If a ship of the line should be propelled with the efficiency obtained in galleys, the surface of the blades of the oars should be increased. If in particular paddle wheels («revolving oars») of the type proposed by earlier authors[23] should be used, it would require twenty of them. This is extremely cumbersome and it is moreover far from clear by which mechanisms human muscular power could be applied to such revolving machinery. On these grounds paddle wheels are dismissed. No detailed study of their mechanics is made.

1.5.3 Vertical oars

Bernoulli envisages long poles suspended outside the vessel from horizontal beams and with large blades at the ends. By a system of ropes and pulleys men would move those in a vertical plane, alongside the vessel, from fore to aft, pushing the blades against the water. During the return movement the poles are raised by tilting the beam (see his *Fig. 3*). The system could be fitted to large warships, normally rigged for sail. Their efficiency is comparable to that of conventional oars. Bernoulli merely mentions them but he clearly thinks his next invention is superior.

1.5.4 Side panels

Large panels suspended from beams near the upper deck and moved towards and away from the sides of the ship by ten men each would replace advantageously the smaller oars. The panels would make an angle of 70° degrees with the side of the ship; this angle is maintained during the pulling stage and is changed – by a clever device – to $-70°$ during the pushing stage. Detailed calculations are made of the power required, the dimensions, the speed to be obtained when applying the system to a battle-ship. This system is well fitted to such a ship, is more efficient than any other device and makes an optimal use of the muscle-power of the men; that at least is what Bernoulli, the inventor, claims.

23 E.g. J.M. de Chazelles, *Remarques sur la différente maniere de voguer des rames ordinaires & des rames tournantes, nouvellement proposés par le Sieur du Guet*: Mém. Paris 1702 (1743), pp. 98–100; Euler in his competing memoir (cf. *supra* p. 36, note 6) devotes many pages to them but limits his examples to rather small ships with $A = 25$ square feet which should attain a speed of $2\frac{1}{2}$ feet/second.

1.5.5 Jet power

In the *Hydrodynamica*, Section XIII, Bernoulli had investigated at length a system whereby water would be poured in a reservoir and a spout, at the bottom of the reservoir and facing aft, would expel this water under pressure. The force of reaction would propel the ship forward. He comes briefly back to it, but now his enthusiasm is less and he claims the efficiency cannot exceed 25% and is therefore not competitive with systems where the men apply their muscles to oars rather than to pumps to push the water up the reservoir; he thinks specifically of his own system of side panels for which he computed an efficiency of 55%. The argument is flawed, as explained in Part 4, p. 65 h. v.

1.5.6 Gunpowder engines

Also this natural source of energy was investigated in the *Hydrodynamica* (in a special subsection after Section 10) and the energy content (as we would now say) of an ounce of gunpowder was stated. Bernoulli takes this over here. He is thinking of a kind of cannon that would fire masses (water?) aft and the reaction on the cannon would propel the ship. A cubic foot of gunpowder consumed in an hour would – according to his estimates – give to a ship of the line a speed of 6.8 feet/sec.

1.5.7 Steam

Bernoulli is aware of the existence, in London, of a steam engine which «distributes water to the entire town» as he writes in the *Hydrodynamica*, Section X, §43. In this essay he goes further and brings an estimate of its power which would mean (if it were used in a naval vessel and no work would be wasted) a speed for the ship of the line of 3.7 feet/sec.

1.6 *Bernoulli's conclusion*

In Part II, §34 p. 247 h. v. Bernoulli reflects on the big principles which underlie his essay viz. the principle of most efficient use of power and the conservation of energy (as we now would say).

1.6.1 Efficiency

Bernoulli stresses that his proposed method to look for the essential part of the work required by a process is as general and as important in mechanics as the problem of finding the curve that includes the maximum area in a given perimeter is in geometry. In addition the argument is completely general and does not depend on the kind of forces put into effect, which brings him to the second point.

1.6.2 Announcing the energy concept

Bernoulli, as said in the introduction, thinks in terms of kinetic energy («living force»). His problem is one of pure motion, and dissipation, but in this restricted frame he states properties which we now recognize as belonging to the wider concept of energy:

- Energy can be stored, transmitted, consumed and its amount can be measured independently of the form the source takes.
- Its general properties (and especially its efficient use) can be stated independently of the form the source takes.

He gives an interesting list of sources of energy (i.e. of *sources de force vive*):

- human muscles,
- fire,
- compressed air,
- hot air,
- steam,
- gun powder,
- wind,
- wound-up springs.

To stress the point he takes up, in his penultimate paragraph, the cases of gun powder and steam.

1.6.3 Final Words

Bernoulli writes literally:

> «... nevertheless I am convinced that those two examples [gunpowder and steam] computed according to infallible principles will be enough

> to make us abandon all hopes to substitute ...the motive powers enclosed in the natural things for human labour ...»[24]

There are two reasons for his pessimism. The first is that he realizes that all such heat engines work with an appalling loss of efficiency because of lost steam and lost heat (the discovery of the intrinsic inefficiency of thermal processes as shown in the Carnot-cycle was yet to come). The second reason is that he does not envisage any other way to propel a ship by using «the elasticity of the air» [air stands for gases from gunpowder or steam or hot air] than a kind of reaction motor: the motive power is used to expel mass (i.e. water, universally available) to the rear and it is the reaction force that pushes the ship forward. This by itself is wasteful; in the *Hydrodynamica* he computed (albeit wrongly) the low efficiency of the process and this argument is repeated here. His final conclusion is then that the most effective use of human power is automatically the most effective use in all cases. And that – he does not state it explicitly, but he has led his readers to conclude for themselves – is precisely what the author of the memoir has achieved.

1.7 *General conclusion*

The basis for all the scientific arguments used in the memoir can be found in earlier publications of Daniel Bernoulli, mainly in the *Hydrodynamica*. The central theory is build around three concepts:

- The conservation of energy, including chemical and muscular energy, in an early and imperfect form, with the emphasis on the creation and consumption of kinetic energy.
- The concept of power, i.e. work done per unit of time or force putting an object in motion at a given speed. (Only constant – terminal – speeds in a resisting medium are considered in the memoir).
- The distinction between useful energy, unavoidably required by the laws of nature in performing a task, and the wasteful energy which one can hope to restrict by clever use of devices.

The waste consists in movements of the water masses which are not those induced by the displacement of the hull; in case of the jets in a reaction motor these are ordered movements but in case of oars (or of modern propellers) these are disordered movements. They are the precursors of the wasted energies in mechanical devices of Lazare Carnot and, indirectly, of the heat

24 Cf. p. 250 h. v.

losses in thermodynamic processes. It is therefore an important seminal idea. It is precisely this insistence on reducing the waste energy that made Bernoulli bet on the wrong technology; with conventional oars one could obtain an efficiency of some 45% (and even 55% with his own system, at least on paper); with the motors he knew, he expected a few percent. But of course this is not the only criterion and soon coal fired steam engines, even working with a dismal efficiency, would prove more docile, more readily available and more economical than the muscular power provided by a crew of a few hundred men.

2 **Part 2: sources of power and efficient use of power**

2.1 *Human power (I, § 4 p. 187 h. v.)*

Bernoulli, erstwhile professor of anatomy, is aware of the contemporary theories – inspired by Descartes – on the origin of muscular power[25]. Apparently the muscles are considered as simple machines that transform into work the animal spirits which are produced at a fixed rate in the body and fed to the muscles through the nerves. Bernoulli has an utterly mechanistic view of the animal spirits: they are like little tensioned springs (i.e. as pure reservoirs of energy). The present phrasing denotes a concern with energy conservation in the relation between the animal spirits and the contraction which is absent in the earlier work.

From earlier experiments, mentioned in the *Hydrodynamica*, Section IX, Bernoulli learned that on average a man can produce power at the rate of 20 pounds feet per second. The idea of the energy reservoir is prevailing here, because he admits that in a working day of 8 hours this is exactly three times as much: 60 pounds feet/second (i.e. 93.6 Watts); a sizeable but not unreasonable performance. His belief in the conservation law and the mechanistic view of muscles tempts him to postulate an extension: that every man, be he sinewy and short or stout and tall produces about the same amount of work in a day; for the subject of the essay this bold hypothesis is not really essential.

Our physicist-anatomist mentions that, besides the work done by arms and legs, the body should produce a comparable amount of power, just for living. The muscular efforts of the heart are an important part of this, which

25 Cf. D. Bernoulli St. 10, *De motu musculorum*: CP I, 1726 (1728), pp. 297–313 – *Werke* 1, pp. 92–103 and the Introduction to the Medical Writings by V. Zimmermann, esp. p. 30.

was measured and equals one twelfth of the normal rate of 20 pound feet per second.

2.2 *The resistance of a ship (I, § 10 p. 190 h. v.)*

Bernoulli follows Newton in supposing that, at the velocities considered, the drag forces due to viscosity (they call it «tenacity») are negligible compared to those caused by the inertia of the waters being displaced. He is aware of the careful calculations of the area of the bow by Bouguer[26]; but is quick to point out that in practice the effective area, to be used in (2), may be different due to effects of shape. The only safe method is to do an experimental determination of A. This was a most wise and farsighted conclusion, because a century later W. Froude and R. E. Froude discovered[27] that a ship's resistance is due to both viscosity effects and the generation of waves; the first are proportional to the surface area of the hull, the second to the section; eddies formed at the stern are a third effect.

Actually, the main effect is given by a formula similar to (2)

$$F = C'_D S \rho v^2 \,, \tag{4}$$

where S is the «wetted surface» of the ship. C'_D is the coefficient for viscous drag, a function of the Reynolds number and therefore of the product vL, where L is a characteristic length of the object (i.e. the ship)[28]. Inasmuch as C'_D is slowly varying or one compares situations with similar Reynolds numbers, most of the conclusions Bernoulli draws from (2) are still acceptable, as they rely on proportionality of F with v^2.

For Bernoulli and his contemporaries the unit of force is the pound (i.e. the weight of a one pound mass). Comparing the modern forms of some formulae and those in the memoir one has to put $g = 1$, especially in equation (1).

26 P. Bouguer, *Traité du navire, de sa construction et de ses mouvements*, Livre III, Paris: Jombert, 1746, Sect. I, Ch. III–VIII).

27 a. o. W. Froude, *Report to the lord commissioners of the admiralty on experiments for the determination of the resistance of a full-sized ship, at various speeds by trials with H. M. S. Greyhound*, London: Taylor and Francis, 1874.

28 Cf. e.g. V. L. Streeter (ed.), *Handbook of fluid dynamics*, New York: Mc Graw-Hill, 1961.

2.3 *The law of the resistance against flow (I, § 11 p. 191 h. v.)*

Per second a mass of $A\rho v$ will impinge on the area A, with a velocity v, hence the area meets a momentum of $p = A\rho v^2$ per second. If the fluid would consist of tiny elastic balls with sufficient space between them they would rebound (without hitting oncoming balls) and the rate of change of the momentum and hence the force would be $2p$. This is the first argument Bernoulli tried out in *Dissertatio de actione fluidorum*[29] and it is also the argument Newton used. Its experimentally demonstrable failure leads him to assume the liquid leaves parallel to the surface, the force along the direction of flow is then p, but this is still not satisfactory. To be correct one has to take into account the pressure and the (turbulent) motion behind the surface. (The argument used in *Dissertatio de actione fluidorum* is very questionable). If the liquid is viscous many more water particles will be deviated from their path, leading to viscous drag, but Bernoulli – and all his contemporaries[30] – believe that in water this effect is negligible[31]. A modern calculation, using the accepted values of the drag coefficient and their scaling laws, by A. Lederer for the frigate used as an example by Bernoulli gave a viscous drag force of 138 kg; the residual drag (mostly inertial effects as in (2)) was found to be 37 kg.

The formula (2) was related to a hydrostatic phenomenon[32]: if one thinks of the flow as issuing through a hole with section A in the bottom of a cylinder filled with fluid up to a height h then, according to the law of Torricelli, the velocity v of flow is equal to the velocity of a mass falling from h,

$$v = \sqrt{2hg}\ . \tag{5}$$

Hence

$$\frac{1}{2} A\rho v^2 = \rho A h g\ , \tag{6}$$

29 Cf. D. Bernoulli St. 14a, *Dissertatio de actione fluidorum*: CP II, 1727 (1729), pp. 304–342 – *Werke* 4.

30 Euler has given due attention to it, but much later: L. Euler E 520, *Essai d'une théorie de la résistance qu'éprouve la proue d'un vaisseau dans son mouvement*: Mém. Paris, 1778 (1781), pp. 597–602 – Euler, *Op. Omnia* II/21, pp. 223–229.

31 *«A l'époque on ignorait la notion de résistance de frottement qui, pour les navires de commerce, représente 60% à 75% de la résistance totale et ne peut se calculer comme la résistance de rencontre, la seule envisagée par Bernoulli.»* (A. L.)

32 Newton introduced it (*Principia* (1726), Book II, Section VII, Corollary II, p. 319) but taking the height $2h$ and an «effective» hole that is smaller than the actual hole (Case 1 of the same Section).

i.e. the force of flow on the area A is equal to the weight of a vertical cylinder of fluid erected on the base A and of height h; which is the force required to plug the hole. Bernoulli uses (6) to define an «equivalent height h for the velocity v» and computes hence, by a hybrid method, the power needed to move a ship at this velocity; he proceeds as follows[33]:

- the force is 70 Ah pounds because the density of water is 70 pounds per cubic foot (of fresh water, pounds and feet of Paris);
- the velocity is $\sqrt{2gh} = 2\sqrt{\frac{1}{2}gh}$ and $\frac{1}{2}g = 15$ feet/sec, 15 feet being the distance a mass in free fall from rest covers in one second.

The power is the product of the two: 140 $Ah\sqrt{15h}$. (The notation of Bernoulli is here $c = v$, $n = A$ and $a = h$). Reversing the formula (5) he obtains $h = v^2/2g$ and therefore, in the units used:

$$h = \frac{v^2}{60} \tag{7}$$

and substituting in (2)

$$W = \frac{7}{6}Av^3 \tag{8}$$

is the power expressed in the units pounds feet/sec. Similarly, the force – in pounds – needed to keep the ship sailing at velocity v is

$$F = \frac{7}{6}Av^2 \, . \tag{9}$$

If a man can apply 60 pounds feet/sec (average for a period of 8 hours), it is easy to compute how many rowers are required to propel a ship at velocity v and inversely what velocity N men will impart to the ship:

$$v = \sqrt[3]{\frac{360N}{7A}} \, . \tag{10}$$

This is an upper bound, because energy will certainly be wasted somewhere.

33 We keep the same notation; Bernoulli has changed his (see further).

2.4 *The geometry of oars*

2.4.1 Standard oars (I, §16 p. 199 h. v.)

An oar extends 12 feet inside the ship, where it is handled by five men; it extends outside 24 feet and the last 5 feet make out the blade which is dipped in the water. The problem is that the tip of the blade moves faster through the water than the other parts nearer the thole; as the force exerted on the water is proportional to the square of the velocity this makes a sizeable variation of force density over the length of the blade. Let us call the length variable x, running from 0 (tip of the blade) to d (upper end of the blade). Given the velocity of the ship and the velocity by which the oars are moved relative to the ship, we can compute the force $F(x)\,dx$ that an element dx at x exerts on the water.

One would customarily define a «centre of force» at a distance x_0 from the tip by

$$x_0 = \frac{\int_0^d xF(x)\,dx}{\int_0^d F(x)\,dx} .$$

This is not what Bernoulli does. Let us call the mean force

$$f = \frac{1}{d}\int_0^d F(x)\,dx ;$$

according to Bernoulli the *centre des efforts* is at the point x_e such that

$$F(x_e) = f ,$$

i.e. the point where the local force is equal to the average.

In practice the difference is negligible: at a ship's speed of 7.2 feet/sec the effective length (from the thole) of the oar is 21.609 feet (Bernoulli computation) against 21.929 feet (customary definition). In the *Hydrodynamica* he took simply the middle of the blade as the centre of the efforts which would give here 21.5 feet.

2.4.2 Other kinds of oars (II, §16 p. 226 h. v.)

For any kind of oars, the essence of the argument remains the same: a ship requires a force which is uniquely given by its section and the prescribed speed. The oars have to transmit this force and, by formula (2), there is a choice: one can make A (the blade) larger and the speed v (of the blade) smaller or vice-versa. For a constant force, making v larger increases the power needed (formula (3)). Bernoulli discusses from this point of view the oars used in galleys (the blades should be larger) and older proposals to use water-wheels.

But the ratio of the useful energy to the wasted energy is proportional to the square root of the surface of the blades to the surface of the ship, see (13). This means that if one wants to propel a ship of the line ($A = 150$ feet squared) – by oars of any type – with the efficiency obtained in galleys ($A = 16$ feet squared), the surface of the blades has to increase in the same proportion. If «rotating oars» or waterwheels are to be used, as discussed already in 1702 by Chazelles, Bernoulli calculates that a ship of the line would need 20 of those. If driven by 500 men she could reach a speed of 4.25 feet per second. He is clearly not in favour; this dismissal does not rest on any detailed consideration of the water set in motion by rotating devices but only on their cumbersome nature and the difficulty to adapt the muscular efforts of men to a rotation movement.

2.5 *The experiments on the Lake of Geneva*

2.5.1 Measuring the speed of a ship (I, §20 p. 199 h. v.)

The traditional way of the seamen is to use the log. But Bernoulli advocates another instrument, which allows a continuous measurement. One measures the speed of the rowing boat, as illustrated in *Fig. 1* of Bernoulli's text, by determining the angle α between the thread which tows the sphere and the vertical. This sphere is subject to a vertical force, the weight, and a horizontal one, the drag due to the velocity v. The towing force is in equilibrium with those, the tangent of the angle is the ratio of the drag to the weight.

In a formula:

$$\alpha = \arctan\Big(\frac{3}{8}\frac{\rho_0}{\rho_1 - \rho_0}\frac{1}{rg}v^2\Big) \qquad (11)$$

where ρ_0 is the density of the liquid, ρ_1 of the sphere, r = radius of the sphere, g = acceleration of gravity, v = velocity. For $r = 1/2$ foot, ρ_1 equal three times the density of the water and $v = 3.5$ feet/sec the angle α is 8.7 degrees. A change of half a foot/sec changes the angle by 2.5 degrees.

Poleni[34] described it extensively in his prize essay to the *Académie* in 1733, but it was known to Bernoulli before.

34 Cf. G. Poleni, *De la meilleure manière de mesurer sur mer le chemin d'un vaisseau, indépendamment des observations astronomiques*: Prix Paris III (pièce VIII) 1733 (1752).

2.5.2 The results of the experiment (I, §21 p. 201 h. v.)

Bernoulli had his theory tested on a small scale, with two rowing boats; the first – manned by two men (and subsequently four) – towing the second; two observers in the second boat would measure the tension in the cable and the speed; see *Fig. 1* of Bernoulli. He reports four runs of the experiment and he analyzes those in two different ways: in the first he takes run nr. 1 as exact, in the second he takes the mean of all four runs.

In §21, experiments 1 to 4, p. 201 h. v. he applies the formula

$$n = \frac{6P}{7c^2} \tag{12}$$

equivalent to (9), where

- n is the resistant area of the boat in square feet (our A),
- P is the weight in the basin in pounds (our F),
- c is the velocity in feet/sec (our v),

with four different velocities. He estimates that the first run has been done with particular care, giving by hypothesis the correct area: $n = 0.394$ square feet. He then *corrects* the experimental velocities, using the formula. He obviously takes the weights to be exact and the measurements of the velocities open to errors. We summarize in the following table; the last column gives the values of n obtained from the raw data (remark – with Bernoulli – that run 4 coincides nearly with run 1, for him an additional justification of the corrections).

N°	velocity c	weight P (ounces)	corrected velocity	$n_{\text{exper.}}$ from (12)
1	3.50	90	definition	0.394
2	3.80	100	3.69	0.371
3	4.00	122	4.073	0.408
4	4.30	138	4.302	0.400

In I, §22 Bernoulli comes back to those results, but this time he computes the average of the four values of n: $\langle n \rangle = 0.393$, giving implicitly the same weight to each. As this is the same as the result of run 1 he takes this as justification of the «little corrections» he made in the first analysis.

In a fifth experiment the towing boat was manned by four men, rowing in a normal fashion as in run 1; the speed was 4.50 feet/sec. Bernoulli assumes apparently that each of these men is as his standard man and confidently assumes that in the fifth experiment the power applied was double of the power in run 1. The velocities should then compare as 1 to $\sqrt[3]{2}$ and indeed they almost do. Bernoulli draws the conclusion that the mechanical power is indeed as the 3rd power of the velocity.

In a sixth experiment the two boats are interchanged; it turns out that one of them shows a somewhat greater resistance viz. $n = 0.481$ square feet. The combined resistance is therefore known, the velocity at a normal pace also (cf. run 1) and from (8) the power needed to propel the two boats is computed: 43.75 pounds feet/sec. Assuming once more the two standard men who produce a steady 60 pounds feet/sec each, dividing the power needed by the power produced one obtains an efficiency of 36.5%.

Bernoulli considers that this is not too different from the efficiency obtained by the galley slaves, viz. 44.6% and he marvels that under those very different circumstances the results are very similar. It comforts him in his hypothesis that the production of power at a steady pace by a man is a fairly unvarying quantity.

2.5.3 The Friend of Bernoulli (I, §20 p. 199 h. v.)

The «friend who carried out the experiments»[35] is the mathematician Gabriel Cramer from Geneva. In a letter[36] Bernoulli asked him to perform experiments in the river Rhône in order to measure the force experienced by a boat in a given current and the power required to keep the boat still by rowing up-stream. He suggested in fact a series of experiments with from one to six rowers in as many tracts of the river with increasingly stronger currents. He advocated the use of the hydrometric pendulum to determine the speed of the flowing river; this seemed to him the best method «which was then taken up by Poleni[37] who wrote an essay on it which won a prize».

Cramer replied in a long letter[38]. He performed some fifteen experiments, in part in the river Rhône, as instructed by Bernoulli, but also on the lake of

35 Cf. p. 200 h. v.

36 Letter of D. Bernoulli to G. Cramer of May 14th, 1738.

37 G. Poleni, *De la meilleure manière de mesurer sur mer le chemin d'un vaisseau, indépendamment des observations astronomiques*: Prix Paris III (pièce VIII) 1733 (1752). There is a similar remark in the letter that Bernoulli wrote to Euler on September 22nd, 1733, while on his European tour after he left St. Petersburg. But there Bernoulli tells also that he himself used the method to measure the speed of the ship sailing from Kronstadt to Danzig and that it was quite accurate.

38 Letter of G. Cramer to D. Bernoulli of September 23rd, 1738.

Geneva (i.e. in still waters). The system of the two boats and the measurement of the force on the towing cable seem to be his invention. More than a hundred years later, William Froude would use exactly the same setup, this time with steamships, in order to measure the resistance of ships[39]. It was those experiments on the lake which Bernoulli chose to include in his essay. He simplified them somewhat: Cramer gave the weight as «118 to 127» ounces in the third experiment, «136 to 140» ounces in the fourth and Bernoulli cited those resp. as 122 and 138 ounces. To obtain the speed Cramer relied exclusively on the hydrometric pendulum; he used a formula for the drag of a sphere in a liquid which Bernoulli had derived in *Dissertatio de actione fluidorum*[40] and which is equivalent to equation (11).

In a letter[41] of the end of 1738 Bernoulli gives Cramer all his remarks on those experiments: 17 points in all. Bernoulli reveals himself an astute proposer and critic of experiments; he remarks a. o. that the hydrometric pendulum should have been checked against a current of known speed.

At the same time Bernoulli sends a copy of a letter of Cramer and of his remarks to Euler[42]. Euler replies immediately on September 15th: «The transmitted experiments on the force of oars have pleased me beyond measure. I hope to derive good use from them for the second part of my *Scientia navalis*[43] on which I am still working»[44]. Euler confirms this in one of his first letters to Cramer[45].

2.6 *Traditional propulsion and its powerneeds*

The important question is: how much power is wasted through the water that the oars set in motion and how does this relate to the power needed to keep the hull moving?

39 Cf. W. Froude, *Report to the lord commissioners of the admiralty on experiments ...*, London: Taylor and Francis, 1874.

40 Cf. D. Bernoulli St. 14a, *Dissertatio de actione fluidorum*: CP II, 1727 (1729), pp. 304–342 (part II, § II).

41 Letter of D. Bernoulli to G. Cramer, of 1738 (undated).

42 Letter of D. Bernoulli to Euler of August 29th, 1739.

43 L. Euler E 110, *Scientia navalis seu tractatus de construendis ac dirigendis navibus*, St. Petersburg 1749 – Euler, *Op. Omnia* II/18 & 19.

44 Letter of Euler to D. Bernoulli of September 15th, 1739. «*Die übersandten Experimente circa vim remorem, haben mir über die massen wohlgefallen, und hoffe ich daraus in meiner Scientia navalis parte altera, woran ich noch arbeite grossen Nutzen zu schöpfen.*»

45 Letter of Euler to G. Cramer of August 3rd, 1743.

2.6.1 Oars (I, §27 p. 204 h. v.)

The calculation is in principle straightforward: compute the velocity of the blade relative to the water, verify the total surface of the blades (correcting for the fraction of the time the oars are in the water) and apply formula (3) to obtain the power absorbed by this motion. If we use a suffix «$_o$» to indicate variables related to the oars and «$_s$» for the ship, the formula for the force exerted by the oars is (see equation (9)) $F_o = \frac{7}{6} A_o v_o^2$. But by the principle of reaction this is also the force needed to keep the ship moving and hence

$$A_o v_o^2 = A_s v_s^2 .$$

This relates the velocity v_o of the oars to the velocity of the ship and hence the ratio of the power required by the oars to the power absorbed by the ship is

$$\frac{W_o}{W_s} = \sqrt{\frac{A_s}{A_o}} . \tag{13}$$

2.6.2 Floats (I, §38 p. 210 h. v.)

In I, §38 to I, §40 (pp. 210–213 h. v.) a purely hypothetical state of affairs is considered: instead of pushing against flowing water (i.e. rowing) the mariners are supposed to push against massive floating objects which we call henceforth floats. The action against these masses will impart to them an accelerated motion, whilst the (equal) reaction will propel the ship; those masses slide freely on the water, without any braking action. Only their inertia is offering a resistance. Bernoulli makes in parallel a discussion of this imaginary system and the conventional mode of rowing. It is the mass of the floats which determines the efficiency: from each push they acquire a given velocity (i.e. a kinetic energy) which is counted on the waste side. The more massive the floats the more efficient is the system; infinitely massive floats would give a 100% efficiency.

2.6.3 Agreement (I, §40, III p. 211 h. v.)

Comparing the last discussion, of the system of floats, with the results for the conventional oars, Bernoulli finds a remarkable agreement between both. This conclusion rests on the equality (within the limits of accuracy) of the two results for the loss of power: 29.8% for the poles pushing against floats and 29.7% for the oars pushing against water. This is less impressive than presented. In the case of rowing Bernoulli assumed as given: the number of oars, their dimensions, the rhythm of the strokes and the speed of the galley. With the help of the theory just formulated he deduces: the effective section of the ship, the power dissipated by the hull, the power dissipated by the oars (and the ratio of the last two numbers), the total power expended (given by the number of men) and finally the fraction of that power wasted by the oars: 0.297. For the system with floats he takes for granted the effective section of the ship – as computed above – and that of the oars and he fixes the mass of the float such that the dissipation of energy by the floats is equal to the loss by the oars in case of rowing. By this last hypothesis he has linked both systems. The rather involved calculations mask the *idem per idem* character of the argument.

2.7 *Theoretical and practical bounds on the speed*

2.7.1 The upper bound (II, §3 p. 215 h. v.)

The limit for a ship's speed is given by formula (10). The section A for a vessel of the first line is taken from Bouguer: A is 150 feet squared. This gives at once the second column of the *Table fondamentale*, which lists $\sqrt[3]{\frac{12}{35}N}$. It is supposed that each man rows at a steady pace, producing power at 60 pound feet per second. In combat, during short spells, an intense effort will produce about double the power. To quantify this Bernoulli falls back on his experiments with the rowing boats: at a normal pace they reached 3.50 feet/second and under effort 4.33 feet/second. The ratio of their power should therefore be $\left(\frac{3.50}{4.33}\right)^3$; as Bernoulli likes to work with fractions rather than decimals, he takes this as

$$\left(\frac{3\frac{1}{2}}{4\frac{1}{3}}\right)^3 = \left(\frac{21}{26}\right)^3 = \frac{9261}{17576} \approx \frac{92}{175}\,.$$

To obtain his third column he multiplies the second by $\sqrt[3]{\frac{175}{92}}$ and obtains, by striking out common factors,

$$v = \sqrt[3]{\frac{105}{161}N}$$

which could in fact have been further simplified to $v = \sqrt[3]{\frac{15}{23}N}$.

2.7.2 Real velocities (II, §7 p. 218 h. v.)

Bernoulli sets out to compute a correcting factor which might be applied to the numbers in the previous *Table fondamentale* to give the actual velocity of a ship of the line propelled by N oarsmen. He takes the basic quantities from the analysis of the galley: 260 men acted as if they pushed continuously an area of 36 square feet against the waters. Consequently N men on the ship will push $\frac{36}{260}N$ square feet. Recalling formula (13) this gives for the power absorbed by the oars $W_o = W_s\sqrt{\frac{150.260}{36N}}$, which is almost $W_s\frac{33}{\sqrt{N}}$. Assuming that additional effects will increase the total power with $\frac{1}{3}$ Bernoulli finds that the total power needed is, as a multiple of W_s (i.e. the basic power required by the ship):

$$W_s\left[\frac{4}{3} + \frac{44}{\sqrt{N}}\right]. \tag{14}$$

As the velocities are as the cubic root of the power one has to divide the numbers in both columns of the *Table fondamentale* by the factor $\sqrt[3]{\frac{4}{3} + \frac{44}{\sqrt{N}}}$ in order to obtain the *Table des vitesses réelles*. Numbers in parentheses are those of Bernoulli, when they differ by more than 0.020.

3 Part 3: the inventions of Bernoulli

3.1 *The Bernoulli propulsion system*

The pet idea of Bernoulli, to which he devotes 19 pages, (see Part 2, §22 p. 233 h. v.) is inspired by the movements of fishes and birds: they move their fins (or wings) sideways and not, as in all the proposals just reviewed, from fore to aft[46]. He discusses first the geometry and the forces connected with a vertical plane making an angle with the keel and being swiftly pulled towards the ship, in a movement perpendicular to the keel. He describes thereafter how to build the mechanism to put this idea to work and he finally devotes many pages in defence of his invention and works out practical examples.

46 One might add that the same principle is put to work when sculling in a small boat.

He makes an unhappy choice of variables: every surface is denoted by a letter «S» but as there are three of them this brings in Greek sigma's (in the end-of-word form) in addition to lower and upper case letters «S»; this is begging for misprints. We shall try to be more discerning and make the pure geometrical language transparent with some trigonometry; the arguments are, however, strictly Bernoulli's.

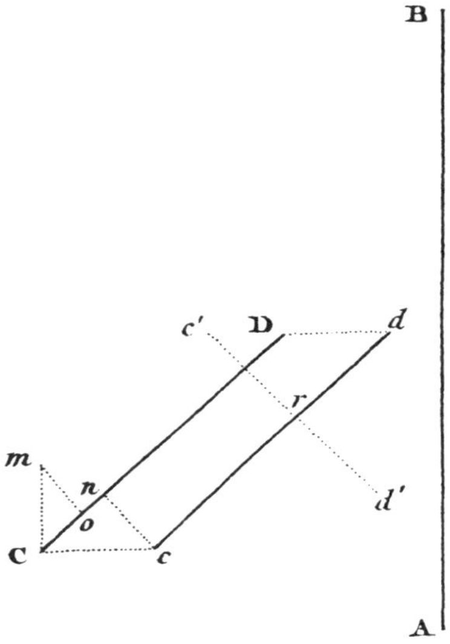

Figure 1: The figure of Bernoulli explaining his invention of a new kind of fin-like oars.

3.1.1 Geometry of oblique blades

We refer to *Fig. 4* of the memoir, which is reproduced in our Fig. 1. AB is the direction of the keel (and of movement), CD is a vertical board which is moved by the workmen along Cc. We prefer to use the following symbols: $\vec{v}_s$ is the velocity of the ship, $\vec{v}_0$ the velocity of the board drawn towards the ship, S_0 the area of the boards; cf. Table 1 and Fig. 2.

The board has a total velocity $\vec{v} = \vec{v}_s + \vec{v}_o$; we can write it as the sum of a velocity parallel to the board and a velocity $\vec{v}_\perp$ perpendicular to it. The force exerted by the moving board is

$$f_\perp = \frac{1}{2} S_o v_\perp^2 \ .$$

item	our notation	Bernoulli
main section ship	A	n
area of board	S_o	
total area of boards	S	S
length of stroke	r	r
angle DCc	α	
angle ncC	$\beta = 90° - \alpha$	
	$r \sin\alpha$	s
	$r \cos\alpha$	ς
speed of ship	v_s	c
speed of board vs. side of ship	v_o	v
resistance of ship	F	P
force on boards	F_o	π

Table 1: List of symbols related to the propulsion system.

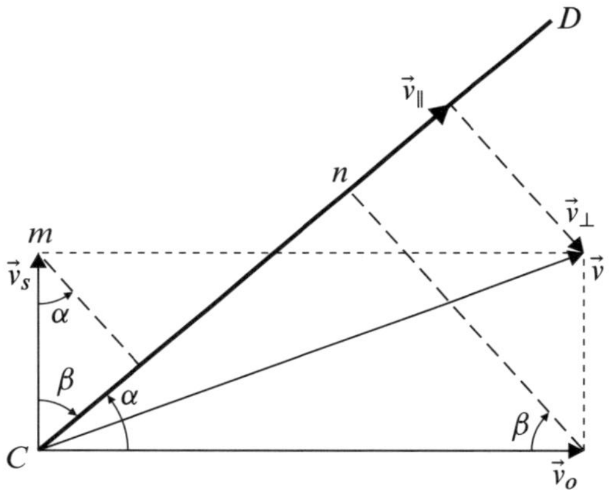

Figure 2: The geometry of oblique rowing boards.

By projecting $\vec{v} = \vec{v}_o + \vec{v}_s$ on the perpendicular to the board:

$$v_\perp = v_o \sin\alpha - v_s \cos\alpha \ .$$

We shall exert a force in the direction aft-fore on the ship if

$$v_o \sin\alpha > v_s \cos\alpha \ .$$

The force exerted by the board is:

$$f_\perp = \frac{1}{2} S_o (v_o \sin\alpha - v_s \cos\alpha)^2 \ .$$

Only the component in the direction aft-fore contributes to the ship's propulsion:

$$f = \frac{1}{2} S_o (v_o \sin\alpha - v_s \cos\alpha)^2 \cos\alpha \ .$$

The force produced by the men pulling the board towards the ship is

$$f_o = \frac{1}{2} S_o (v_o \sin\alpha - v_s \cos\alpha)^2 \sin\alpha \ .$$

We suppose a number of boards are simultaneously pulled on either side of the ship; let their total surface be S. The sum of their forces, say F, equals the resistance of the ship, hence:

$$\frac{1}{2} A v_s^2 = F = \frac{1}{2} S (v_o \sin\alpha - v_s \cos\alpha)^2 \cos\alpha \ .$$

Solving for v_o:

$$v_o = v_s \left(\frac{\cos\alpha}{\sin\alpha} + \frac{1}{\sin\alpha} \sqrt{\frac{A}{S \cos\alpha}} \right) . \qquad (15)$$

The minimum power required (the «essential power») is:

$$F v_s = \frac{1}{2} S \cos\alpha (v_o \sin\alpha - v_s \cos\alpha)^2 v_s \ ,$$

and the power expended (F_o = sum of all f_o):

$$F_o v_o = \frac{1}{2} S \sin\alpha (v_o \sin\alpha - v_s \cos\alpha)^2 v_o \ .$$

The ratio of the total power over the essential power is, using (15):

$$\frac{F_o v_o}{F_s v_s} = \frac{v_o}{v_s} \tan\alpha = 1 + \frac{1}{\cos\alpha} \sqrt{\frac{A}{S \cos\alpha}} \ .$$

The ratio of the essential power (*travail utile*) to the waste power (*travail inutile*) is therefore[47]

$$(\cos\alpha)^{\frac{3}{2}}\sqrt{\frac{S}{A}}\,, \tag{16}$$

or, in Bernoulli's notation:

$$\left(\frac{\mathsf{S}}{r}\right)^{\frac{3}{2}}\sqrt{\frac{S}{n}}\,.$$

In the discussion of this formula Bernoulli remarks that the best efficiency is obtained if

- primo: $S \gg A$: area of boards much larger than area of ship's section,
- secundo: $\cos\alpha \approx 1$ (i.e. $\alpha \approx 0$): boards almost perpendicular to the axis of the ship.

But of course $\alpha = 0$ requires $v_o = \infty$, so one will have to compromise.

The most rational solution consists of making S as large as possible: bigger boards in larger numbers; this allows to neglect the second term in (15). One takes then v_o as large as the pulling by men allows; from the prescribed speed v_s follows the angle by

$$\frac{v_o}{v_s} \approx \cot\alpha\ .$$

For smaller S equation (16) shows that one should take $\cos\alpha$ as large as possible and hopefully close to 1; the waste power is then to the essential power as $\sqrt{\frac{A}{S}}$ and this is precisely the formula (13) we found for the conventional oars.

3.1.2 Examples

The example taken is of the same large ship with a section $A = 150$ square feet; moving at 4 feet/sec it needs a force of 2800 pounds (cf. equation (9)).

47 This is expressed by Bernoulli as: *«en raison composée réciproque de la raison sesquipliquée des cosinus de l'angle nCc et de la raison sousdoublée de la surface S.»* (Part 2, §25(d) p. 236 h. v.)
There is a discrepancy between this text and his formula; *réciproque* should be deleted.

A number of the previous formulae are evaluated for angles $\alpha = 10$, 20 and 30 degrees. Supposing the total area of the boards to be 600 feet squared one obtains an efficiency of 0.645 for an angle $\beta = 90° - \alpha$ or 70 degrees[48]. Even reducing this arbitrarily to 0.550 in order to account for sundry losses, it is still substantially more than the 0.446 for the galleys at cruising speed. If 500 men were driving the boards one could expect a theoretical speed of 5.55 feet/sec, as appears from the *Table fondamentale*. At an efficiency of 0.550 this speed is reduced by $\sqrt[3]{0.550}$ giving a speed of 4.55 feet/sec.

In §25(f) (p. 237 h. v.) Bernoulli compares this to the performance of galleys, using formula (10): a galley at that speed would need 53 rowers. One should remark here that the comparison is not fair; also the losses incurred by the galley should be taken into account. From formula (14) this is a factor $\left(\frac{4}{3} + \frac{44}{\sqrt{N}}\right)^{-\frac{1}{3}}$ which would give 146 rowers; in view of the $\frac{16}{150}$ ratio of the resistances this is a much more favorable figure in defence of the Bernoulli system.

3.1.3 The mechanism

The whole discussion is only valid if, unlike with the conventional oars, there are no idle times. For this Bernoulli suggested a mechanism that, at the end of the pulling stroke inverses the angle of the board with the ship so that the pushing stroke is equally effective. This can be effected by mounting the board on a hollow cylinder which rotates around a shaft; a side-hole in the cylinder and a plug in the shaft moving in the hole will allow the board to rotate e.g. from 70° to −70° at the end of the pulling stroke and be stopped in this position ready for the pushing (cf. the *Fig. 5* of the memoir)[49].

The shafts are hinged to beams strutting out from the ship and are constrained to move in a plane perpendicular to the axis of the ship. Below the hinge, at one quarter of the length of the shaft a rod is attached to it which extends into the ship. At the end of the rod a transverse bar is fixed which

48 Here a transcription error has apparently occurred. The result is stated for 60 degrees but it corresponds to 70 degrees. On February 16th, 1771, in a letter to Johann Albrecht Euler, Bernoulli mentioned the printed version of his memoir was full of errors and in his proposal for experiments he does propose an angle of 70 degrees. In a second letter, of June 8th, 1771, he proposes – without explanation – 60 degrees.

49 *«Le système préconisé par Bernouilli eut été sujet à usure rapide à la suite des chocs répétés entre la rainure et la clavette, que ces pièces eussent été en bois ou en métal.»* (A. L.)

is gripped by 10 men who are standing upright and impart a see-sawing motion to it. They push and pull with a force of twenty pounds each and their strokes are three feet wide. The large warship taken as example would carry 25 of those new-style oars on each side.

3.2 *Experiments*

After the memoir came out in print, in 1769, Bernoulli suggested to Johann Albrecht Euler, then secretary of the Academy in St. Petersburg, to try out this new style of oars in a project to be funded by the Academy[50]. In his letter of May 7th, 1771, J. A. Euler proposed improvements on the mechanism and mentioned he has submitted plans to count Orlow, president of the Academy and recently promoted chamberlain, who showed great interest. On June 8th, 1771, Bernoulli wrote back with detailed proposals how to fit his invention to the royal yacht[51]. Bernoulli wrote also to another friend in Geneva, the astronomer J. A. Mallet, in 1771 to do small-scale experiments on the lake; two men in a small boat should perform a preliminary trial before launching the expensive Russian experiment. In the correspondence of Bernoulli with Mallet during 1771–1772 we read that the idea worked but that the makeshift equipment was too different from the original design to allow any conclusions as to the efficiency. The experiment was not pursued, mainly because the St. Petersburg project was abandoned; count Orlow had left St. Petersburg on a tour of Germany and Switzerland which he intended to last three years; already in his letter of June 8th Bernoulli had aired his misgivings about such a project without a mighty protector at court.

3.3 *Alternative schemes (II, § 32 p. 244 h. v.)*

In cases where speed is not the prime concern, but the only aim is to induce some motion in the ship one could adopt other tactics. A few mariners in a long-boat would tow a large board, floating horizontally on the sea, ahead of the ship; a cable is fixed to the board which is paid out from the ship. A cable's length away the board is tilted vertically, presenting a large area towards the ship. The crew would then pull with full strength on the cable, hauling the ship towards the board. The manœuvre would then start all over.

50 Letter of February 16th, 1771, which was read to the Academy by J. A. Euler on March 22nd (March 11st, Julian calendar).

51 This letter was read on July 1st (June 20th).

The principle can be refined by using folding boards or a kind of umbrella's (even a whole string of them), being towed ahead in a state of low resistance but offering a high resistance to the hauling[52]. The final numerical example is merely an order-of-magnitude estimate.

4 Part 4: jet propulsion and its efficiency

Bernoulli recalls that in his *Hydrodynamica* he proposed to use the reaction force on a water container, spouting water through a lateral opening, for ship propulsion. We shall review the physics of this system and then comment on this proposal. But between the writing of the *Hydrodynamica* and the present memoir there is a whole development as testified in the correspondence of Bernoulli.

4.1 *Experiments and correspondence*

In the same letter to Cramer in which he commented on the experiments with oars (end(?) of 1738) he gives definite instructions to experiment with jets, produced by pumps forcing a jet towards the rear, under the waterline. He had already written to Euler[53] that he had asked Cramer and on November, 8th, that Cramer had done experiments. We are not sure what those experiments were, but it seems that somewhat later a junior colleague in Geneva, Jean Jallabert, takes over. On July 6th, 1740, Bernoulli writes instructions to Jallabert[54], very similar to those he sent Cramer. We have two undated letters from Jallabert, probably written in the summer of 1741; one mentions the experiments, the other reports extensively the results[55]. They are somewhat disappointing because the mariner actioning the pump reached not even half the speed he obtained by rowing. Similar experiments were set up by a Mr Rivaz on the lake and he reached 3/4 of the rowing speed[56] but on January 10th, 1742, Bernoulli[57] writes again saying he can't see any difference between the two experiments and perhaps Mr Rivaz had only been flattering him. In the correspondence with Euler from 1738

52 *« Ces corps devraient être en toile assez forte … donc leur surface serait assez rugueuse, d'où un grand effort de frottement lorsqu'ils sont tirés par la chaloupe. »* (A. L.)
53 Letter of D. Bernoulli to L. Euler of May 24th, 1738.
54 Letter of D. Bernoulli to J. Jallabert of July 6th, 1740.
55 Letter of J. Jallabert to D. Bernoulli, (draft), undated.
56 Letter of D. Bernoulli to J. Jallabert of November 3rd, 1741.
57 D. Bernoulli to J. Jallabert of January 10th, 1742.

to 1741 the subject is regularly mentioned and Euler had approached the Chamberlain von Korff in order to have such experiments carried out in St. Petersburg[58]; nothing came of this project as far as we know.

4.2 *The jet as conceived by Bernoulli*

Bernoulli proposes a two-stage mechanism to produce the jet. In the first stage water from the sea (or rain) is taken aboard and stored, more or less at sea level, in an auxiliary tank (b). In the second stage water from (b) is pumped up to an upright tank (a) which is kept filled to a level of A feet[59] above sea level; near the bottom of this tank a hole of area S points aft and from this hole the jet proceeds with velocity c (measured from the ship). The hole is small compared to the diameter of the tank, such that the law of Torricelli applies, i.e.

$$c = \sqrt{2gA} \; . \tag{17}$$

Bernoulli remarks that taking the water aboard, from the sea to tank (b), requires a force because the mass of water has to be accelerated from rest to the velocity $-v$ of the ship. In this Section XIII of the *Hydrodynamica* he is mainly concerned with forces; the notions of velocities and of power are seldom used. We present first the calculation in a modern form and will compare later with the results of Bernoulli in the *Hydrodynamica* and the present memoir. Still describing all movements in the reference frame of the sea, we can distinguish the following forces and sources of power.

4.2.1 Jet

We assume the hole in the tank has an area S; in one second a mass $S\rho c$ leaves the hole. Accelerating this mass $S\rho c$ from tank (a), with velocity $-v$ to the jet with velocity $c - v$ in one second, we need a force $S\rho c^2$ (the reaction force on the ship is then $-S\rho c^2$). The acceleration is c. (Force and acceleration are constant.) In order to find the work done we have to integrate the force times the velocity over the time, i.e.

58 Letter of L. Euler to D. Bernoulli of September 15th, 1739.

59 We adopt here the notation of our author in the *Hydrodynamica*: capital letters A, B, ... are used for the height of water columns.

$$\int_0^1 S\rho c^2(-v+ct)\,dt\;,$$

$$W'_{\text{jet}} = S\rho c^2\Big(\frac{1}{2}c - v\Big) = \frac{1}{2}S\rho c^3 - S\rho c^2 v\;. \tag{18}$$

4.2.2 Water aboard

Accelerating water from zero to $-v$ needs a force whose reaction on the ship is $S\rho cv$, which pushes towards the rear. The power needed is

$$W_{\text{water}} = \frac{1}{2}S\rho cv^2\;.$$

4.2.3 Ship

From those two effects the ship experiences a force

$$-S\rho c(c-v)$$

which balances the reaction force of the sea; maintaining the velocity $-v$ requires the power

$$W_{\text{ship}} = S\rho c(c-v)v\;. \tag{19}$$

4.2.4 Total power

The sum of those three terms gives the total power:

$$W_{\text{total}} = W'_{\text{jet}} + W_{\text{water}} + W_{\text{ship}} = \frac{1}{2}S\rho c(c^2 - v^2)\;. \tag{20}$$

This result could have been obtained more simply by using just the conservation of water, momentum and energy without referring to two reservoirs (a) and (b), but Bernoulli did not mention this.

4.2.5 Efficiency

The efficiency is:

$$\frac{W_{\text{ship}}}{W_{\text{total}}} = \frac{2\frac{v}{c}}{1+\frac{v}{c}}\;, \tag{21}$$

a function of $\frac{v}{c}$ which grows monotonically from 0 to 1 as v increases from 0 to c.

A further discussion shows that this efficiency depends finally on the ratio $\frac{S}{A_s}$ (A_s is the effective section of the ship, formerly denoted by A), i.e. how large a nozzle is allowed either by the mechanical design or by the condition that the water should have sufficient momentum to leave the nozzle as a jet. Here we shall only mention that if $S = A_s$ we have $\frac{v}{c} = \sqrt{3} - 1$ and the efficiency (21) is then 84.5%.

4.3 *The Bernoulli mechanism as seen from the ship*

If we choose to describe the same phenomena in a reference frame fixed to the ship, we shall say that the jet has velocity c and the ocean comes rushing in with velocity v.

4.3.1 Jet

The jet exerts a reaction force on the ship:

$$p_j = -S\rho c^2 \tag{22}$$

and requires a power $\frac{1}{2}S\rho c^3$.

4.3.2 Water aboard

Stopping the water that comes in to fill tank (b) requires a force and the reaction on the ship is

$$p_w = S\rho cv \ . \tag{23}$$

This water brings kinetic energy to the ship at the rate of

$$\frac{1}{2}S\rho cv^2$$

reducing its power requirements by the same amount.

4.3.3 Ship

The sum of the two previous forces is the force on the ship

$$p_s = -S\rho c(c - v) \tag{24}$$

which cancels the force $\frac{1}{2}A_s\rho v^2$ of the flowing water on the ship. As the ship is at rest this force is not displaced, no work is done and no power is required.

4.3.4 Total power

The total power is then the sum of the two first terms:

$$W_{\text{total}} = \frac{1}{2}S\rho c(c^2 - v^2)$$

as found by the other method.

4.4 *The argument in the* Hydrodynamica

Bernoulli uses consistently «equivalent heights» instead of velocities, related to each other by a relation of the type of (17). We shall represent his arguments exactly, but translate all formulae into the equivalent formulae with velocities, so as to make the comparison with the previous results more obvious. Here then is how Bernoulli, in the *Hydrodynamica*, treats the problem of finding the optimal conditions for a reaction motor.

The force on the ship is due to the reaction force caused by the jet (pushing the ship forward) and the acceleration of the water taken aboard, slowing the ship down. He obtains the formula:

$$p_s = Sc(c - v)$$

equivalent to (24), because he takes $\rho = 1$. Under this influence a velocity v is reached, such that

$$v^2 = \frac{2S}{A_s}(c^2 - cv) \tag{25}$$

which is equivalent to (2).

Suppose now we have a workforce on board with a power of N. This relates, writes Bernoulli, to the velocity c by

$$N = \frac{1}{2} S c^3 \, . \tag{26}$$

Eliminating S between (25) and (26) he obtains:

$$v^2 = \frac{4N}{A_s}\left(\frac{1}{c} - \frac{v}{c^2}\right) . \tag{27}$$

Suppose we are given N, we then ask for the value of c that will give the maximal value to v (i.e. the largest obtainable speed of the ship for the power given). Differentiating (27) one obtains:

$$2v\, dv = \frac{4N}{A_s}\left(-\frac{dc}{c^2} + 2\frac{v}{c^3}\, dc - \frac{dv}{c^2}\right) .$$

At the maximum $dv = 0$ and the condition for this is

$$-\frac{1}{c^2} + 2\frac{v}{c^3} = 0$$

or

$$v = \frac{1}{2} c \, . \tag{28}$$

Bernoulli remarks that at this optimal value the force driving the ship, p_s is equal to the force taking in the water, p_w and (in absolute value) each of them is half of the reaction force p_j of the jet:

$$|p_j| = Sc^2 \quad |p_w| = \frac{1}{2} Sc^2 \quad |p_s| = \frac{1}{2} Sc^2$$

and this ends the theoretical part of the problem in the *Hydrodynamica*.

The error is clear when one compares (26) with (20): there is a term $-\frac{1}{2} S \rho c v^2$ missing[60]. Probably Bernoulli had been reasoning as if on the ship, visualizing the workmen pumping up water into tank (a) and forgetting that

60 The Bernoulli formula in the *Hydrodynamica* (D. Bernoulli St. 31, *Hydrodynamica*, Strasbourg 1738), Section XIII, §25

$$M = \frac{N}{8A\sqrt{A}}$$

the onrushing waters carried energy with them, or he simply could not see how to recuperate it. Or he reasoned as if on shore, forgetting that on a ship sailing with velocity $-v$ the same power has to be divided into two parts: one part pushing the ship through the sea, the other accelerating the water in the jet which reaches then a lower velocity than when lying still in the harbour.

The Bernoulli solution can still be seen as the optimum for a defective system where part of the energy is systematically lost.

4.5 *The remarks in the memoir*

In I, §41 (p. 213 h. v.) Bernoulli recalls this result and uses it to obtain the ratio of useful power to total power. He gives no details of his argument and states simply that, taking the most favourable choice of variables, the power pushing the ship is three times smaller than the power expended elsewhere (i.e. the efficiency is at most 25%) much lower than in a galley with oars.

He probably remembered his final remark in the *Hydrodynamica* that the force pushing the ship is – in the best circumstances – a third of the sum of the other two (i.e. of the forces that accelerate waters). Bearing in mind the rule that power = force × velocity he may have been tempted to apply the same ratio to the power. This conclusion cannot be drawn, however, because in those other terms the work is done at different velocities (from zero to v or c).

4.6 *Sources from nature: gunpowder and steam*

Bernoulli considers extensively these two sources of work. The references for the values he quotes are listed in the notes to Part II, §36 (p. 249 h. v.) of his text; we comment briefly on his methods.

4.6.1 gunpowder (II, §36 p. 249 h. v.)

To measure the amount of *vis viva* in, say, two ounces of gunpowder, this quantity was put at the bottom of a gun pointed vertically and loaded with an iron cannon-ball. The time T elapsed between the firing and the ball

should have been:

$$M = \frac{N}{8\sqrt{A}(A - B)} .$$

See also G. Mikhailov, *Werke* 5, pp. 416–417, note 24.

hitting the ground was measured: 34 seconds. Bernoulli had solved the differential equation for vertical motion in a constant gravitational field with a braking force proportional to the square of the velocity as is (2). By his formulae T was given as a function of V_0, the initial velocity of the ball. By numerical solution through an iteration algorithm this could be solved for V_0 and hence various parameters of the shot determined. Especially the culminating point and the height H a similar shot would have reached in vacuum; the last number meant that the ball had started with a *vis viva* equivalent to the work done raising the ball by H. Bernoulli, in the *Hydrodynamica*, worried over the errors which beset these measurements and he tried to estimate them. He calculated that due to the imperfections of the system most gases escape uselessly from the cannon. His guess was that there is from 10 to 100 times more «living force» in those gases than is finally found in the cannon ball.

When stating in the present memoir that a cubic foot of powder could raise 200 million pounds by a foot, he is taking the results of those experiments, and correcting upwards by a conservative factor of five.

It should be pointed out that Bernoulli never mentioned an internal combustion engine: he is only thinking of cannons firing masses in the back direction and pushing the ship forward through their recoil.

4.6.2 Steam (II, §36 p. 249 h. v.)

Bernoulli considers that such a steam engine, as described by Weidler, will be used to raise water in a reservoir which feeds a jet and propels the ship by reaction. There is no indication that the engine could drive a paddle wheel although the idea was in the air and Denis Papin had already advocated it around 1690.

The critique that this engine has a low efficiency (in Bernoulli's sense) is only partly justified: the steam is used to lift the piston, the cylinder is then cooled with cold water and the atmospheric pressure drives the piston back which activates the pump. In principle no kinetic energy of the steam is lost; in practice the cylinder does not fit perfectly and steam escapes, just as the hot gases in a cannon escape between the barrel and the ball. There is, in 1753, not yet a notion of thermal efficiency, only the mechanical energy draws the attention of the author.

Principes Hydrostatiques et Méchaniques sur la question proposée pour la seconde fois par l'Académie Royale des Sciences: *Quelle est la meilleure maniere de diminuer le Roulis et le Tangage d'un Navire, sans qu'il perde sensiblement par cette diminution aucune des bonnes qualités que sa construction doit lui donner (St. 48)*

1 **Historical perspective**

1.1 *The history of the prize*

The question put for the prize of 1755 was:

> «The ways to lessen as much as possible the rolling and pitching of a ship, without losing appreciably any of her good qualities her construction should provide».

The appointed judges were Bouguer, Clairaut, Le Monnier, Camus and Nicole[1]. The prize of 2000 *livres* was awarded to M. Chauchot, *sousconstructeur des vaisseaux du Roy au département de Brest*[2]. The memoir, by order of the *Académie*, was immediately printed[3]. The *Académie* went on and stated that the matter being very important it should be looked into at the deepest level, especially as pitching is concerned; consequently the same question was asked again for 1757.

The same judges were retained. The prize of 2000 *livres* was given to Daniel Bernoulli[4], the tenth and last time he won a prize of the *Académie*. At the same session the question for 1759 was put, this time asking for a discussion of the stresses induced in a ship's structure by rolling and

1 P. V. Acad. Sci. Paris of September 6th, 1754.

2 P. V. Acad. Sci. Paris of April 9th, 1755.

3 M. Chauchot, *Sur les mouvemens de roulis et tangage des navires*, Paris: G. Martin, H. L. Guerin, L. F. Delatour & A. Bourdet, 1755. The memoir counted 55 pages and 3 plates of engravings. The copy in the Archives de l'*Académie* is bound at the end of vol. VIII of the *Prix Paris*. The other memoirs in that volume were printed by Pancoucke in 1771 and in an *Avertissement* on p. 2 Chauchot's paper is mentioned as «printed separately by Delatour». In the Basle copy of vol. VIII it is absent.

4 P. V. Acad. Sci. Paris of April 20th, 1757.

pitching and how to assemble the parts so as to withstand those stresses[5]. The memoir was printed in the *Prix Paris*[6]; it was reported in the *Histoire de l'Académie, année 1757*[7].

1.2 *The style*

In the first half of the memoir, Chapitres 1 to 4 (pp. 254–303 h. v.) Bernoulli exploits the formulae he published in *De statu aequilibrii corporum humido insidentium*[8] and *De motibus oscillatoriis corporum humido insidentium*[9]. He does not, however, cite those but derives them anew in a form adapted to the symmetry of the problem. Compared with those articles the present essay is much more wordy and eschews mathematically involved arguments. For example there is no derivation – nor any citation – of the formulae for the driven pendulum; instead there is an attempt at an intuitive explanation and ultimately the reader is invited to do himself an elementary experiment. The essay seems to have been written under time pressure, because there subsist many inconsistent notations. We tried to be consistent in this introduction, but this meant that our version of a formula will differ from any of the Bernoulli versions.

1.3 *Visibility and impact of the memoir.*

The memoir is in fact composed of two parts: the first dealing with the equilibrium and oscillations of ships on level water, the second part being devoted to the ships on a wavy sea. The basic ideas for part one were fully developed in the two articles which Bernoulli sent to St. Petersburg[10] in 1738 and 1739. It was a stroke of bad luck that they would appear in print so

5 That prize was divided between Euler E415, *Examen des efforts qu'ont à soutenir toutes les parties d'un vaisseau dans le roulis & dans le tangage ou recherche sur la diminution de ces mouvemens*: Prix Paris VIII (pièce V) 1759 (1771) – Euler, *Op. omnia* II/21, pp. 1–30 and A. Groignard, *L'examen des efforts qu'ont à soutenir toutes les parties du Vaisseau dans le roulis & dans le tangage, & la meilleure maniére de procurer à leur assemblage la solidité nécessaire pour résister à ces efforts, sans préjudicier aux autres bonnes qualités du Vaisseau*: Prix Paris VII (pièce IV) 1759 (1769).

6 Prix Paris VIII (pièce IV) 1757 (1771).

7 Hist. Paris 1757 (1762), p. 184.

8 D. Bernoulli St. 30, *De statu aequilibrii corporum humido insidentium*: CP X, 1738 (1747), pp. 147–163 – *Werke* 6.

9 D. Bernoulli St. 32, *De motibus oscillatoriis corporum humido insidentium*: CP XI, 1739 (1750), pp. 100–115 – *Werke* 6.

10 Cf. notes 8, p. 74 h. v. and 9, p. 74 h. v.

much later[11], in 1747 and 1750. Because in the meantime Pierre Bouguer had published the *Traité du navire*[12] and Leonhard Euler the *Scientia navalis*[13].

Bouguer had arrived, apparently independently, at similar results; he presented them in a geometric language, emphasizing his discovery of the metacentre, a point everybody could visualize. It had been believed by a number of authors (a. o. Witsen, following Stevin) that a ship is only stable if the centre of mass is lower than the centre of buoyancy. Hoste[14] had shown on a special example, using geometrical arguments, this was not true. Bernoulli, using moments of forces, demonstrated the general condition for stability and in addition determined the rolling period of a ship. Bouguer reached the same conclusions. Also Bouguer showed how the natural period of the ship is related to the moment of inertia. Those discoveries were connected with his name, because the book was widely read and Bouguer himself was well introduced in naval circles; there was obviously no reference to Daniel Bernoulli in this work. After the article on stability appeared[15], Bouguer wrote to Bernoulli, on April 28th, 1751: «I think I forgot to tell you that I read your memoir in volume 10 of the memoirs of Petersburg and I could only congratulate myself that my results are so perfectly in accord with yours»[16].

11 There are numerous letters of complaint by Bernoulli, to the Academy in St. Petersburg and to other correspondents, concerning this delay:
- August 19th, 1743, to Moula: he counts ten papers on the waiting list;
- August 15th, 1744, to Schumacher: *«...elles perdraient les graces de la nouveauté si on différait de les imprimer.»* («...they would loose all the grace of novelty if one would delay their printing»);
- November 12th, 1746, to the Academy: Bernoulli insists that his articles should be printed with the date of reception accurately marked, because *«...elles contiennent des nouvelles découvertes, que d'autres à qui je les avais communiquées ont trouvé digne pour s'en parer»* («...they contain novel discoveries that others, to whom I communicated them, have found worthy to adorn themselves»).

12 P. Bouguer, *Traité du navire, de sa construction et de ses mouvements*, Paris: Jombert, 1746.

13 L. Euler E 110–E 111, *Scientia navalis seu tractatus de construendis ac dirigendis navibus*, St. Petersburg 1749 – Euler, *Op. Omnia* II/18 & 19.

14 P. S. J. Hoste, *Théorie de la construction des vaisseaux, qui contient plusieurs traitez de mathématique sur des matières nouvelles et curieuses*, Lyon: Anisson et Posuel, 1697.

15 Cf. D. Bernoulli St. 30, *De statu aequilibrii corporum humido insidentium*: CP X, 1738 (1747), pp. 147–163 – *Werke* 6.

16 Letter of P. Bouguer to D. Bernoulli on April, 28th, 1751: *«Je crois que j'ai oublié de vous dire que j'ay lu vôtre mémoire sur les innatans dans le 10e volume des memoires de Petersbourg et que je n'ai pu que m'aplaudir que mes résultats s'accordassent si parfaitement avec les vôtres».*

Euler knew the articles of Bernoulli. In fact it is a recurring subject in their correspondence from May 1739 to April 1740. Both agreed that the results on stability and on free oscillations were obtained by either independently. But the *Scientia navalis* is a textbook, with almost no references. This book too was widely quoted. A more popular version appeared in French in 1773, without any references either[17].

When Bernoulli wrote his memoir for the 1757 prize he could have mentioned his earlier discoveries. But he knew very well that Bouguer was going to be on the reading committee. Would it be wise, by citing, to give himself openly away as the author of the anonymous text and furthermore to remind that examiner he was not the first with his discovery? Bernoulli, in the first part of the memoir, simply rederives his former results in a way which avoids the use of the metacentre, at the same time putting in a number of flattering remarks for Bouguer. On the whole his relations with Bouguer were amiable. E.g. there is a letter from Bouguer, of the 6th of August 1756, to Bernoulli; the latter had obviously enquired whether his submission for the *Prix Paris* had arrived. Bouguer answers very diplomatically that the secretary of the *Académie* up to now has only received three pieces: *«L'une a pour devise* plenis trajiciens *&c, la seconde* insequitur clamorque *&c et la troisieme* Qui dubiis ausus committere flatibus alnus *&c»*[18]. By citing the motto of Bernoulli with seven words against the two for either the first and second he gives a signal to Bernoulli that he has recognized his text, but he still acts according to the rules that impose anonymity on the contestants.

As a second stroke of bad luck the printing of this memoir itself was delayed by 14 years[19]. Consequently when in 1758 Duhamel du Monceau[20] published a treaty of naval architecture no mention of Bernoulli was made. Vial du Clairbois, who had translated Chapman[21], wrote in 1776 the first

17 L. Euler E426, *Théorie complette de la construction et de la Manœuvre des vaisseaux...*: Acad. Imp. des Sciences, St. Petersbourg 1773 – Euler, *Op. omnia* II/21, pp. 82–229.

18 Letter of P. Bouguer to D. Bernoulli of August 6th, 1756.

19 In a letter to an unknown addressee but obviously meant for the *Académie* on August 22nd, 1755, Bernoulli repeats his earlier lament: *«L'excessive lenteur avec laquelle on imprime les autres pieces qui remportent les prix, leur fait perdre toute la grace de la nouveauté...»*. (The excessive slowness with which the other pieces who obtained prizes are being printed makes them lose all the grace of novelty...). And in his letter of July 1st, 1768, the secretary of the *Académie*, Dortous de Mairan, had to admit that indeed the memoirs take too long to appear.

20 H.-L. Duhamel du Monceau, *Élémens de l'architecture navale, ou Traité pratique de la construction des vaisseaux*, Paris: Jombert, 1752, 2nd ed. Paris 1758.

21 F. H. Chapman af, *Tractat om Skepps-Byggeriet, tillika med Förklaring och Bevis öfver Architectura navalis mercatoria*, Stockholm: J. Pfeffer, 1775.

treatise on ship's construction[22] based on geometric principles, he cited Bouguer and Euler but ignored Bernoulli. So did he when writing the hundred pages on stability in the *Encyclopédie méthodique* (Marine) in 1787 which is exclusively and avowedly based on Bouguer. In the same book is a list of suggestions for a naval library: Euler and Bouguer appear prominently, Daniel Bernoulli only through his *Hydrodynamica*[23].

When Dupin, in 1814[24] developed the analogy between a floating body and a solid with a curved bottom resting on a horizontal plane, he unknowingly took up a suggestion by Bernoulli (cf. his § 3, p. 254 h. v.). Neither did Carnot, who presented the communication, notice this.

None of those other authors, however, had tackled the movements of a ship riding the waves, which was obviously a very hard task[25]. When Bertin[26] in 1877 put into a book his mathematical theory on the subject he paid due respect to the pioneering efforts of Bernoulli; but only for the second part of the memoir. The first part he considered «the things everyone expected to see»; he obviously did not know the 1738–1739 articles and thought, as every one else, that the elementary theory was due to Bouguer.

2 Note on symbols and units

2.1 *List of symbols*

In Table 1 are listed the symbols used in this introduction and the notes, their definition and the equivalents most often used by Bernoulli.

22 H. S. Vial du Clairbois, *Essai géométrique et pratique sur l'Architecture navale, à l'usage des gens de mer*, Brest: R. Malassis, 1776.

23 D. Bernoulli St. 31, *Hydrodynamica*, Strasbourg 1738 – *Werke* 5, pp. 91–424.

24 P. V. Acad. Sci. Paris of August 30th, 1814.

25 In an earlier letter to Bouguer – of November 11th (?), 1754 – Bernoulli announced Euler and he himself had found the free oscillations for a general floating object, «which was a rather difficult problem». He then goes on: *«cependant la solution de notre probleme est encore extremement insuffisante parceque la surface de l'eau n'est rien moins qu'unie et horizontale.* («Our solution, however, is still extremely insufficient because the surface of the water is nothing less than even and horizontal»). Further in the letter he mentions the dangers of resonance and the possibility to alter the frequency of rolling by redistributing the masses in the ship.

26 L. E. Bertin, *Les vagues et le roulis*, Paris: Berger-Levrault et Cie, 1877. This expert detailed his praise and criticism in an earlier article: E. Bertin, *Notes on waves and rolling. Historical introduction: essay on Daniel Bernoulli*: Naval Science 2 (1873), pp. 344–351.

symbol introd.	definition	symbol Bernoulli
A	centre of gravity	s...
B	centre of buoyancy	f...
B'	idem, in tilted position	f′...
σ or θ	angle of heel	σ
$T(\sigma)$	restoring torque at angle σ	Pr
x	distance from stern	x
a	half width of slice	
	or of ship (midships)	a or $a/2$
q	width of ship at water line	q
s	distance AB	s or d
τ_0	period of free oscillation	t
τ	period of driven oscillation	θ
M	immersed surface of slice	M...
α	thickness of slice	
	or slope of sea surface	α
P	weight of ship	P
F	stability	π

Table 1: List of symbols used in the introduction and the notes.

2.2 *The unit of length*

The unit used is the foot, but which one? A clue can be found in §62 (p. 322 h. v.) where Bernoulli states that «a beam drawing 18 feet» will oscillate with a period «somewhat less than $2\frac{1}{3}$ sec»[27]. Turning the problem around one finds this means a foot equals somewhat less than 30.033 cm. This coincides almost with the Basle foot.

On the other hand in §22 (p. 276 h. v.) a pendulum with (semi)period of 2 seconds has a length «somewhat in excess of 12 feet». We find that 12.35 french «pieds du roi» would do, whereas we would need almost 13 Basle feet. In §56 (p. 315 h. v.) we learn that a cubic foot of water weights $\frac{2000}{28}$ pounds; this fits the french system: 1 pied du roi = 0.3248...m and 1 pound = 489.5 g.

27 Cf. p. 323 h. v.

So it would seem that Bernoulli was fickle in his system of units. Of course, in view of the practical problems he was handling and the enormous uncertainties on some of the parameters this had really no importance.

3 The equilibrium of a ship

3.1 *The mechanics of floating*

The first four chapters of the memoir rely on two formulae in *De motibus oscillatoriis corporum humido insidentium*[28]. In that article he investigated the rocking and bobbing of a floating object. His method is a synthesis of three of his own lines of research: the statics of a floating object[29], the movements of a rigid body[30] and normal vibrations of a complex system[31]. The article was purely theoretical, was concerned in the end with the floating mechanics of a body of the most general shape and emphasized approximations where only movements close to an equilibrium position were investigated. In the present memoir, devoted to a most practical subject, he retains only the results that apply to ships as they are commonly build; where possible he goes beyond the neighbourhood of equilibrium. In the whole memoir there is but one reference to his own publications viz. to a text in the mechanics of a solid body.

The essential simplification is that a ship has a plane of symmetry, defined by bow, keel and stern. A second simplification serves more a didactic purpose: consider a vertical section through the ship, orthogonal to the plane of symmetry, or rather a slice of thickness α. In *De statu aequilibrii corporum humido insidentium*[32] it was shown how to find the equilibrium positions of such a slice which we constrain to remain vertical but which is free to move in this vertical plane. As a ship can be conceived as a stack of such slices the formulae for the slices will lead to the formulae for the ship.

28 Cf. D. Bernoulli St. 32, *De motibus oscillatoriis corporum humido insidentium*: CP XI, 1739 (1750), pp. 100–115 – *Werke* 6.

29 Cf. D. Bernoulli St. 30, *De statu aequilibrii corporum humido insidentium*: CP X, 1738 (1747), pp. 147–163 – *Werke* 6.

30 D. Bernoulli St. 13, *De mutua relatione centri virium, centri oscillationis et centri gravitatis*: CP II, 1727 (1729), pp. 208–216 – *Werke* 3, Mechanik p. 24 (Introduction) and p. 136 (text).

31 D. Bernoulli St. 23, *Theoremata de oscillationibus corporum filo flexili connexorum*: CP VI, 1732/33 (1739), pp. 108–122 – *Werke* 6.

32 Cf. D. Bernoulli St. 30, *De statu aequilibrii corporum humido insidentium*: CP X, 1738 (1747), pp. 147–163 – *Werke* 6.

3.1.1 The floating slice (§§5–6 pp. 258–259 h. v., & 20 p. 275 h. v.)

According to the argument which goes back to Archimedes, the weight of a solid object is a downward vertical force which acts on the centre of mass A (see Fig. 1); the pressure of the water on the hull results in an upthrust which

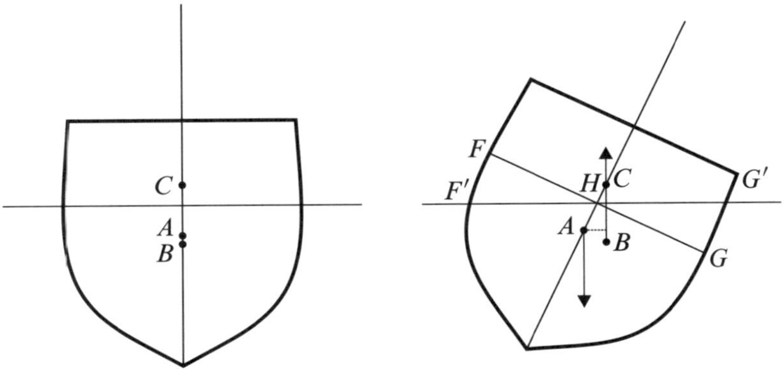

(a) (b)

Figure 1: A section midships in the normal position (a) and heeling (b); A: centre of mass; B: centre of buoyancy; C: metacentre.

is an upward vertical force acting in the centre of mass of the displaced water (or, as Bernoulli states more accurately, «the centre of mass of the immersed part of the hull considered as a homogeneous body»); this is now commonly called the centre of buoyancy B. At floating equilibrium the law of Archimedes states that weight and buoyancy are equal in magnitude; in position (b) they will produce a torque which tends to rotate the vessel back to its vertical position. That is, provided the point A lies below the point C, the so-called metacentre. For a ship to be stable this is the essential condition.

This restoring torque can be calculated: simple geometric considerations on the centres of mass of the equal triangles $F'HF$ and $G'HG$ allow to determine the position of B and hence the torque[33]. If we call:

- $s =$ the vertical component of the distance AB (positive if A is above B),
- $q =$ the width FG at the water line in the reference position (usually the equilibrium position),

33 The argument is given in detail in the introduction to D. Bernoulli St. 30, *De statu aequilibrii corporum humido insidentium*: *Werke* 6.

– M = the weight of the vessel's slice (= surface of immersed part times thickness times density of the water),
– σ = the angle of heel,

we find for the restoring torque:

$$T(\sigma) = \left(\frac{1}{12}q^3 - Ms\right)\sin\sigma\ . \tag{1}$$

Provided the left and right sides of the vessel are straight and parallel this is an exact formula. If not, q and s are functions of σ and the formula (1) is only valid up to order σ^2. By a slight variation of the argument Bernoulli obtains the differential form of the formula, stating the increase $dT(\sigma)$ of the torque when the vessel is tilted from σ to $\sigma + d\sigma$:

$$dT(\sigma) = \left(\frac{1}{12}q(\sigma)^3 - Ms(\sigma)\right)d\sigma\ . \tag{2}$$

The ratio between the effect (the torque dT) and the cause $d\sigma$ is aptly called the *stability* by Bernoulli; we shall denote[34] it by F.

Integrating over σ from the equilibrium position up to a given angle, one obtains a formula for the restoring torque at any angle:

$$T(\sigma) = \int_0^\sigma \left(\frac{1}{12}q(\sigma)^3 - Ms(\sigma)\right)d\sigma\ . \tag{3}$$

3.1.2 The floating ship (§§ 5–6 pp. 258–259 h. v., & 13 p. 268 h. v.)

The result (1) can be carried over for a «prismatic vessel», i.e. a vessel in the form of a prism, where all sections are equal, and especially the centre of mass of each slice is in a similar position.

For real ships, where q, M and s are functions of the position of the slice along the length of the vessel, and assuming this vessel is undeformable and for each slice the weight equals the buoyancy, one can write a double integral, integrating (2) over the angle σ and then over the length x from

34 In modern texts (and for $\sigma \approx 0$) one finds $F = M(r - s)$ where $r \equiv CB$ is called the metacentric radius; this formula is easily seen to be equivalent with (1). The length $r - s = CA$ is called the metacentric height.

stern to bow (i.e. for x from 0 to $2b$); this is:

$$T = \int_0^{2b} dx \int_0^{\sigma} \left(\frac{1}{12} q^3(x;\sigma) - M(x)s(x;\sigma) \right) d\sigma \ . \qquad (4)$$

This formula, by Bernoulli's own account, is completely impractical because there is no way to actually find those functions for a real ship. In the very practical problem he is addressing he will develop a strategy of encirclement: try out different approximations and learn from them what they can tell – with more or less confidence – about rolling and pitching. The formulae can be calculated exactly for floating rectangular boxes and for circular cylinders; both are a case of «prismatic vessels». Ellipsoids too can be calculated, or hulls made of two ellipsoids with the same width but different lengths put together. One could think of a ship as having almost vertical sides build on such a bottom of two half-ellipsoids; comparing with contemporary drawings of naval vessels this seems a close approximation.

The formula can be simplified by introducing lumped variables: it is a technique that appears in *De motibus oscillatoriis corporum humido insidentium*[35], but neither here does Bernoulli refer explicitly to this publication. Assuming that a ship is symmetric with respect to a plane through its keel, bow and stern and that its sides, at equilibrium, are vertical at the water line, we can write:

$$T = T_1 - T_2 \ , \qquad (5)$$

$$T_1 = \int_0^{\sigma} 2p(\sigma)S(\sigma)q(\sigma)\, d\sigma \ , \qquad (6)$$

$$T_2 = P \int_0^{\sigma} s(\sigma)\, d\sigma \ , \qquad (7)$$

where:

- S is the area on the horizontal plane at the water level between the left part of the hull and the median along the length of the ship,
- q is the distance from the median of the centre of mass of this area (in the sense of a plane surface),
- p is the distance from the median of the centre of mass of the wedge with the area S and the angle $d\sigma$ (the wedge cut out by the part of the

35 Cf. D. Bernoulli St. 32, *De motibus oscillatoriis corporum humido insidentium*: CP XI, 1739 (1750), pp. 100–115 – *Werke* 6.

ship that dips into the water when the heeling angle changes by $d\sigma$); this wedge is considered homogeneous,
- P is the weight of the ship,
- s is the vertical component of the distance between the centre of mass of the ship and the centre of buoyancy.

The term T_1 is related to the area the ship defines on the sea's surface; T_2 depends on the height of the centre of mass above the centre of buoyancy and – unless the former is under the latter, when the sign of T_2 is reversed – it is a term which tends to make the ship unstable. For ships as commonly build in the 18th century T_1 is about three times larger than T_2.

In addition if we can restrict the movements to the neighbourhood of an equilibrium position things are considerably simpler.

3.2 *Ballast tanks and circular sides (§§ 6–7 pp. 258–260 h. v.)*

Imagine a «prismatic» ship of arbitrary section. Being tilted over an angle σ it will experience a torque $T(\sigma)$. If a ballast-tank, filled with water, is attached underneath the hull this will change both the centre of mass and the centre of buoyancy of the whole.

The centre of mass A' of the tank coincides, however, with its centre of buoyancy; both the weight and the buoyancy act at the end of the lever arm OA' to produce the torque due to the tank (see Fig. 2): as they are equal

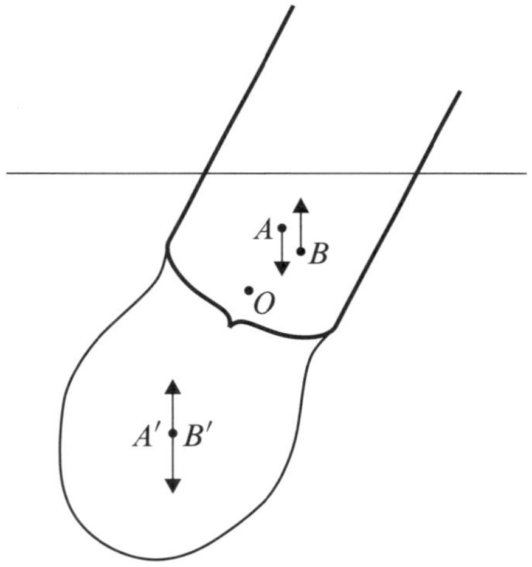

Figure 2: Hull with ballast tank, tilted with respect to the sea level.

and opposite this added torque is zero. Therefore, as stated by Bernoulli, an additional ballast-tank does not change the stability. If a ship has a circular cross section, the form of the submerged part is always a circular segment and therefore the restoring torque is easily computed once the distance of the centre of mass A from the centre O of the circle is known.

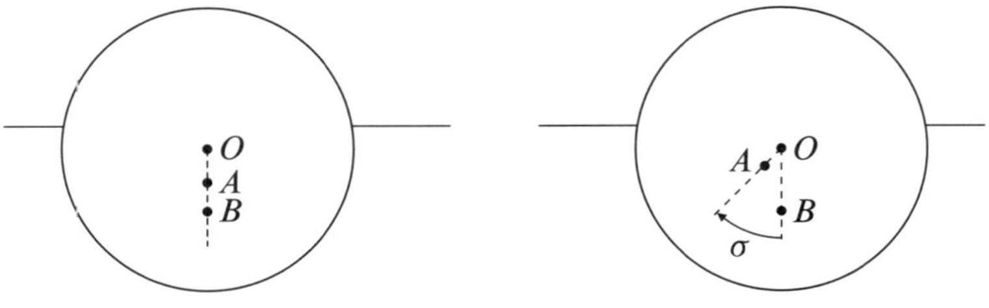

Figure 3: For circular cylinders the buoyancy centre is independent of the angle of heel.

We consider a slice of thickness α and m_c the area of the submerged part of such a circular hull; let A be the centre of gravity (which depends of course on the distribution of masses in the cylindrical vessel), B is the centre of buoyancy and its position in space is independent of the tilting angle σ. We call $f = OA$, $g = AB$. Our formula (3) reads then (in units[36] where $\alpha\rho g = 1$)

$$T = \int_0^\sigma \Big(\frac{1}{12}q^3 - m_c(f + g - f\cos\sigma)\Big)\, d\sigma \ ,$$

or

$$T = \Big[\frac{1}{12}q^3 - m_c(f + g)\Big]\sigma + m_c f \sin\sigma \ ,$$

it is easily verified by using the well-known formulae from the geometry of segments of a circle that the square bracket is zero and hence

$$T = m_c f \sin\sigma \tag{8}$$

and in differential form

$$dT = m_c f \cos\sigma \, d\sigma \ .$$

But because a ship with circular sides and any form of bottom could be converted to a ship with circular section by adding a suitable tank whose

36 In this instance, unlike in the rest of this section, g means the acceleration of gravity.

lower shape is a circular cylinder, the above formula gives at once the stability of ships with circular sides (A is then the centre of mass of the ship plus the ballast). By a slight extension this will allow to compute almost exactly the stability of flat boats (see § 3.4.2 p. 88 h. v.).

3.3 *Floating slabs*

3.3.1 Rectangular slabs (§ 9 p. 262 h. v.)

We take a slice with vertical sides (and hence parallel), in equilibrium, and assume furthermore that it has a vertical symmetry axis. Suppose we tilt the slice over an angle θ, keeping the immersed surface constant (which insures that the law of Archimedes is still obeyed).

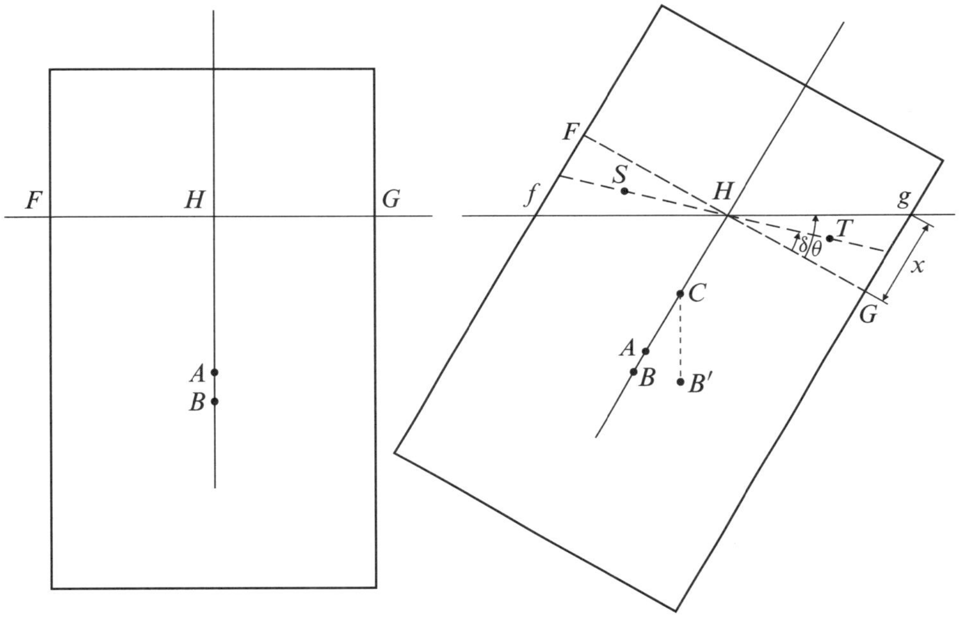

Figure 4: A floating rectangular slab. The following abbreviations are used in the text: $\overline{FG} = 2a$, $\overline{AB} = s$.

We refer to Fig. 4. The immersed part loses the triangle FHf and gains the triangle GHg. The surface of each triangle is $\frac{1}{2}GH \cdot Gg$ and their centres of mass, S and T are on the median line, at $\frac{2}{3}$ from the apex H. Let B' be the centre of buoyancy in the tilted position. The line BB' is therefore parallel to ST and by the definition of centre of mass (m is the immersed area)

$$m\,.\,\overrightarrow{HB'} = m\,.\,\overrightarrow{HB} + \frac{1}{2}\,.\,Gg\,.\,HG\,.\,\overrightarrow{HT} - \frac{1}{2}fF\,.\,HF\,.\,\overrightarrow{HS}\,, \tag{9}$$

which is equivalent to

$$m\,.\,\overrightarrow{BB'} = Gg\,.\,HG\,.\,\overrightarrow{HT}\,. \tag{10}$$

From this we can compute the projection of $\overrightarrow{AB'}$ on the horizontal, which multiplied with the force gives the torque (the force is equal to m times the thickness of the slice α, times the density of the water ρ, times the acceleration of gravity g). As shown in the introduction to *De statu aequilibrii corporum humido insidentium*[37] in *Werke* 6 this leads to the torque:

$$T(\theta) = \frac{1}{12}\overline{FG}^3\,\frac{\tan(\theta)\cos(\theta-\delta)}{\cos(\delta)} - ms\sin(\theta) \tag{11}$$

where the angles θ and δ are shown in Fig. 4 and $s = AB$ and we chose α, the thickness of the slice, such that $g\alpha\rho = 1$.

Remark that if A would be lower than B, the sign of the second term should be changed. Bernoulli, in §9 (p. 262 h. v.), sets

- $FG = 2a$,
- distance from bottom to water level $= c$ and hence $m = 2ac$,
- $a\,.\,\tan(\theta) = x$, the distance of the watermarks on the side of the ship,
- $AB = s$ is called d by Bernoulli, but we shall still use s in this note.

The torque for a rectangular slab can then be written as

$$T(\theta) = m\alpha\rho g\Big(\frac{a^2}{3c} - s + \frac{x^2}{6c}\Big)\sin\theta\,. \tag{12}$$

The published formula is faulty; probably due to a transcription error the third term is printed twice as $\frac{xx}{2c}$; the error appears in all formulae derived from (12).

37 Cf. D. Bernoulli St. 30, *De statu aequilibrii corporum humido insidentium*: *Werke* 6.

3.3.2 Rectangular versus round (§ 10 p. 263 h. v.)

In order to show that straight sides give more stability than round sides Bernoulli produces a numerical example in which he compares the values of the torques computed by formulae (12) and (8) for angles from 0 to 45 degrees. He takes his example such that both are the same for $\sigma = 0$, i.e.

$$m_c f = m\Big(\frac{a^2}{3c} - s\Big) ,$$

and both those terms are multiplied by $\sin\sigma$ in the expressions of the torque; to order σ there is no difference then. It is the term proportional to x^2 which will make the rectangular shape more stable at larger angles. The ratio of the torques for the rectangular slab and for the circular one is then

$$1 + \frac{1}{2 - 6\frac{cs}{a^2}}(\tan\sigma)^2 .$$

σ	circular	rectangular
0	0	0
1	1.454	1.455
2	2.908	2.915
3	4.361	4.385
4	5.813	5.870
5	7.263	7.374
10	14.471	15.371
15	21.568	24.665
20	28.502	36.053
25	35.218	50.534
30	41.667	69.444
35	47.798	94.668
40	53.566	128.995
45	58.926	176.777

Table 2: Torques for a circular and a rectangular section, showing that their difference increases with the angle of heel.

The value in the table for 40 degrees, in the third column, should read 279.86 instead of 274.70. Unfortunately Bernoulli used the formula as printed in his text which has its coefficient of x^2 too large by a factor three; the effect is certainly there, but less spectacular than he thought. In his example the ratio is equal to 7 at 45° whereas it is in fact only equal to 3. This is shown in Table 2, computed with the correct formula.

3.4 *List of formulae for special cases*

Bernoulli has treated various cases of floating and derived special formulae for each, changing often notation. We give here a systematical survey of those formulae.

We follow in this section the use of Bernoulli that $g = 1$, i.e. force is measured in units of weight.

3.4.1 Hull in the form of a circular cylinder (§7 p. 260 h. v. & §15 p. 271 h. v.)

Cf. Fig. 3; let $f = \overline{OA}$, the distance of the geometric centre of the circle to the centre of mass, then we have:

$$T = Mf \sin\sigma \ .$$

3.4.2 Boats with a flat bottom and (approximately) circular sides (§8 p. 262 h. v.)

Cf. *Fig. 2* of Bernoulli and his §7. A virtual circular ballast tank *cgd*, filled with water, is appended to the flat bottom; *o* is the centre of the circle, *m* is the centre of mass of boat plus ballast. Let μ be the immersed surface *beg* and ϕ the distance $\overline{om}$, then:

$$T = \mu\phi \sin\sigma \ .$$

In the limit for a very shallow boat, when $be = cd \ll bc$ and *m* is practically the centre of mass of the ballast:

$$T = \frac{1}{12}\overline{be}^3 \sin\sigma \ .$$

The stability of shallow boats with a flat bottom can therefore be computed almost exactly (whether the sides are circular or not will in that case make almost no difference).

3.4.3 The floating rectangular prism (§9 p. 262 h. v.)

Cf. Fig. 4. Let the waterline $\overline{FG} = 2a$,
the length of the prism: $2b$,
the distance from the bottom to the water line: c,
the distance $\overline{AB} = s$ between centre of mass and centre of buoyancy,
the distance between the watermarks on the side: $\overline{Gg} = a \tan\sigma = x$

$$T = M\Big(\frac{a^2}{3c} - s + \frac{x^2}{6c}\Big)2b \sin\sigma \ .$$

3.4.4 The elliptical hull (§13 p. 268 h. v.)

In the limit of a small angle of heel σ, when the form of the hull at the waterline is an ellipse with semimajor axis b and semiminor axis a:

$$T = \Big(\frac{\pi}{4}a^3 b - Ps\Big)\sigma \ , \qquad (13)$$

P is the weight of the ship,
s is the distance $\overline{AB}$ between centres of mass and buoyancy.

3.5 *Pitching (§14 p. 271 h. v.)*

Up to now we considered only the stability against rotation around a longitudinal axis, which will be used when discussing rolling. For pitching the axis is transverse. Mathematically this is entirely similar: just exchange b and a, the length and the width, and any formula related to rolling is converted to the corresponding formula for pitching.

3.5.1 Pitching for ellipsoidal hulls

In §13 (p. 268 h. v.) the torque for a half-ellipsoidal hull was computed, (cf. formula (13)) and the exchange between length and width being duly made, we obtain for the first term: $\frac{\pi}{64}ab^3\sigma$ (a = width[38], b = length, σ = pitching angle); because $b \gg a$ the first term is now much larger and the second term can be neglected. With $\pi = \frac{22}{7}$ this gives the formula $\frac{11}{224}ab^3\sigma$. Also relying

38 In order to obtain the same formulae as Bernoulli we follow – in this section only – his changed notation, changing a and b by a factor 2.

on §13 we find that the corresponding first term for a rectangular hull is: $\frac{1}{12}ab^3\sigma$. The ratio of the two terms (i.e. for ellipsoidal and for rectangular hull is $\frac{\pi}{64}12 = 0.58905...$ which is approximated by $\frac{22}{7}\frac{12}{64} = \frac{132}{224} = 0.58929...$; within a sensible accuracy $\frac{3}{5} = 0.6$ is almost as good.

Bernoulli computed first the formula for the elliptical hull; for the rectangular one he forgot he had already computed it, but he remembered they are almost in the ratio of $\frac{3}{5}$. Therefore he simply stated that the coefficient is $\frac{3}{5}\frac{11}{224} = \frac{55}{672}$ i.e. $\frac{1}{12.22}ab^3\sigma$ instead of the exact value $\frac{1}{12}ab^3\sigma$. In order to illustrate the formula for pitching, in the ellipsoidal case, Bernoulli computes the weight that, applied to the bow, would induce an angle of σ degrees. He takes the same dimensions for the ship as in the earlier example. Of course it is most revealing to display it as a fraction of the total weight. For a hull that under the water line has the shape of a half-ellipsoid, the volume of the displaced water is $P = \frac{\pi}{6}abc$ (c = depth); this is approximated by $\frac{1}{2}abc$. The lever arm is $\frac{1}{2}b$. The required weight m must be such that (cf. formula (13))

$$\frac{1}{2}bm = \frac{\pi}{4}\left(\frac{b}{2}\right)^3\frac{a}{2}\sigma ,$$

which gives the weight m as a fraction of the weight of the ship:

$$m = \frac{3}{16}\frac{b}{c}\sigma P .$$

Bernoulli's result is derived from his approximate formula for the torque:

$$\frac{1}{2}bm = \frac{11}{224}ab^3\sigma ,$$

combined with the approximation $P = \frac{1}{2}abc$ and the values $b = 140$ and $c = 18$ of the previous example, which gives:

$$m = \frac{55}{36}\sigma P , \qquad (14)$$

which, in the original text, was misprinted as $\frac{556}{36}P$.

4 The oscillating ship on a level sea

4.1 *Oscillators: pendulum and ship (§§ 18–26 pp. 274–281 h. v.)*

The oscillations of a ship are discussed by analogy with those of a pendulum. we recall those briefly and indicate the formulae Bernoulli takes as point of departure of his theory.

4.1.1 The free pendulum (§ 22 p. 276 h. v.)

We recall the equation of the simple pendulum expressing that the mass times the (tangential) acceleration equals the tangential force:

$$ml\ddot{\theta} = -mg \sin\theta$$

where m is the mass, l the length, θ the angle from the vertical, g the acceleration of gravity (i.e. $9.81\,\mathrm{m/s^2}$). For small oscillations we can take $\sin\theta = \theta$ and a solution is then $\theta = A \sin\sqrt{\frac{g}{l}}t$ which means a period $\tau_0 = \pi\sqrt{\frac{l}{g}}$ (in the sense to be defined, see footnote 40). A two-second pendulum has therefore a length of $l = 9.81(\frac{2}{\pi})^2 = 3.976$ meters or 12.24 *«piés du roy»* which, as Bernoulli writes, is indeed «slightly more than 12 feet».

4.1.2 The driven pendulum (§ 23 p. 277 h. v.)

Suppose now a periodic force with period τ is acting on the pendulum; its equation of motion – for small amplitudes – is:

$$ml\ddot{\theta} = -mg\theta + k \sin\Big(\pi\frac{t}{\tau}\Big) .$$

A particular solution is

$$\theta = a \sin\Big(\pi\frac{t}{\tau}\Big) .$$

Substituting this in the equation one finds the amplitude

$$a = \frac{k}{m(\frac{g}{l} - \frac{\pi^2}{\tau^2})l}$$

which is uniquely given by the frequency and the intensity of the driving force. Bernoulli measures the parameter k by giving the elongation θ_0 at which the pendulum would be in equilibrium with a static force of magnitude k. This requires $-mg\theta_0 = k$ and hence we can write the elongation of the driven pendulum as:

$$a = \frac{\theta_0}{1 - (\frac{\tau_0}{\tau})^2} \ . \tag{15}$$

Suppose now $\tau = 1$ second, and $\tau_0 = 2$ seconds: it follows that the force which would cause by itself a static elongation over $\theta_0 = 12$ degrees, results in an amplitude of $a = \frac{1}{3}\theta_0 = 4$ degrees, «in alternating cooperation and opposition with gravity». Further on in §22 the analogous example of a driving force having a period of four seconds results in an increase of the elongation by a factor 4/3.

Obviously, when τ equals τ_0, the amplitude blows up. In reality terms we have neglected when setting up the equation – such as friction – will prevent the catastrophe but in the real world the oscillations can still build up to a dangerous level, a fact Bernoulli will point out several times.

4.1.3 The freely oscillating ship (§24 p. 278 h. v.)

If a ship, in the harbour, is tilted sideways and then suddenly released it will rotate around an axis parallel to its length under the influence of the restoring torque $T(\sigma)$; we may assume that the ship will, while rotating, stay in hydrostatic equilibrium[39]. Assuming this, arguments from the mechanics of solid bodies assure us that the axis of rotation will pass through the centre of mass. Let the torque producing a tilt over an angle σ in a floating object be $T(\sigma)$; we admit a linear relation

$$T(\sigma) = F\sigma \ ,$$

where F is called the stability of the floating object. We assume the classical equation for an oscillating system (I = moment of inertia):

$$I\ddot{\sigma} = T(\sigma) = F\sigma \ .$$

39 This is discussed in §4.4.1 p. 96 h. v.

The solution, up to a phase, is

$$\sigma(t) = a \sin\left(\sqrt{\frac{F}{I}}t\right) \tag{16}$$

and hence the natural period[40] is $\tau_0 = \pi\sqrt{\frac{I}{F}}$. It is inversely proportional to the square root of the stability; in addition the stowing of the cargo, by changing the moment of inertia I, can also influence the natural frequency. This form of rolling Bernoulli calls *balancements libres* (sometimes *balancements naturels*) which we translate by *free oscillations*. He uses freely the analogy of these oscillations with those of a simple pendulum and rather than computing the rolling period in seconds he computes the length of the synchronous pendulum. This is of course only possible in the simple cases where the stability is completely computable, i.e. for round and for rectangular hulls; a comparison of a cylindrical and a rectangular hull of the same dimensions shows that the latter is about twice as stable; the advantage increases moreover with the angle as appears from Table 2.

4.1.4 Driven oscillations of a ship (§25 p. 279 h. v.)

It may happen that at sea the ship is subjected to regular and prolonged buffeting by water masses (e.g. breaking waves). Assume that these produce a periodic torque like

$$T(\sigma;t) = T_0 \sin\left(\pi\frac{t}{\tau}\right),$$

they will force the vessel to oscillate on the waters with this same period τ. Without further mathematical developments Bernoulli remarks this is entirely similar to the phenomenon of the driven pendulum and he produces the relevant formula for the amplitude of such driven oscillations, adapted to the oscillations of a ship. If, following Bernoulli, we take the natural period of the ship to be n times slower than the external torque (due to water masses)[41] (i.e. $\frac{\tau_0}{\tau} = n$), formula (15) becomes:

40 Bernoulli calls «period» what in modern parlance is called half-period, viz. the time between two extreme elongations and not the time to come back to the initial position and velocity. We shall follow Bernoulli's convention.

41 In practice $n < 1$, the ship oscillates faster than the sea.

$$|\sigma_{\max}| = \left|\frac{\theta_0}{1-n^2}\right| , \tag{17}$$

where θ_0 is the angle due to the statically applied maximal torque and $\sigma_{\max}$ is the elongation of the ship's oscillation. One could alternatively express $\theta_0 = \frac{T_{\max}}{F}$ and hence

$$\sigma_{\max} = \frac{\tau^2 T_{\max}}{(\tau^2 - \tau_0^2)F} \tag{18}$$

which is the other formula he exhibits in §25 and discusses in §26. We point out already now that $\tau_0^2 F = \pi^2 I$, i.e. the second term in the denominator is nothing but the moment of inertia which does not depend on either the surface at water level nor on the distance between the centre of mass and the centre of buoyancy, two parameters which govern the stability.

4.1.5 Resonance (§26 p. 279 h. v.)

It is immediately clear that trouble is in store as soon as $\tau = \tau_0$, which is commonly called resonance. In almost all circumstances the period of the seawaves is much longer than the proper («natural») frequencies of the common vessels, but it is a phenomenon Bernoulli has constantly in mind and to which he keeps coming back. He will never go beyond this formula and does not e.g. investigate the effect of damping on the resonance.

4.2 *A parenthesis on gauging ships (§29 p. 285 h. v.)*

Without really mentioning it, Bernoulli takes part in a discussion on mathematical methods of gauging ships, by citing a result of Bouguer and putting forward his own, simpler, estimate.

Bernoulli refers to the chapter in Bouguer[42] on gauging a ship. This is done by calculating the volume between the sections at sea level for a charged and for an empty ship and involves measuring the area of the sections. Bouguer, as an example, gives the widths at fifteen feet intervals of a ship 120 feet in length:

1, 16, 26, 28, 29, 28, 27, 22, 1

42 P. Bouguer, *Traité du navire, de sa construction et de ses mouvements*, Livre II, Paris: Jombert, 1746, Sect. I, Ch. VI.

and approximates the area by a sum of trapezes; he finds it measures 2655 square feet. Bouguer is in fact carrying on a debate with Varignon[43]. Both were involved in a request to the *Académie* by the *Conseil de Marine* to ascertain the best method for gauging ships amongst the many in use in various French and foreign ports. Varignon proposed to approximate the hull by an ellipsoid and then to compute its sections by the exact geometrical formulae. The semimajor axis would be the width of the vessel midships. Bouguer did measure a ship very carefully and found Varignon's method to be off by $\frac{1}{7}$. His thesis was that one should simply resort to some sort of numerical integration, e.g. by trapezoids[44]. Bernoulli's remark is both a criticism of the method of trapezes (whose results fall always short of the true area) and a rescue of the simplified method by ellipses. This can be improved simply by approximating the hull by two half ellipsoids having common minor and semimajor axes. One should take those at the ship's largest (vertical) section, which is not midships but commonly at about $\frac{5}{12}$ of the length from the bow. The formula of Varignon is not changed by this. Stating just the area obtained that way, as Bernoulli does, is of course no proof. The curious reader can convince himself that the numbers of Bouguer can be better approximated by two half ellipses of half major axes of 70 and 50 feet than by a single ellipse having half a major axis of 60 feet. Evaluating the area by Simpson's method of integration, which should be better adapted to the curves of a ship, one finds 2710 square feet, quite close in fact to the 2734 of Bernoulli.

4.3 *Scaling laws and applications (§§27–33 pp. 281–290 h. v.)*

The scaling laws follow at once from a dimensional analysis of the formulae.

1. The stability F, i.e. the ratio of the torque to the angle of heel: from (4) the stability is proportional to the 4th power of the dimensions (called the *raison biquarrée* by Bernoulli).
2. The mass is proportional to the volume (we assume obviously constant densities) and hence to the 3rd power of the dimensions.
3. The moment of inertia, under the same assumptions, is proportional to the 5th power.

43 Hist. Paris 1721 (1723), pp. 43–51.

44 The full story can be read in P. Varignon, *Jaugeage d'un navire ellipsoide*: Mém. Paris 1721 (1723), pp. 44–56, and in J.-J. Dortous de Mairan, *Remarques sur le jaugeage des navires*: Mém. Paris 1721 (1723), pp. 76–107.

4. According to (16) τ_0 is proportional to $\sqrt{\dfrac{I}{F}}$ and hence to the square root of the dimension (*la raison sous-doublée*).
5. As important as the period is the instrument to check it, i.e. the pendulum. The length of a pendulum is proportional to the square of its period. Hence a pendulum isochronous to the oscillations of a ship has its length proportional to the dimensions of the ship.
6. Doubling the dimensions increases the surface of the sails fourfold, and also the force exerted on the masts. Those are the lever arms for the torque induced by the wind, which therefore increases eightfold, i.e. as the cube of the dimensions. The stability, however, increases as the fourth power; the same wind will therefore incline the larger ship to a lesser degree, i.e. inversely proportional to the dimensions.
7. The position of the centre of mass, if we would increase all dimensions by the same factor, would be found in the homothetic point. But in practice the builders do not simply scale up a smaller boat to construct a larger one. The larger vessels carry correspondingly more and larger guns and the form of the hull is not the same. The remark by Bernoulli that the larger vessel has its centre of gravity above the homothetic point is a purely factual one and not related to scaling.

4.4 *The rotating slab*

Where is the axis of rotation of an oscillating ship? Both Euler and Bernoulli are groping for a clear view of the problem and the point is constantly argued over in their correspondence. We show the theorem of Bernoulli, viz. that the axis goes through the centre of mass, first by giving a modern derivation, showing the limits of the theorem, and subsequently we present the arguments he used.

4.4.1 The rotation axis

The Lagrangian solution. If a floating object – and Bernoulli considers here again a vertical slab, cut from the broadest part of the ship – is turned over an angle θ_0 from its equilibrium position and suddenly released, it will make angular oscillations. There is no a priori fixed axis, such as occurs in a pendulum or the escape wheel of a watch. The object is so to say free to pick its own axis. We shall first formulate the complete problem of the floating slab in modern terms, i.e. by using Lagrangian mechanics. The position of a figure in the plane can be given by the cartesian coordinates q_1 and q_2 of a given mark O on the figure and the angle θ a line through O makes with e.g. the vertical direction. Any motion can be split up in a translation of the figure and a rotation around O. The kinetic energy is given by:

$$E_{\text{kin}} = \frac{1}{2}\int \rho \dot{R}^2 \, d^2R \ ,$$

where $\vec{R}$ is the position coordinate of a point of the slab, referred to a fixed origin outside and ρ is the mass density of the slab. The potential energy is a function $V(q_1, q_2, \theta)$ depending on the shape of the slab. For a ship oscillating freely and acted on only by weight and buoyancy, V will not depend on the horizontal coordinate q_1, and the Lagrangian equation related to q_1 is exclusively determined by the kinetic energy. The velocity $\dot{\vec{R}}$ of any point can be decomposed as

$$\dot{\vec{R}} = \dot{\vec{q}} + \dot{\vec{\theta}} \times \vec{r} \ ,$$

where $\dot{\vec{\theta}}$ is the angular velocity around O and $\vec{r}$ is the position coordinate referred to O. The final form of the kinetic energy can be obtained by standard methods, as explained e.g. in Sommerfeld[45]:

$$E_{\text{kin}} = \frac{1}{2}m(\dot{q}_1^2 + \dot{q}_2^2) + \frac{1}{2}I_O\dot{\theta}^2 + m(x_A\dot{q}_2 - y_A\dot{q}_1)\dot{\theta} \ , \qquad (19)$$

where x_A and y_A are the coordinates of the centre of mass A relative to the origin O and m is the mass. The potential energy, $V(q_2, \theta)$, is independent of the horizontal coordinate q_1 because all forces are vertical. The moment of inertia around O is related to the moment of inertia around A, the centre of mass, by the theorem of Steiner:

$$I_O = I_A + a^2 m \ , \qquad (20)$$

where a is the distance OA. We write the Lagrangian equations of motion and simplify them by using (20) and noting that $\dfrac{\partial V}{\partial q_1} = 0$; this gives:

45 A. Sommerfeld, *Vorlesungen über Theoretische Physik I, Mechanik*, 4^e Auflage, Wiesbaden 1949, Ch. IV, art. 22.

$$I_A\ddot{\theta} = x_A\frac{\partial V}{\partial q_2} - \frac{\partial V}{\partial \theta}\,, \tag{21}$$

$$m\ddot{q}_1 = my_A\ddot{\theta}\,, \tag{22}$$

$$m\ddot{q}_2 = -mx_A\ddot{\theta} - \frac{\partial V}{\partial q_2}\,. \tag{23}$$

From (22) follows that O will perform an horizontal oscillation synchronous with the angular oscillation, and having the acceleration $y_A\ddot{\theta}$, i.e. an oscillation with its axis in A.

In addition a vertical oscillation will occur; but a vertical movement of A, combined with a rotation around A is equivalent with a rotation around a point O', to the right (or the left) of A; the argument can be found in §35 and is illustrated by *Fig. 4* of Bernoulli. This instantaneous axis of rotation O' will wander, unless θ and q_2 would oscillate in unison.

Let us presently simplify the problem by assuming that O is on the axis of symmetry, i.e. $x_A = 0$. The movement of the slab will in general be a complicated one because the potential energy V contains cross terms between q_2 and θ (i.e. between heave and roll) and cannot be described by a simple oscillation; a remark Bernoulli made already in *De motibus oscillatoriis corporum humido insidentium*[46]. But as remarked there, those terms vanish for small movements, and in that case we obtain a simple superposition of an harmonic oscillation in θ and one in q_2.

Observe that by tilting a ship very slightly we do not raise perceptibly its centre of mass and hence do not excite the vertical oscillation; the movement will be exclusively a (small) oscillation around an axis through the centre of mass A.

The Bernoulli solution (§34 p. 290 h. v.). We now turn to the arguments of Bernoulli. In the present memoir he has tacitly assumed there is no vertical movement, drawing the conclusion that if A has no horizontal movement it must be on the axis of rotation; that is his first argument.

His second argument can be summarized as follows. We neglect anew any translational movement of O and «assume» a torque acts on the ship, i.e. weight and buoyancy are equal, meaning that at all times an Archimedian equilibrium is maintained and no extra coupling terms between θ and q_2 arise in the potential energy. The torque of course does not depend on the

46 Cf. D. Bernoulli St. 32, *De motibus oscillatoriis corporum humido insidentium*: CP XI, 1739 (1750), pp. 100–115 – *Werke* 6.

axis. A second assumption is that the torque is proportional to the angle of heel θ. The third assumption is that the kinetic energy is of the form:

$$\frac{1}{2}(I_A + a^2 m)\dot{\theta}^2 ,$$

where a is a parameter to be determined. Comparing with (19) we see that this is tantamount to assume that $\dot{q}_1$ and $\dot{q}_2$ are negligible, i.e. no horizontal or vertical movements of O are present; the equation of motion is then:

$$(I_A + a^2 m)\ddot{\theta} = -F\theta \tag{24}$$

and, as Bernoulli remarks, F is independent of the choice of O. The equation is easily solved:

$$\theta(t) = \theta_0 \sin\left(\sqrt{\frac{F}{I_A + a^2 m}}t\right) \tag{25}$$

which is an oscillation with amplitude θ_0 and period (in the modern meaning) $\tau_0 = 2\pi\sqrt{\frac{I_A + a^2 m}{F}}$.

Bernoulli thereupon uses the argument that, as we displace O, the r.h.s. of (24) stays the same; therefore if a decreases the variation of $\dot{\theta}$ increases, for the same time interval we obtain a larger angular velocity $\dot{\theta}$ or, alternatively, a smaller inertia. He then appeals to a principle that forces the movement to maximize the angular velocity or to minimize the inertia: «*... il est donc clair que le centre de rotation se placera de lui-même là où le plan* [i.e. the slab] *aura la moindre inertie pour recevoir le même degré de vitesse angulaire.*»[47] Next a theorem is invoked, on a solid pivoting around a fixed axis, showing that this acquires the largest angular velocity in a given time if it pivots around its centre of mass. It is not clear that the theorem – meant for a fixed axis and one degree of freedom – applies to the present case with two degrees of freedom where we are looking for a «spontaneous axis».

This principle of minimal inertia is one of the many that were proposed at the time in the quest for the foundation of mechanics; see e.g. the work

47 Cf. p. 291 h. v.

of Maltese[48]. It was already used in *De percussione excentrica*[49] and is discussed in the Introduction[50] to that work[51]. As remarked there, we shall see that also in the present problem it can be related to the principle of minimal kinetic energy of Maupertuis (published eleven years later). The hypothesized solution will be like

$$\theta(t) = \theta_0 \sin\left(\sqrt{\frac{F}{I_A + a^2 m}}\, t\right)$$

and the kinetic energy, averaged over a period, will therefore be

$$E_{\text{kin}} = \pi \theta_0^2 \sqrt{F} \sqrt{I_A + a^2 m} \ .$$

The principle of Maupertuis would require to select the mode of oscillation that, at constant total energy, makes this expression a minimum. That the total energy is constant entails that θ_0 is constant, only a can be varied and the minimum is obtained for $a = 0$.

4.5 *Movement under impact (§§ 35–36 pp. 292–295 h. v.)*

As the argument mixes concepts from the theory of floating objects with those of a physical pendulum, we shall derive the result in a way more familiar to the reader and show the connection of both problems afterwards. Imagine a freely floating object subject to a horizontal force $\vec{F}$ during a short time interval dt. Let the centre of mass be A; we assume $\vec{F}$ lies in the vertical plane through A. Let the line of the force $\vec{F}$ meet the vertical through A in D (see Fig. 5) and set $\overline{AD} = y$. We consider the point D' on AD, on the other side of A, such that also $\overline{AD'} = y$. Imagine now in D' two opposing horizontal forces of magnitude $\frac{F}{2}$, and in D two equal forces $\frac{\vec{F}}{2}$,

48 G. Maltese, *La storia di « F = ma». La seconda legge del moto nel XVIII secolo*, Firenze: Olschki, 1992, p. 92.

49 D. Bernoulli St. 27, *De percussione excentrica*: CP IX, 1737 (1744), pp. 189–206 – *Werke* 3, pp. 145–159.

50 D. B. *Werke* 3, pp. 58–61.

51 It was used also, as selfevident, by Chauchot on p. 5 of his memoir of 1755. (Cf. M. Chauchot, *Sur les mouvemens de roulis et tangage des navires*, Paris: G. Martin, H. L. Guerin, L. F. Delatour & A. Bourdet, 1755.)

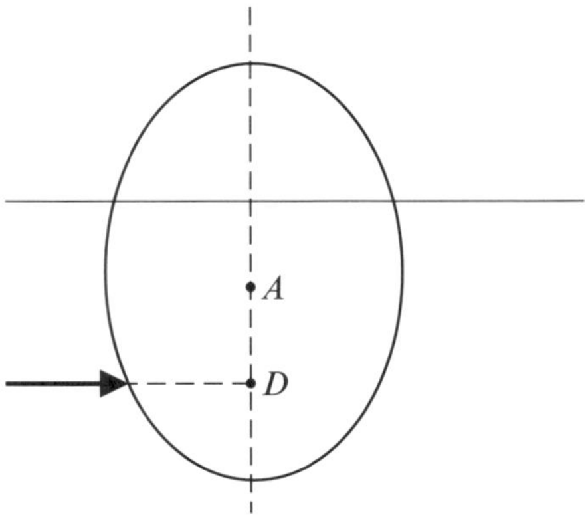

Figure 5: A floating body hit by a force in D.

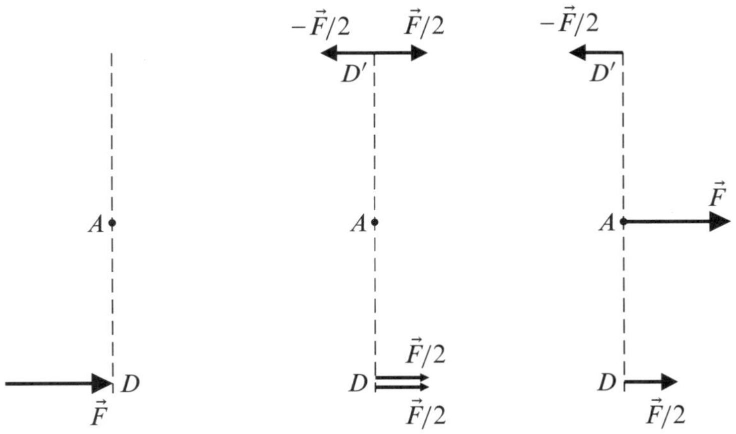

Figure 6: The three setups are equivalent, the third gives manifestly a torque and a force on the centre of mass.

which is strictly equivalent with the previous hypothesis, see Fig. 6. This argument was known to Bernoulli because it was used (in a more involved way) by Chauchot.

By the well-known laws of combining forces on a solid this is equivalent with a torque of magnitude yF and a force $\vec{F}$ applied in A. Consequently the body will be set in motion: a horizontal translation over x, combined with a rotation over θ around A will be the result:

$$x = \frac{1}{2}\frac{F}{m}(dt)^2 ,$$
$$v = \dot{x} = \frac{F}{m}\,dt ,$$
$$\theta = \frac{1}{2}\frac{yF}{I_A}(dt)^2 ,$$
$$\omega = \dot{\theta} = \frac{yF}{I_A}\,dt ,$$

where m is the mass and I_A the moment of inertia around A. We shall find the line DD' has moved to FH, which cut each other in B: the line has made an infinitesimal rotation around B, *the spontaneous rotation axis*, as Bernoulli calls it. From Fig. 7 we see that

$$\overline{AB}\,.\tan\Big(\frac{yF}{2I_A}\,dt^2\Big) = \frac{F}{2m}\,dt^2$$

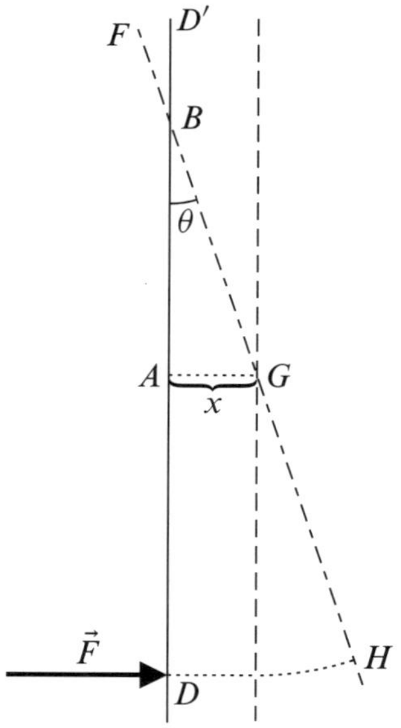

Figure 7: The effect of the torque and the force on the line DD'.

and for small *dt* this simplifies to

$$\overline{AB} = \frac{I_A}{ym} \, .$$

Introducing the radius of gyration a by $I_A = a^2 m$ and calling $\overline{AB} = b$, we find:

$$b = \frac{a^2}{y} \, . \tag{26}$$

Consequently, the closer to the centre of mass we hit the object, the further away is the spontaneous axis of rotation and the smaller is the rotation. Let us now turn to the arguments used by Bernoulli in §§ 34–38 (pp. 290–296 h. v.), which are derived directly from his pioneering work on the collision of a bar with a sphere, *De percussione excentrica*. We refer the reader to the extensive Introduction[52] in *Werke* 3. Already in the article on floating objects[53] the same reasoning was used. We point out the main points. By comparing the Fig. 7 with *Fig. 4* of Bernoulli it is clear that he distinguishes clearly the translation and the rotation movement. He does not obtain it, as we did, from the equations of motion of the solid but by arguments related to the collision laws of Huygens; although he does refer to the angular velocity he still uses the ratio of *cg* to *cb* to define the angle, as he did in *De percussione excentrica*[54]. He keeps coming back to the mechanics of the compound pendulum and interprets the formula (26) – which is the only relevant one in the present context – as the relation between the distance from a hypothetical suspension point D to the centre of mass (our y, his x) and the length of oscillation ($\overline{DB} = y + b$). This was explained in the comments to *De motibus oscillatoriis corporum humido insidentium*[55]. His parameter m is what we call the radius of gyration a; he cites two properties of this parameter (see note to § 37, p. 295 h. v.) but it is clear that neither of them plays any role in the present problem. As it was through considerations of the compound pendulum that he found his way to the solution of the problem at hand we find those preoccupations back in the text.

52 D. B. *Werke* 3, pp. 40–61.

53 Cf. D. Bernoulli St. 30, *De statu aequilibrii corporum humido insidentium*: CP X, 1738 (1747), pp. 147–163 – *Werke* 6.

54 Cf. D. Bernoulli St. 27, *De percussione excentrica*: CP IX, 1737 (1744), pp. 189–206 – *Werke* 3, pp. 145–159.

55 Cf. D. Bernoulli St. 32, *De motibus oscillatoriis corporum humido insidentium*: CP XI, 1739 (1750), pp. 100–115 – *Werke* 6.

4.6 *Increasing the load (§ 44 p. 301 h. v.)*

Provided the walls of the ship are approximately straight and parallel an increased load will not change the area at the water line (term (6) in the expression for the stability), it only affects the second term (7) which is negative and depends on the distance between the centre of mass and the centre of buoyancy. The purpose of the calculation is to find out how this distance is changed. The relevant points are sketched in Fig. 8.

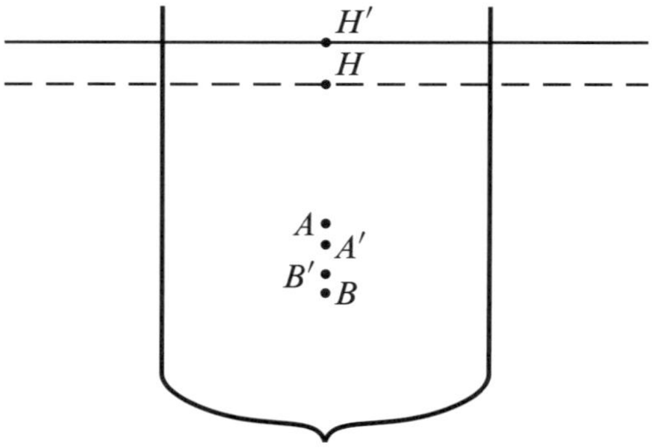

Figure 8: Effect of increased load: A = centre of mass, B = centre of buoyancy, H = water level; the letters with an accent denote the positions after the load has increased.

Bernoulli introduces the following variables:

- $b = \overline{AH}$,
- $c = \overline{BH}$,
- S = horizontal area at the water level,
- aS = volume of the displaced water = weight of the ship,
- αS = increase in the volume of the displaced water due to the increased load = added weight,
- β = distance between the centre of gravity A and the added mass (positive if lower than A).

We have to compute $\overline{A'B'}$.

- The position of A': by the definition of the centre of mass («law of the lever»)

$$\overline{A'A} \,.\, aS = (\beta - \overline{A'A}) \,.\, \alpha S \ ,$$

hence $\overline{A'A} = \frac{\beta\alpha}{a+\alpha}$ and

$$\overline{A'A} + \overline{AH} = \overline{A'H} = b + \frac{\beta\alpha}{a+\alpha} \ .$$

– The position of B': similarly, by the definition of centre of mass:

$$\overline{BH} \ . \ aS - \frac{1}{2}\alpha \ . \ \alpha S = \overline{HB'}(a+\alpha)S \ ,$$

hence

$$\overline{HB'} = c + \frac{-\frac{1}{2}\alpha^2 - \alpha c}{a+\alpha} \ .$$

– Distance $\overline{A'B'}$:

$$\overline{A'B'} = \overline{HB'} - \overline{HA'} = \overline{AB} - \frac{\alpha c + \frac{1}{2}\alpha^2 + \beta\alpha}{a+\alpha} \ .$$

Comparing the second term (7) in the stability before and after the load is added, we have $-aS(c-b)$ before and $-(a+\alpha)S\left(c-b-\frac{\alpha c + \frac{1}{2}\alpha^2 + \alpha\beta}{a+\alpha}\right)$ after, whose difference is:

$$S\left(\alpha b + \frac{1}{2}\alpha^2 + \alpha\beta\right)$$

which is the increase in stability. The last formula published by Bernoulli in §44 is easily simplified to give the above result.

4.7 *Recommendations*

Bernoulli's recommendations appear somewhat scattered in the memoir, we gather them here for the convenience of the reader. They can all be linked to the distinct factors in his formula for the stability (5). Increasing the stability will make the ship less sensitive to perturbations by air and sea in the first place; it will also shorten the period of the free oscillations[56].

56 *«Si la stabilité est très grande les accélérations peuvent être excessives, rendant le bateau inconfortable et pouvant aller jusqu'à des dégâts sur le bateau.»* (A. L.)

There is no difference of principle in the treatment of pitching as compared to rolling. A pitching ship rotates around an axis through the centre of mass, orthogonal to the length; the moment of inertia is much larger, the period longer and the maximal angles smaller. The damping of these oscillations is much more pronounced than for rolling and at the end of the essay Bernoulli will come back to this. Looking closely at (4) or (5), the various terms suggest the following means to increase stability.

4.7.1 Increasing the area at the water line (§26 p. 279 h. v.)

This can simply be done by building bigger ships. At the same time the size of the ship will have consequences for other properties; in §26 Bernoulli states a number of scaling laws and investigates in all the necessary detail how the formula (18) will be affected if the size is increased n-fold.

Keeping the size of the ship constant, as well as the ratio of width to length, a rectangular imprint of the hull on the sea surface has a larger area than an elliptical one. Along the water line then a ship should be like a rounded rectangle, and not only in the upright position but for all practical angles of heeling. This pleads for straight vertical sides for some distance above and below the water line.

4.7.2 Increasing the moment of inertia (§38 p. 296 h. v.)

This is related to $p(\sigma)$ in (6) and means putting on any deck the heavier cargo towards the sides. It also means putting the heavy loads further down, but this affects mainly the second term T_2. This rule, however, must always be considered together with the rule on resonance.

4.7.3 Lowering the centre of mass (§44 p. 301 h. v.)

This is related to the parameter s in the second term, (7). For maximal stability the centre of mass should lie below the centre of buoyancy because then the sign of this term is reversed and it would increase the stability. But ships in the 18th century had it the other way[57] and the negative sign means this term stands for instability. As it is, all means should be used to lower the centre of mass and increase the moment of inertia: use high-density ballast and put it close to the keel, put the heavy cargo down, build light superstructures (including masts not too high). Bernoulli gives a handy

57 Modern ships as well.

means of computing the distance of the centre of mass from the centre of buoyancy by simple measurements in the harbour[58].

4.7.4 Avoiding resonance (§§24–43 pp. 278–301 h. v.)

We should measure accurately the free oscillations of every ship, it is a fundamental property every pilot should be aware of. More observations and research should be devoted to the movements of the sea; this way one will be able to detect the onset of periodic movements in the vicinity of the natural frequency of the vessel and take countermeasures. A suggested measure is to change the moment of inertia of the ship and the position of the centre of gravity. Another is to change course so as to meet the impact of the waves at a different frequency. As far as our present experience shows, Bernoulli stresses, the periodic movements of the sea are slower than the natural frequencies of our ships; in order to remain on the safe side of the resonance one should try to keep the natural frequencies fast. This can be obtained by lowering the moment of inertia, e.g. by concentrating the cargo towards the middle of the ship (i.e. towards the centre of mass); but this of course decreases stability.

4.7.5 Increase damping (§§38–41 pp. 296–299 h. v.)

Structures which protrude from the hull and run along the length of the vessel will tend to damp the rolling[59]. A sharp keel works this way. Bernoulli thinks furthermore of sidebands slightly above the water line which, during rolling, would dip into the water and brake the oscillations. In pitching, damping occurs naturally as the fore and aft parts of the ship are alternately lifted out of the water but in the next instant hit it with their full surface.

4.7.6 The ideal form of a ship (§33 p. 289 h. v.)

A ship should have vertically straight sides at all points which are likely to dip in and out of the water. Its imprint on the sea surface should be

58 See §56 p. 315 h. v. The dimensions and the displacement of the ship should be known. By installing beams protruding from the side of the ship and suspending weights on them one imparts a known torque on the ship; the resulting angle of heel should be measured. Assuming the hull of the ship is a half ellipsoid (or two half ellipsoids with a common minor axis) the term T_1 can be evaluated. Its difference with the torque is an estimate for T_2 and hence for s.

59 This is exactly what bilge keels do in modern ships.

rectangular[60] (bow and stern excepted). Below this region the vertical section should fast grow narrow, then for the lowest part stay narrow with a strong extruding keel. A high-density ballast should be loaded in the narrow part above the keel. Bow and stern should be higher because in pitching they plunge into the water. The top of a ship can have volume, the bottom must have weight.

5 The oscillating ship on a wavy sea

5.1 *Ocean waves*

5.1.1 The model for surface waves (§46 p. 304 h. v.)

The main idea of Bernoulli could be summarized as follows. The water in the ocean moves as if it were contained in an array of tubes; in each tube the water oscillates. The ocean therefore is a collection of oscillators. This system should then behave as that paradigm of all coupled systems: the composite pendulum. Through it Bernoulli discovered the phenomenon of the normal modes[61]; he now states that the waves at the surface of the sea result from the normal modes of the oscillating watercolumns. He supposes these modes are excited by the wind, rather like blowing across a pipe excites the normal acoustic modes of the pipe. We shall presently consider this model in detail.

5.1.2 Oscillating water columns (§47 p. 306 h. v.)

U-tubes and Newton's model. The oscillations in a simple U-tube are well known[62]: it is synchronous with a pendulum of length $\frac{1}{2}L$, and hence has a frequency $\omega = \sqrt{\frac{2}{L}}$, where L is the length of the watercolumn in the U-tube, which is easily proven: in a U-tube, water at rest will have the same level in both vertical branches of the tube. Away from rest the level in e.g. the left branch will move up and down; we call $x(t)$ its distance from the equilibrium position (see Fig. 9). The difference between the levels in one branch and the other is $2x(t)$. Hence there is a force of $-2x(t)S\rho g$ on the

60 *«La forme rectangulaire à coins arrondis provoquerait immanquablement des vagues, d'où une augmentation de résistance à l'avancement inadmissible en pratique.»* (A. L.)

61 D. Bernoulli St. 23, *Theoremata de Oscillationibus corporum filo flexili connexorum*: CP VI, 1732/33 (1739), pp. 108–122.

62 Newton, *Principia* (1726), Book II, Proposition XLIV, Theorem XXXV (in the first edition of 1687, it is known as Theorem XXXIV).

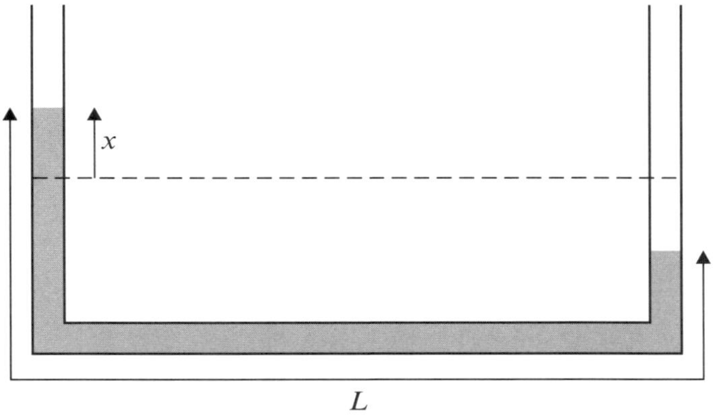

Figure 9: The water in a U-tube of length L which joins points at equal distances from two nodes (and hence half a wavelength apart) will oscillate with the same frequency, whatever its position or the amplitude.

water column (S = surface area, ρ = density, g = acceleration of gravity). This column has a mass of $LS\rho$ (L = total length of the column). The column moves as a whole, its velocity is identical to the velocity of its levels. Hence, by Newton's equation:

$$LS\rho\ddot{x} = -2xS\rho g$$

with solution

$$x(t) = x_0 \cos\left(\sqrt{\frac{2g}{L}}t + \phi\right)$$

(ϕ is an arbitrary phase angle, x_0 is the maximal elongation). Comparing this with the oscillation of a mathematical pendulum of length l:

$$\theta(t) = \theta_0 \cos\left(\sqrt{\frac{g}{l}}t + \phi\right)$$

we see that the liquid in the U-tube oscillates as a pendulum of length $\frac{L}{2}$.

Note that this oscillator is isochronous: the period is independent of the amplitude of the oscillation. Newton proposed[63] that the wave's movement

63 Newton *Principia* (1726), Book II, Proposition XLVI, Problem X (in the first edition known as Problem XI).

could be understood if any two points near the surface and half a wavelength apart would be joined by a U-tube (see Fig. 9), the frequency is then exactly the same near a node or near a maximum. In order to explain the movement of the waves we need admit the existence of an array of crossing tubes, as shown in Fig. 10.

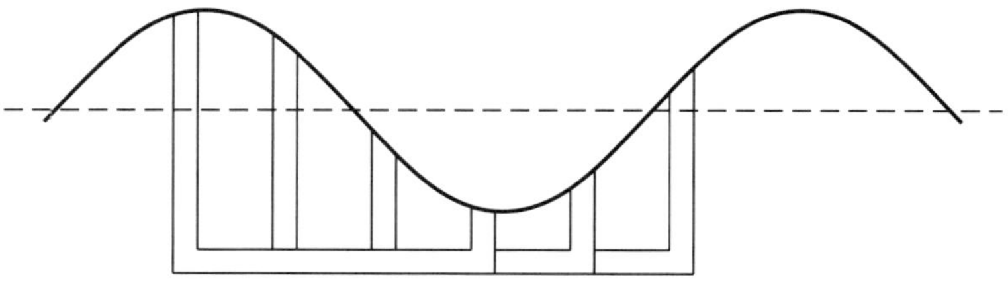

Figure 10: The standing sinusoidal wave can be explained by an array of crossing tubes, which share a common frequency, independent of the amplitude.

That those tubes have to cross each other makes the model hard to believe, claims Bernoulli, and he presents a much better solution.

V-tubes (Johann Bernoulli). In his *Hydrodynamica* (St. 31) Bernoulli refers to an article by his father Johann[64] on oscillations of a watercolumn in a V-tube. Suppose we replace the U-tube of the previous section by a tube with sloping ends, making an angle α with the horizon; water in such a tube will oscillate with a frequency $\omega = \sqrt{\dfrac{2\sin\alpha}{L}}$.

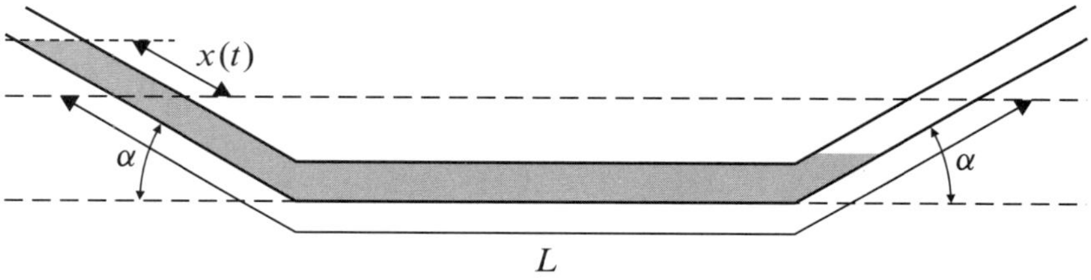

Figure 11: In a V-tube the vertical difference in levels, the so-called head, is $2x\sin\alpha$.

64 Joh. B. Op. CXXXVII, *Theoremata selecta pro conservatione virium vivarum demonstranda et experimentis confirmanda. Excerpta ex Epistolis datis ad filium Danielem*: CP II, 1727 (1729), pp. 200–207.

We measure $x(t)$, the distance of the level from the equilibrium position, along the sloping tube (see Fig. 11). The force of the excess column (length $2x$) on the underlying column must be taken in the direction of motion, i.e. along the slope. The equation of motion of Newton gives now

$$LS\rho\ddot{x} = -2x\sin(\alpha)S\rho g$$

with the solution

$$x(t) = x_0 \cos\left(\sqrt{\frac{2\sin(\alpha)g}{L}}t + \phi\right),$$

i.e. «tautochronous» with a mathematical pendulum of length $\dfrac{L}{2\sin\alpha}$.

Bernoulli's model. Imagine a tube joining two points which lie symmetrically around a node, and make the slope of the tube (i.e. $\sin\alpha$) near the surface proportional to the length of the tube. Think then of an array of nested tubes, as in Fig. 12. A set of tubes with the sines of their slopes pro-

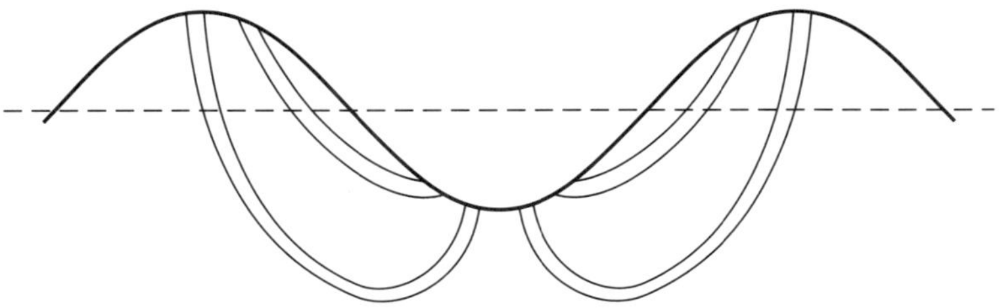

Figure 12: The standing sinusoidal wave can be better explained by an array of nested tubes, which also share a common frequency.

portional to their total lengths will have:

$$\frac{L}{2\sin\alpha} = \text{constant}$$

and the water columns in all the tubes of the set will vibrate with the same frequency; the water in all such tubes can therefore oscillate in phase: all the liquid at the surface will then move together and simulate a standing wave.

5.1.3 Wind and waves (§46 p. 304 h. v.)

Bernoulli has no doubts that the waves on the sea are created by the wind, blowing over the surface. There is no hint as to a mechanism for this interaction. But one remark, which he presents as self-evident, is important: it takes the wind exactly one period of the standing ocean wave to move from one crest (or through) to the next. I.e. the wind travels one wavelength in one period[65].

5.1.4 Seawaves, the Airy theory

The model of Bernoulli explains the motion of the waters in a swelly sea as if they would oscillate in a system of nested tubes. A hundred years later this hypothesis was vindicated by the progress of hydrodynamics. A mathematical model for standing waves, based on Euler's equations for hydrodynamics, was made by Green and developed by Airy. Details can be found in Lamb[66]. It follows that the surface waves can be described by the following formula (in Lamb's notation):

$$\eta(x,t) = a\cos(kx)\sin(\sigma t + \varepsilon) \ .$$

It is here assumed that the water is incompressible and the flow is irrotational, consequently there is a velocity gradient. The geometry is two-dimensional: x and z are in the horizontal plane, but it is assumed that no variables depend on z; $y = 0$ is the plane equilibrium surface of the water. That surface is a distance h above the level seafloor; $\eta(x,t)$ is the height of the water surface at the point x and the moment t, referred to the equilibrium plane. The relation between time and space frequencies is, according to the model

$$\sigma^2 = gk\tanh(kh) \ .$$

The ensuing relation between period and wavelength is displayed in Fig. 13.

65 There is a similarity with Newton's remark: «... a wave will describe its own length in the time in which the pendulum oscillates twice» (Newton, *Principia* (1726), Book II, Proposition XLVI, Problem X). This, however, refers to standing waves and uses a notion of velocity adapted to those; it is **not** understood as the velocity of travelling waves.

66 H. Lamb, *Hydrodynamics*, 6th ed., New York: Dover publ., 1945, Ch. IX.

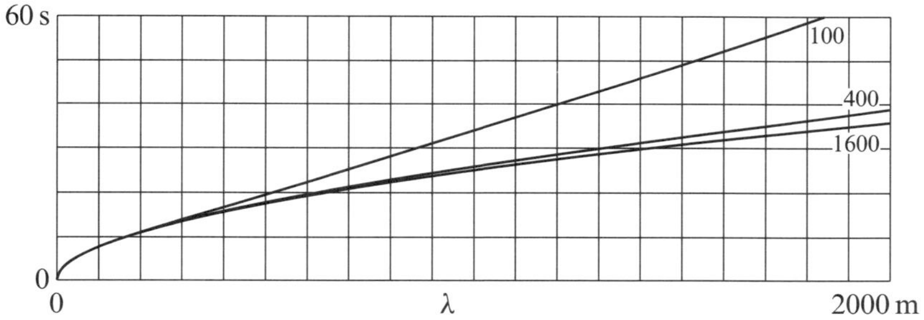

Figure 13: The relation between wavelength (in meters) and period (in seconds) for standing waves in the sea; the sea is resp. 100, 400, 1600 m deep.

The same theory allows to draw the lines of flow along which the water moves from crest to hollow; this is displayed in Fig. 14.

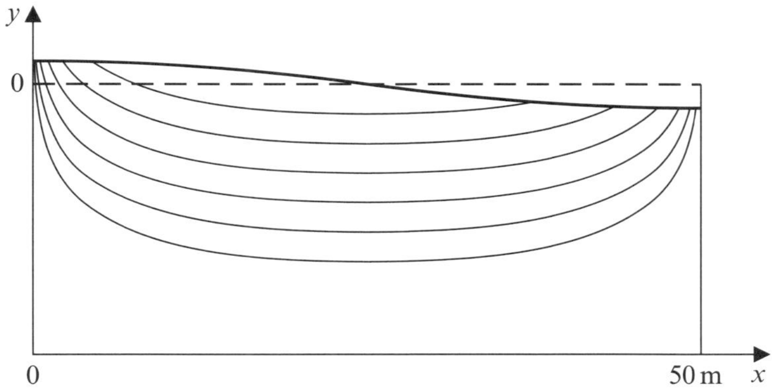

Figure 14: Lines of flow for standing waves from crest to hollow, the seafloor is at the bottom of the figure. The amplitude of the movements decreases rapidly with the depth.

From this figure one can read off the truth of Bernoulli's hypothesis; although the tubes are not strictly of the same width all over their length, they do show the general relation between length and terminal slope he postulated.

As long as we are on the open sea there are no boundary conditions which could impose fixed wavelengths and frequencies. The depth of the seafloor influences only the dispersion of the waves, i.e. the relation between frequency and wavelength. His analogy between acoustic waves in a pipe and surface waves on the ocean breaks down in this respect. That the periods of waves are proportional to the square roots of the wavelength, as «is sufficiently admitted», is strictly true for a deep sea (when $h \gg \lambda$) only.

Superposing two standing waves with phases $\pi/2$ apart we can obtain, in the same theory, a model of running surface waves. Their phase velocity is

$$v_{\text{phase}} = \frac{\sigma}{k} = \sqrt{\frac{g\lambda}{2\pi}\tanh\left(\frac{2\pi h}{\lambda}\right)}\,.$$

Looking at the movements of the water molecules under the surface, these are, for the running waves and far enough from the surface, small circles whose radius[67] is ae^{ky}. When Bernoulli assumes that the waters are perturbed to a depth proportional to the wavelength he is essentially correct. His hunch, that this depth could be six times the wavelength is a gross overestimate. Taking his example of a wind speed (and hence wave speed) of 30 feet/second one would obtain with the present theory and for a deep sea a wavelength of 64 m or about 200 feet, against his result of 1950 feet.

5.1.5 Successes and pitfalls of an analogy, hindsight

The streamlines. The Fig. 14 shows the streamlines. Comparing the theory of Bernoulli with the formulae of Airy we remark that indeed:

- the streamlines form an array of nested tubes,
- the slope at the surface varies from horizontal at the node to vertical at the through and at the crest.

Progressive waves. The model of Bernoulli refers exclusively to standing waves. But from there it is a short way to travelling waves: by combining two standing waves a quarter wavelength out of phase we can make two progressive waves. Their phase velocity makes them travel a wavelength in one period. It is clear that a gusty wind will excite exactly those waves which travel at the same speed: a travelling wave will be able to pick up energy by being pushed along by a moving local variation in the air pressure. This way the cryptic remark by Bernoulli is seen to be correct[68].

67 Remember that for points lower than the equilibrium plane $y < 0$.

68 But still an oversimplification. Swell is caused by a storm in a region thousands of kilometers away and days ago, where wind and water interact in a most chaotic manner. In spreading, the longer waves move faster away from the disturbance. At a given place and time this gives a swell with a fairly defined wavelength. These

The delusion of normal modes. In his comparison of ocean waves with acoustic waves in a pipe Bernoulli goes so far as to predict that the periods of the waves will be in the ratio's of the uneven numbers 1, 3, 5, 7, ... depending on the depth of the water (analogous to the tones of a semiclosed pipe). This proved to be wrong; the depth of the ocean influences the dispersion and the velocity of the waves, but not the frequency. Normal modes do occur, but only in a fairly small confined sea, due to the reflections on the shores.

5.2 *Ships in a swell*

In Chapters I to III, (pp. 253–290 h. v.) Bernoulli examined the equilibrium of a vessel, its free oscillations, its response to incidental or periodic forces; in Chapter IV (pp. 290–303 h. v.) he gave a theory of surface waves. From now on he will again take up the questions of stability, but this time for a ship riding the ocean's swell. He considers this as a very extended system of surface standing waves whose wavelength is much larger than the size of a ship.

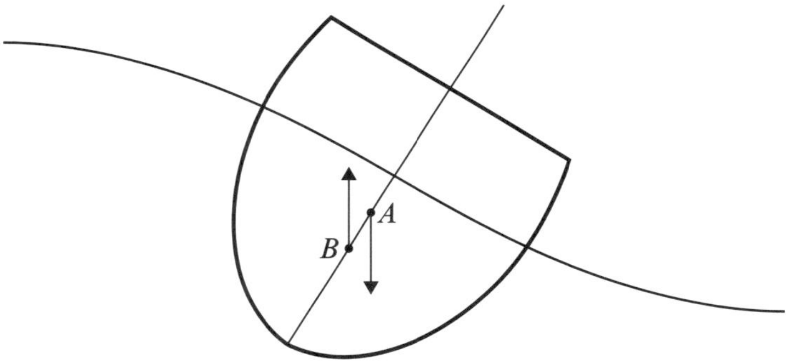

Figure 15: A ship with its deck parallel to the sloping sea experiences a torque tending to incline it further.

5.2.1 Ships in equilibrium on a sloping sea (§§ 51–56 pp. 311–315 h. v.)

Suppose a ship at a position of a node, e.g. point *e* in *Fig. 5* of the essay. Suppose further its deck would be parallel to the surface of the water.

waves do pick up some energy from winds with the right velocity, but this cannot be considered to be the cause of the swelly sea.

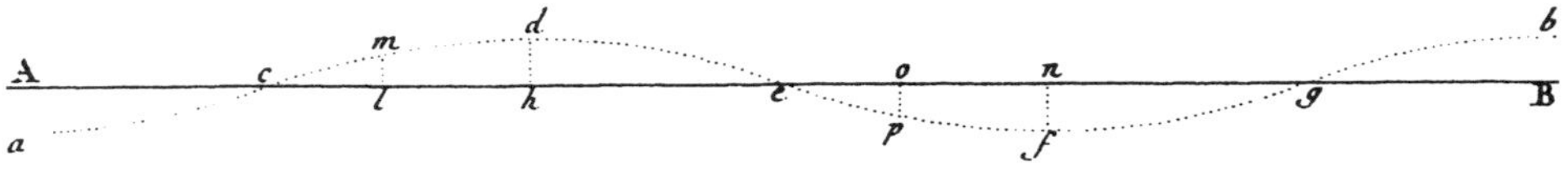

This would not be a position of equilibrium: the points A and B (cf. Fig. 15) are both in the symmetry plane of the vessel, but because this is now tilted with respect to the horizon, there is still a non-zero torque on the ship. It is easily seen (the argument goes as in § 1.2.1) that if the sea has a slope of α degrees, the ship's deck should have a slope of β, where

$$\beta = \frac{q^3}{q^3 - 12Ms}\alpha \tag{27}$$

and q, M, and s are as in (1). If the centre of mass is higher than the centre of buoyancy (s positive), the ship will slope more than the surface of the sea. On an example taken from practice Bernoulli shows the angle could be doubled.

Consequently in a swell the ship will tend to oscillate with the frequency of the waves over an angle which is larger than the maximal slope of the waves. Bernoulli calls these *balancements d'équilibre*, i.e. equilibrium oscillations, because it is the movement a ship would have if she were on a wavy sea and free of other torques.

In practice the ship will experience in addition some amount of free oscillation induced by incidental water masses, and both movements will add, which may lead to quite large angles. Bernoulli draws a twofold consequence:

- try to make $s = 0$, i.e. make the two centres coincide, in order to reduce the amplitude of the equilibrium oscillations;
- here again avoid at all costs the resonance with the free oscillations.

5.2.2 A new law of Bernoulli (§48 p. 307 h. v.)

In the *Hydrodynamica* Bernoulli published his famous law, relating the pressure, velocity and potential energy of a liquid in a steady flow. In the derivation he made use of the law of Torricelli and of the discovery of Huygens that for a system of masses the sum of the potential energy and the kinetic energy is constant. In the present essay he computes the pressure in a tube as used in his model of ocean waves, i.e. for a flow with velocity varying with time and he relies on the equation of motion (Newton's second law)[69]. We summarize the argument.

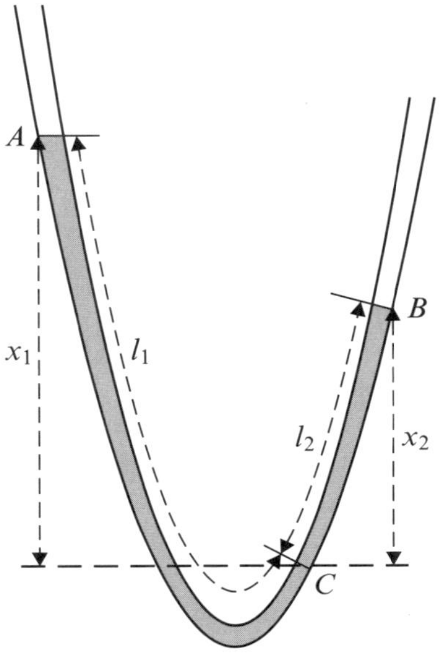

Figure 16: In the gravitational field there is a relation between the pressure at a given point, its height and the masses of the liquid in motion.

We refer to Fig. 16, which represents a curved tube of uniform section (area S) containing a liquid in the part ACB. Let ρ be the density, g the acceleration of gravity and p the pressure at the point C. An element of the liquid around C will experience a force from the left composed of the weight of the column AC less the pressure p at C; the mass of this column is proportional to its length l_1; the acceleration due to this force is therefore:

$$\frac{Sx_1\rho g - Sp}{S\rho l_1} .$$

69 It was Johann Bernoulli who first applied the principle that $F = ma$ to a problem of flow, see letter nr 22 from Joh. B. to L. Euler of March 7th, 1739; in L. Euler, *Op. Omnia* IV/A, p. 275. The results were published in the *Hydraulica*, see Joh. B. Op. CLXXXVI, *Hydraulica nunc primum detecta ac demonstrata directe ex fundamentis pure mechanicis, Anno 1732*: CP IX, 1737 (1744), pp. 3–49, CP X, 1738 (1747), pp. 207–260 – *Opera* IV, pp. 387–488, 1742.

The same argument can be made for the column CB and both accelerations are the same, because both columns always move together with the same velocity. One obtains then the equality:

$$(x_1\rho g - p)l_2 = (p - x_2\rho g)l_1$$

and solving[70] for p:

$$p = \rho g\left[x_2 + (x_1 - x_2)\frac{l_2}{l_1 + l_2}\right] . \tag{28}$$

5.2.3 The altered pressure on the hull (§§49–50 pp. 309–310 h. v.)

Because the ship can be considered as a small object compared to the wavelength, the pressure, due to the law expressed by equation (28), is the same on the whole of the hull. If we call x the vertical distance from a point on the hull to the surface of the ocean (and if we neglect the curvature of the tubes near the surface), it is easily seen that this pressure is proportional to x:

$$p = nx ,$$

where n is a constant nearly equal to one; Bernoulli estimates that n could at most be like $1 \pm \frac{1}{6}$. Therefore the change in pressure due to the moving waters should not play a major role in our problem. The above estimate is not supported by any argument in the text.

In fact, from Fig. 14 it is clear that a ship of sizeable draught, riding a wave near a node, will experience a torque inclining it towards the nearest crest, because the undercurrents at the site of a node are stronger at the keel than at the surface; those will counteract the hydrostatic torque tending to increase the angle of heel beyond the slope of the surface, as demonstrated in § 53 (p. 313 h. v.) of the essay.

Taking the formulae of Lamb[71], we find that the horizontal velocity at a depth y, in the vicinity of a node is – in his notation –

$$v_x = \frac{ga}{\sigma}\frac{\cosh(k(y+h))}{\cosh(kh)}k\cos(\sigma t + \varepsilon)$$

70 Bernoulli – as he usually does – expresses p (the *compression*) by the equivalent height, noted as π which is $p/\rho g$.

71 H. Lamb, *Hydrodynamics*, 6th ed., New York: Dover publ., 1945, § 228.

(h = depth of the sea; a = amplitude of the surface wave; σ = frequency; k = wave number; y = vertical position, counted negative below sea level).

At the depth y the hull will experience a lateral force $\frac{1}{2}\rho v_x^2$ per unit area. At maximum value, for a deep sea and small angles this is:

$$\frac{1}{2}\left(\frac{gak^2}{\sigma}\right)e^{ky} .$$

Knowing the form and size of the hull (especially its length and draught) and the height of the centre of mass, the torque caused by this flow can be computed. As remarked by Bertin[72] the effect, for large ships, is not negligible, and tends to reduce the angle given by (27).

5.3 *The last chapter: pitching and summing up (§§ 58–63 pp. 318–323 h. v.)*

This last chapter is devoted to pitching, in the presence of the ocean's waves, and is obviously in response to the wish of the academy which insisted on pitching when repeating its question of 1755 in 1757. The main difference with rolling is that pitching movements will tend to damp out much faster, due to the difference in resistance to the movement of the ship. Longer ships would pitch less, if we consider the angles only; but there are two drawbacks to longer ships. The first is that a long ship will have large movements up and down at its ends; both stern and bow must be build higher and are still at risk to get under water. The second reason is that in a long ship the stresses on the keel and the hull grow larger, especially due to the pitching and the occurrence of breaking waves.

5.3.1 An example: the floating beam (§ 61 p. 321 h. v.)

Suppose a wooden beam, floating lengthwise. Its length is b, width and height are a and d, respectively; it is immersed over a depth c. How does it oscillate around an axis perpendicular to its length? We take formula (2), adapt it to pitching and find for the stability against pitching:

$$F = \left(\frac{1}{12}b^3 - bc\frac{d-c}{2}\right)a .$$

72 Cf. L. E. Bertin, *Les vagues et le roulis*, Paris: Berger-Levrault et Cie, 1877.

The moment of inertia around the axis is $I = \frac{1}{12}abc(b^2 + d^2)$ and consequently we obtain the frequency $\sqrt{\frac{F}{I}}$. Comparing this with the frequency of a pendulum of length l, we find that a pendulum with this frequency has length:

$$l = g\frac{c(b^2 + d^2)}{b^2 - 6c(d - c)} .$$

If, as assumed by Bernoulli, b is much larger than either a or d or c, we obtain simply that $l = gc$; in the units he uses $g = 1$ and the length of the equivalent pendulum is simply c, the draught of the beam, independent of the length b. The numerical example he presents gives us a clue as to the unit of length he used, as was mentioned in §2.2 (p. 78 h. v.).

5.3.2 Final Words

As his final recommendation Bernoulli repeats the advice to lower the centre of mass at the level of the centre of buoyancy. He ends with a plea to consider his text without prejudice, as an element to be incorporated in the knowledge of pilots and shipbuilders.

Expositio theoretica singularis Machinae hydraulicae Tiguri Helvetiorum extructae *(St. 65)*

1 Introduction

The «singular machine» is a spiral pump, constructed and working in Zurich. It is, says Bernoulli, a most remarkable machine from a scientific point of view and it will not work unless mathematically well planned and build with due respect for the right size and the right ratios.

The machine had been described extensively in a learned journal[1]; it was obvious already there that careful considerations of hydrostatics were essential but deemed very difficult. The author put forward a basic principle for the design of such pumps. In the present essay Bernoulli takes over this description of the pump and of the way it functions but improves on it in three decisive ways:

- He first borrows an astute invention, made in Sevilla but perfected in Paris, which allows to boost the maximal pumping height of a suction pump by 60%, and adapts it to the spiral pump (§6 to §7, pp. 328–330).
- He analyzes the behaviour of the water and the air in the spiral and shows that the general principle of Ziegler does insure a working pump but does not make an efficient one. His analysis does (§8 to §15, pp. 330–335).
- He proposes an alternative design which enjoys the full efficiency of his theoretical model but avoids some of the technical snags which plagued the original machine (§16 to §18, pp. 335–337).

It is on the whole a straightforward article: a non-trivial practical problem is solved by clear thinking and elegant – even if simple – mathematics.

1 J. H. Ziegler, *Vorläufige Anzeige eines neuen Schöpfrades erfunden und verfertigt von Hrn. Andreas Wirz, Zinngiesser, des grossen Raths, Inspector der Gesellschaft der Constablern und Feuerwerkern. Mit Vorwissen und gütiger Bewilligung des Erfinders beschrieben*: Abhandlungen der Naturforschenden Gesellschaft in Zürich 3 (1766), pp. 431–463.

2 The Spiral Pump, Problems and Solutions

2.1 *The principle*

A spiral pump consists of a cylindrical metal drum inside of which is a spiral metal blade, very much as a spring coil in a portable watch. There is a long open slit in the drum where the spiral meets the outer mantle; the drum is half immersed in water, its axis along the water level. Upon turning the drum around its axis it will scoop in water, when the slit is in the water; turning further, the water will proceed to the next winding of the spiral and air will be sucked in. This procedure repeats itself at each revolution of the drum, until all the coils are filled with water in their lower half and air in their upper half. A hole in the centre of a circular side of the drum allows the water and the air to escape. Up to now the water is simply displaced horizontally and in principle no work is done in turning the drum (except for frictional losses).

A short tube is then adapted to the central hole, which rotates with the drum. A similar but fixed tube abuts the first and a bulbous structure (the *choenicium*) encircles both making an airtight junction of the moving and the fixed tube. This last one then curves vertically upwards to a height H. Suppose a column of water stays in the upward tube. It will exert a pressure on the air in the innermost coil of the spiral. This pressure will force the water in the same coil to move up from the bottom of the coil; assuming we look at the spiral along the axis and watch it moving clockwise we should see the water moving up the left side of the coil. This water will thereby compress the air in the previous coil, which in turn will displace the water in that coil upwards to the left. This repeats itself up to the outer coil of the spiral where the fluids are subject to the atmospheric pressure A (which Bernoulli expresses conveniently – and traditionally – as an equivalent height of water). This is illustrated in his *Fig. 1*.

As the drum is rotated and scoops water in at the outer coil, the succession of water and air is moved towards the centre where it is pushed into the vertical tube and streams out at the top, provided the height is less than the maximal obtainable height, shortly to be determined. The turning of the drum requires exactly the power needed to raise the amount of water, neglecting frictional losses. The above is an only slightly modernized version of the explanation given by Bernoulli.

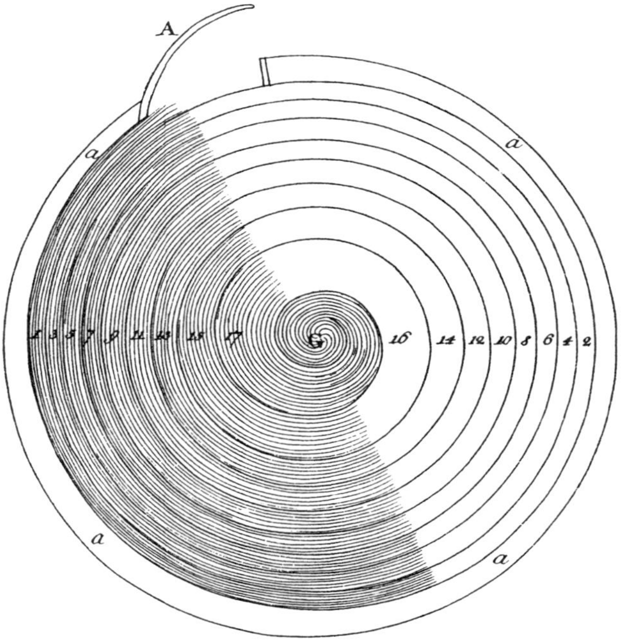

2.2 *The Wirz pump*

The pump in Zurich was constructed in 1746 by Andreas Wirz, a tin-smith and a well known citizen. It worked up to about 1765, raising water from the river Limmat 10 feet higher into a reservoir of dye-works, built on the river bank, but it could raise water up to 18 feet. It was smallish, 3 feet in diameter and 1 foot thick along the axis, but it delivered continuously a jet of bubbling water through a 2 inch pipe. Nine blades, two feet high, were mounted on the rim as on a waterwheel which the current of the river set in motion.

Its inner workings were hidden from the public but excited the curiosity of Johann Heinrich Ziegler, member of the Scientific Society of Zurich. He wrote an article in the Transactions of the Society[2], with the full approval of the inventor, in order to make this marvellous machine known to the world. Ziegler was aware of some of the peculiarities of the pump; he claimed the volume of each circonvolution should be constant, the diminishing radii of the inner coils being compensated by an increasing mutual distance. He es-

2 Cf. *supra* p. 121, note 1.

timated the maximal head the pump could deliver was the sum of the diameters of all circonvolutions. A small experimental model, which he proposed to build, would allow to fix the optimal design and a research programme was suggested. Ziegler admitted that a correct theoretical treatment of the pump would require «no little effort».

2.3 *The Bernoulli Ideas*

The first remark Bernoulli makes is that the total head attainable is smaller than the sum of the diameters, because in the inner coils the air is compressed and the water takes a proportionally larger part. The arcs filled with water he calls «aqueous arcs». The vertical distance between the top and the bottom of such an arc he calls «aqueous height». The aqueous arcs are separated by arcs of compressed air, the more compressed as they are nearer to the centre. It is easily seen that at equilibrium the sum of all the aqueous heights (in the successive windings) balances exactly the height H of the vertical water column. Starting from this remark he will analyse how to reach the maximal pressure at the pump's exit. But this simple mathematical result is not the final answer: a trick, demonstrated in Paris where it was applied to a conventional suction pump, allows to boost the height.

2.3.1 Mixing air and water

To the water sucked up by the Paris pump is added a stream of small air bubbles. This mixed fluid has a lower density than pure water and the air pressure on the reservoir can therefore keep in balance a column of this mixture which exceeds largely the height of 10 meters, the theoretical limit for conventional pumps. This was also demonstrated by the famous Abbé Nollet[3].

2.3.2 Mathematical theory of the pump

Leaving aside for a moment the trick with the injected air, Bernoulli derives an improved structure of the pump. The spiral pump feeds into the vertical

3 Nollet [M. l'Abbé], *Réflections sur quelques nouveaux phénomènes d'hydrostatique*: Mém. Paris 1766 (1769), pp. 431–445.
Nollet [M. l'Abbé], *Sur quelques nouveaux phénomènes d'hydrostatique*: Hist. Paris 1766 (1769), pp. 150–156 (a condensed version of the previous item).

tube alternatively equal volumes (if we could reduce them to standard pressure) of water and air; in a large tube the water, however, will run down the sides of the tube and collect at the bottom and the air will escape upwards. Suppose now the tube had a very small diameter: then indeed one should have that the successive small columns of air and water stay in the tube in the order they were produced. Bernoulli assumes the top little column is made of water; on it presses the atmosphere with a pressure equivalent to a water-column of height A. The little column itself – as all other little columns of water – has a height of, say, α. The underlying little column of air would also have a height of α if it were subject to the (normal) pressure A. But it is subject to the pressure $A+\alpha$ and consequently[4] its height is $\frac{\alpha A}{A+\alpha}$; the subsequent column of air, which supports two columns of water, has a volume of $\frac{\alpha A}{A+2\alpha}$. If there are n coils in the spiral the lowest air column has a height of $\frac{\alpha A}{A+n\alpha}$. Adding the heights of the n water columns and the n air columns we obtain a height of

$$n\alpha + \sum_{k=1}^{n} \frac{\alpha A}{A+k\alpha}$$

with a pump pressure of $a = n\alpha$. Bernoulli asserts that a good approximation can be found to the above sum, which is:

$$\sum_{k=1}^{n} \frac{\alpha A}{A+k\alpha} \approx \frac{\alpha A}{A+\alpha} + A \log_e \frac{2A+2n\alpha+\alpha}{2A+3\alpha} .$$

The result is easily understood if we replace the terms after the first by the integral over the simplest interpolating function; consider the following sum:

$$\sum_{k=1}^{n} \frac{1}{1+k\varepsilon} \approx \frac{1}{1+\varepsilon} + \int_{1\frac{1}{2}}^{n+\frac{1}{2}} \frac{1}{1+x\varepsilon}\,dx$$

$$\approx \frac{1}{1+\varepsilon} + \frac{1}{\varepsilon} \log_e \frac{2+(2n+1)\varepsilon}{2+3\varepsilon} .$$

4 By the law named after Boyle and Mariotte that the product of pressure and volume is constant.

For the values he considers, i.e. $\varepsilon \equiv \frac{\alpha}{A} \approx \frac{1}{30}$ it is a very good approximation. In practice one can install a battery of parallel small tubes which together have the section of the required larger tube, suggests Bernoulli. In his improved design he will include a mixing chamber where foamy water should be produced at the base of the vertical tube and the flow velocity is high such that in the course of the ascent through the tube no significant unmixing occurs. The density of the foam decreases towards the top, as the air in the bubbles is less compressed, but still there will be water actually delivered at an increased height. In the example he considers the sum of all the water-columns is $n\alpha = 32$ feet and the sum of the air columns is almost 22 feet.

2.3.3 The right proportions

At each turn of the drum the same quantity of water is taken in: the quantity that fills exactly half of the outer coil. This is incompressible and one finds exactly the same quantity in each successive coil (provided the spiral is properly conceived); the same volume of air is taken in, but in successive coils this will be more and more compressed, taking less and less volume. The coils themselves get smaller, each one being nested inside its preceding neighbour. There is clearly a limit: the innermost coil must have a volume at least one half of the outermost one; and in this limit the air would have to be infinitely compressed. Bernoulli does not mention this rather obvious condition and it is not clear that his *Fig. 1* conforms to the rule, but his numerical example does.

In order to build the most efficient pump it is clear that the aqueous height should be the largest possible in every winding. This means that the top of the arc filled with water should be at the top of its winding: any other position would have a lower top and a higher bottom i.e. a lower aqueous height. This leaves a given arc for the air which, in the volume left, should experience the right compression to balance the pressure of all the water arcs which lie between itself and the outer mantle of the pump. This right balance can be achieved by a judicious choice of the gap of the winding.

2.4 *The optimal design*

Let us leave for a moment Bernoulli and make use of some computer programming. We shall consider the kth winding, counting inwards, and replace it with a circle with the average radius (cfr. Fig. 1). We neglect the thickness

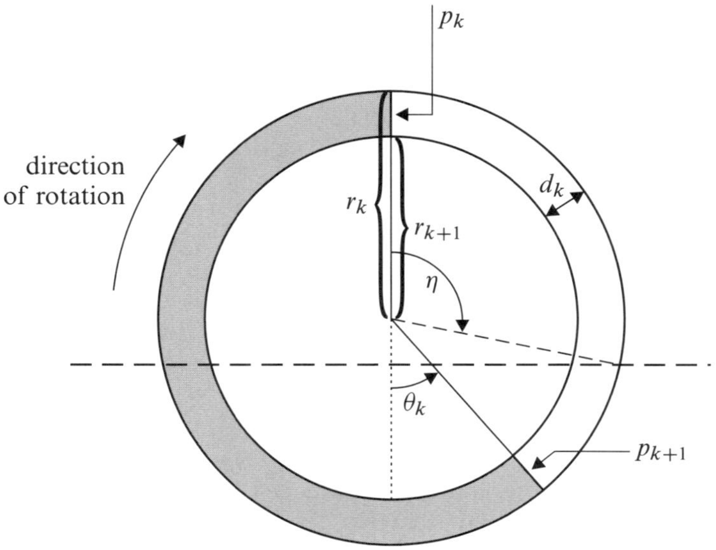

Figure 1: A spire with a maximal aqueous arc to the left.

of the metal blade that separates successive coils. Assume the outer radius of the kth winding is r_k, the gap of the winding d_k, i.e. $d_k = r_k - r_{k+1}$. The water fills an arc from the top to an angle $\pi + \theta_k$. On top the pressure is p_k, at the bottom p_{k+1} (in equivalent water heights). We assume the radius of the drum is r_0, the gap of the outer winding is d. It should be remarked that if the water arc in the outer winding (radius r_0) extends to the top, the slit will scoop in the air only in one quarter of that circonvolution. A better result can be obtained if the drum is less than half immersed. (We are at the moment interested only in the greatest attainable height, not in the discharge of the pump.) We indicate the depth of immersion by the angle η, as indicated in Fig. 1 and therefore the volume of the air in the outer winding is[5] $r_0 d\eta$. We have the following conditions:

The volume of the water in winding k:

$$r_k d_k(\pi + \theta_k) = r_0 d(\pi + \theta_0) \ . \tag{1}$$

5 The length of the drum along its axis is irrelevant for the discussion, as it drops out in the formulae.

The pressure of the air in winding k:

$$p_{k+1} = \frac{A r_0 d \eta}{r_k d_k (\pi - \theta_k)} . \tag{2}$$

The pressure equilibrium:

$$p_{k+1} = p_k + r_k(1 + \cos\theta_k) . \tag{3}$$

From (1) we express d_k as a function of θ_k and r_k. We eliminate p_{k+1} between (2) and (3) obtaining

$$A \frac{\eta}{\pi + \theta_0} \frac{\pi + \theta_k}{\pi - \theta_k} = p_k + r_k(1 + \cos\theta_k) . \tag{4}$$

For each winding r_k and p_k are the characterizing parameters; solving (numerically) (4) first for θ_0 and then for the other θ_k we find from (1) the thickness d_k and we obtain the recursion relation:

$$\begin{pmatrix} p_{k+1} \\ r_{k+1} \end{pmatrix} = \begin{pmatrix} p_k + r_k(1 + \cos\theta_k) \\ r_k - d_k \end{pmatrix} . \tag{5}$$

A computed example. We take the following example: $p_0 = A = 10\,\mathrm{m}$, $r_0 = 0.5\,\mathrm{m}$, $n = 13$, $d = 0.01\,\mathrm{m}$, i.e. a drum of half a meter radius, 13 coils and initial gap of 1 cm and immersed up to 1/3 of its diameter (as was the pump in Zurich). See Table 1.

The important feature we learn from this and other numerical experiments is that the r_n are almost linear functions of n, provided the innermost coil has a volume much larger than half the volume of the outermost coil. The spiral will therefore stop before half the outer radius is reached, and from there feed the perforated axle and the vertical tube, unlike the drawing Nr. 1 on the copper plate.

2.5 *The arguments of Bernoulli*

Bernoulli assumes tacitly there is an optimal construction and he explicitly states that it must ensure that the aqueous arcs extend to the top of their coil. If they do not, because the coils are too large (i.e. their gaps are too wide), each coil will still contain the amount of water and air that was taken it at the first turn, put the pressures will be less and the pump less efficient. This happened in the pump by Wirz.

k	p_k	r_k	θ_k degrees
0	10.00	0.500	76.86
1	10.61	0.490	118.69
2	10.87	0.481	119.80
3	11.11	0.472	120.83
4	11.34	0.463	121.77
5	11.56	0.454	122.64
6	11.77	0.445	123.45
7	11.97	0.435	124.20
8	12.16	0.426	124.89
9	12.34	0.416	125.54
10	12.52	0.406	126.15
11	12.68	0.395	126.72
12	12.84	0.385	127.25

Table 1: Coil number, pressure (waterhead in meter), radius (in meter), water-arc angle (in degrees) for an optimized spiral pump with 13 coils, a diameter of one meter and immersed over $1/3$ of its diameter.

If the inner coils are too small (i.e. the spiral is wound too tightly) the increasing air pressure will make the water at the top spill over into the previous coil, thereby upsetting the balance of that coil as well, and so forth; in some coils the water may find itself in the right part, counteracting the desired effect[6]. In short a tight spiral will ruin the pump: no water will be put through, as the inventor, Wirz, found out at his first attempt, according to Ziegler and echoed by Bernoulli.

Bernoulli does not try to find the exact optimum. He determines how much the air is compressed in the innermost coil, due to the required column of water in the vertical tube. The volume of the compressed air plus the invariant volume of the water give the volume of this coil. The volume of the water is given by half the volume of the outer coil (as is the air in same)[7]. He interpolates linearly between inner and outer coils, but does not specify how the diminishing volumes are distributed between the radius and the gap of the successive coils. From his proposed new design one can guess that he thought of coils with a constant gap; this was a good guess as we saw in the computed model.

6 Bernoulli takes the corresponding «aqueous height» as a negative number.
7 Not strictly true, because of increased air pressure inside.

It is not compulsory – asserts our author – to immerse the pump exactly by half its radius; the present reasoning can easily be extended but in order to avoid prolixity he does not do so.

An example by Bernoulli. In the example proposed the spiral has 32 coils, the section of the tube is the same as the section of the innermost coil and sufficiently small to have a stable superposition of 32 pairs of little columns of air and water; the little columns of water are one foot high, their sum therefore 32 feet, i.e. the atmospheric pressure. It is stated that in the innermost coil (in our notation) $\theta_{32} = 60°$, which follows from the above: this air cushion balances a pressure of $2 \times$ the atmospheric pressure, it must therefore be compressed to half its original volume, which makes the volume of the last (inner) coil three quarters of the outer coil. This is the argument which Bernoulli has made for the general case in § 10, p. 331.

We can now find out the dimensions of his machine (which he did not state). He chose the section of the inner coil to be the same as the section of the (narrow) vertical tube. The spiral is therefore almost such a coiled-up tube. Neglecting the thickness of the walls we have that the outer tube has a circumference of two feet (i.e. a radius of almost exactly 10 cm). The inner radius is then 7.5 cm and between the two there are 30 thin metal walls and 31 gaps: this makes a gap 0.8 mm wide. He mentions that he had a «little machine» (*parvula machina, machinula*) for experimental purposes but we can only guess it had those dimensions.

3 An improved design

Bernoulli voices two criticisms at the Wirz machine:

- «it is very hard to join the spiral metal blade to the circular sides of the enclosing barrel»[8]: leaks between the coils are unavoidable and impair the efficiency of the pump.
- «the decreasing radius of the inner coils makes for a decreased motive force (*vis motrix*)»[9].

One might interpret this remark as follows: the inner coils have a decreasing radius r_k but increasing gaps d_k such that $r_k d_k = r_0 d_0$. This makes the machine less efficient in the sense that the maximal attainable height is smaller and hence the force needed at the base of the watery column to push

8 Cf. p. 335 h. v.
9 Cf. p. 335 h. v.

it up is smaller. If enough force is available, e.g. from the stream pushing the blades of the barrel, this is a bad point for the design.

Bernoulli proposes to take a cylindrical barrel, whose diameter is of one, two or more feet, and to wind around it a lead tube of an inch or so as an helix, one winding touching the next, for sixteen or more turns. One end is open and dips in and out of the water, the other end is connected to the short tube in the axis and this to the fixed vertical tube, as in the old design. It is easily manufactured and no lateral leaks between coils can occur. Of course, here also the volume of the coils has to decrease from the open side to the side connected to the vertical tube. This is most easily done by keeping the section of the helical tube constant but decreasing the radius of the barrel from one side to the other; it is in fact conical and resembles the fusee in a portable watch. He claims that the Zurich craftsmen are now adopting this design.

Another improvement is to fix to the small end of the conical barrel a small cylindrical capsule or chamber; the helical tube empties in this capsule and the central tube drains it (see his *Fig. 3*). Bernoulli is convinced that as the barrel is turning fast and the flow is rapid the alternating flows of water and air will mix in the capsule and make a watery foam which will be pushed up the vertical tube, dense at the bottom and less dense up, which will allow to exceed the limit of the 32 feet corresponding to atmospheric pressure, as explained before.

Texts

DISCOURS
Lû
DANS
L'ASSEMBLEE
PUBLIQUE
de l'Academie des Sciences,

Le 2. Mars 1728.

Par Mr. DE L'ISLE,
Avec
la Reponse
De Mr. BERNOULLI.

A St. PETERSBOURG,
de l'Imprimerie de l'Academie des Sciences.
1728.

W

Discours lû dans l'Assemblée publique de l'Académie des Sciences, Le 2. Mars 1728. Par Mr. De L'Isle, avec la réponse de Mr. Bernoulli

Imprimerie de l'Academie des Sciences, St. Petersbourg, 1728

Question

Si l'on peut demontrer par les seuls faits Astronomiques,
quel est le vrai Systeme du monde?
Et si la Terre tourne, ou non?

C'est une question assez souvent agitée dans le monde, le plus souvent par des ignorans; si la terre tourne, ou non? Ceux qui ne font qu'effleurer cette matiere, sont ordinairement ceux, qui en decident le plus hardiment; mais leur jugement est determiné par deux sortes de prejugés; Prejugé de Religion, par la deference que quelques uns ont, pour le sens litteral de l'Ecriture sainte, qui etablit le repos de la terre: & prejugé d'opinion, en faveur de quelques Astronomes & Philosophes de reputation, lesquels ont soutenu le mouvement de la terre.

Il y a d'autres personnes, qui par un plus profond respect, & par une soumission entiere, pour le texte sacré; ne croient pas qu'il soit seulement permis, de mettre en doute si la terre tourne ou non? pretendant que c'est manquer de foi, de mettre en question, ce que l'Ecriture decide si positivement; au lieu que les autres, sans croire manquer en rien à ce que la Religion exige d'eux, pensent qu'il est permis de proposer cette question, & de l'examiner suivant les principes des sciences humaines; puisque l'Ecriture même nous aprend, que Dieu a donné le Monde, pour un sujet de disputes entre les hommes. Je suis de ce dernier sentiment, & sans croire manquer à ce que je dois à la Religion; je pense qu'il m'est permis d'examiner la question, non pas si la terre tourne ou non; mais si l'on peut demontrer par le raisonnement, & les faits astronomiques, quel est le vrai systeme du Monde? & si la terre se meut réellement? car, quand même j'aurois prouvé, par les principes mathematiques, que la terre se meut: je peux ensuite, n'en rien croire; & n'avoir rien prouvé: en reconnoissant, & avouant, la fausseté, ou de mes principes, ou de mes consequences; puisque la Religion m'oblige de penser le contraire.

Il faut donc regarder la question que je propose ici; comme une question purement mathematique, que je ne discuterai que par les faits d'experiences & de raisonnemens, suivant les principes que les hommes se sont faits, dans les mathematiques & les autres sciences humaines.

Je mets ici en question, ces deux choses à la fois: Quel est le vrai systeme du Monde? & si la terre tourne ou non? parce que le mouvement de la terre ou son repos, ne fait qu'une partie de tout le systeme du Monde; & que l'on ne peut prouver le mouvement, ou le repos de la terre que par la relation qu'il a, avec tout le reste du systeme du Monde.

Par le mot de systeme du Monde, j'entens l'arrangement & la situation des corps celestes entr'eux, & à l'egard de la terre. Ainsi connoitre le vrai systeme du Monde; c'est sçavoir à quelle distance, les corps celestes, & la terre, sont réellement les uns des autres; c'est determiner, lesquels sont en mouvement, & lesquels en repos. C'est enfin sçavoir au juste quelles sont les regles de ces mouvemens.

Suivant cette idée, je dis, que l'on peut bien demontrer geometriquement par les observations & le raisonnement, quel est l'arrangement des corps celestes entr'eux; & à l'egard de la terre, leur grandeur, leur distance mutuelle: comme aussi le changement de distance & de situation respective qu'il leur arrive; mais de sçavoir, par le mouvement réel, de quels corps ce changement arrive? & s'il y a quelques corps celestes en repos? & quels ils sont? C'est ce que je pretens, que l'on ne peut pas demontrer par les seules regles mathematiques; sans faire des suppositions; ou sans admettre des hypotheses physiques, qui quoique vrai semblables, n'ont pas la certitude des demonstrations geometriques; & qui par consequent seront encore long tems disputées.

Pour prouver ce que j'avance, il ne faudroit que parcourrir les progrés que l'on a fait jusqu'ici dans la connoissance des mouvemens celestes; par où, il seroit aisé de faire voir, que, quoi que l'on ait pu determiner les proportions des orbes celestes, distinguer, & marquer au juste les vitesses de tant de mouvemens differens; cela n'a pas encore suffit pour decider laquelle des deux hypotheses du repos ou du mouvement de la terre est la vraie.

Mais, dira-t-on, il nous manque peut etre quelque connoissance dans les mouvemens celestes, d'où l'on pourroit tirer une demonstration geometrique du mouvement ou du repos de la terre.

J'avoüe que quelques progrés que l'on ait fait de nos jours dans l'Astronomie; il reste encore bien des choses à decouvrir, & à perfectioner; mais je pretens; que quand nous aurions cette connoissance exacte des mouvemens celestes, autant qu'il est possible de l'avoir, & de maniére que les Astronomes ne manquassent jamais d'un seul moment dans les predictions qu'ils font des Eclipses, & autres Phenomenes; long tems avant qu'ils arrivent; cela ne suffiroit pas pour demontrer si la terre se meut ou non.

Quelqu'un pourra encore s'imaginer ici que le defaut de nos connoissances, pour decider cette question ne vient que de ce que nous ne pouvons faire nos observations que d'un seul point de vuë, qui est la terre que nous habitons; & que s'il etoit possible de se transporter dans d'autres planetes; & y faire de pareilles observations; & entre autres y observer la terre; on s'appercevroit aisement, si elle se meut ou non.

A cela je repons que les Astronomes, qui ne peuvent faire réellement ce voiage, l'entreprennent quelque fois d'imagination; ils sçavent, & ils se plaisent souvent à se representer de quelle maniere les corps celestes leur paroitroient se mouvoir de dessus chaque planete; si ils y etoient effectivement: il est même certain, que si l'on etoit sur ces planetes, l'on y verroit les mouvemens des autres corps celestes precisement tels que les Astronomes disent; mais cette connoissance, certaine que nous avons, de l'aspect du ciel, de tous les points devuë possibles du Monde entier, dans lesquels nous ne pouvons pas nous transporter réellement, cette connoissance dis-je ne nous avance pas plus pour la decision de notre question. Nous ne deciderons jamais par nos seules observations si ces points de vuë sont fixes ou non.

Pour rendre les preuves de ma proposition à la portée de tout le monde, je me servirai d'une comparaison de ce qui arrive dans le ciel, avec ce que l'on pourroit eprouver sur terre.

Que l'on s'imagine donc etre sur un vaisseau en pleine mer, & que l'on n'apercoive autour de soi, qu'un petit nombre d'autres vaisseaux, qui ne soient pas trop eloignés pour que l'on ne les puisse facilement distinguer les uns des autres, & observer leurs mouvemens; & qu'enfin dans le plus grand éloignement où l'on puisse porter la vüe l'on n'appercoive autour de soi que des côtes de terre, diversement figurées, aux quelles on ne puisse point aborder, non plus qu'aux autres vaisseaux, qui paroissent dans un mouvement continuel; & toujours fort éloignés de nous, tantôt plus, & tantôt moins; tantôt d'un côté, & tantôt de l'autre.

Voila la peinture du Monde visible. Le vaisseau sur lequel nous sommes est la terre. Les côtes que nous voions autour de nous dans la plus grande distance sont les Etoiles fixes: & enfin les autres vaisseaux qui nous paroissent à une moindre distance de nous & dans un mouvement continuel sont les planetes.

Mais pour examiner ce que nous pouvons conclurre, des observations de ces mouvemens apparens, descendons dans un plus grand detail, en suivant toujour la même comparaison: Commençons, par l'examen des côtes: en les regardant avec attention elles nous paroissent tourner toutes ensemble autour de nous, par un mouvement continuel & reglé: C'est à dire que les terres, qui nous paroissoient ci devant repondre à la proüe de nôtre vaisseau; ni repondent plus un moment aprés; mais elles paroissent s'en eloigner peu à peu; jusqu'a se rendre vis à vis la pouppe, comme il seroit arrivé si nôtre vaisseau eut viré de bord; Mais il faut concevoir ce mouvement apparent des côtes qui nous environnent, comme se faisant toujours du même sens; & sans que ces terres nous paroissent changer de figure; ni sans qu'elles nous paroissent plus proches ni plus éloignées de nous, dans un tems que dans un autre.

Que pourrions nous conclurre de ces observations seules; si non, ou que le vaisseau sur lequel nous sommes tourne sur lui même, ou que ce vaisseau restant immobile, les terres qui l'environnent tournent autour de lui, ou enfin que l'un & l'autre sont en mouvement; mais diversement. Voila tout ce que nous en pourrions conclurre si nous n'avions point d'autre moien de juger de ce qui en est effectivement, que les observations que je viens de rapporter.

L'on en voit facilement l'application, dans le mouvement journalier des Cieux. Nous voions tous les jours les Astres se lever & se coucher, rouler continuellement sur nos têtes, sans qu'il ait été possible jusqu'ici de determiner par les seules observations de ce mouvement, si ce sont les cieux qui font ce mouvement autour de la terre immobile, ainsi que les anciens l'ont d'abord pensé; ou bien si c'est la terre qui par sa revolution sur elle même, est la cause de cette apparence; ainsi que ceux qui soutiennent le Systeme de Copernic, le croient.

Passons à l'observation des objets plus proches de nous; je veux dire des planetes; ou de ces vaisseaux que nous pouvons distinguer les uns des autres par leurs pavillons; & qui nous paroissent avoir differens mouvemens; tant entr'eux, qu'a nôtre égard; & à l'égard des côtes, auxquelles nous les rapportons. Mais pour ne point nous embarrasser de trop de mouvemens differens, examinons avec ordre le mouvement de chacun en particulier; & distinguons le mouvement que chacun de ces vaisseaux nous paroit avoir à nôtre égard; & à l'égard des côtes vis à vis desquelles il repond.

A nôtre égard; c'est à dire à l'égard des differentes parties du vaisseau, sur lequel nous sommes; chacun des autres vaisseaux nous paroit tourner autour de nous; de même que j'ai dit ci devant que les côtes nous paroissent tourner; & ils paroissent tous aller du même sens; mais avec différentes vitesses apparentes: de sorte que quand le prémier a fait un tour, un second n'a point encore achevé le sien, & un troisiéme a fait plus d'un tour.

C'est de là que ces vaissaux nous paroissent changer peu à peu de situation apparente entr'eux; car de deux vaissaux qui nous paroissoient joints ensemble; & tous deux repondans precisement à la proüe du nôtre; lors que celui qui aura fait sa revolution plus promptement, nous paroitra de nouveau repondre à la proüe, le second n'y etant pas encore revenu paroitra s'être éloigné de ce prémier, d'un sens contraire à leur mouvement autour de nous.

Avant d'aller plus loin, que pouvons nous conclurre de cette difference de vitesse apparente dans la revolution de chaque vaisseau autour de nous. Nous en pouvons deduire differentes veritez suivant les differentes suppositions que nous ferons.

Si nous supposons par exemple, que nôtre vaisseau n'ait pas remué; nous devons conclurre que tous les autres vaisseaux, ont réellement tourné autour de nous, chacun avec une differente vitesse; mais si nous supposons que ces vaissaux n'ont eu aucun mouvement; & que tout ce que nous avons observé, provient du mouvement du nôtre; nous serons alors contrains de reconnoitre que notre vaisseau a dû avoir deux mouvemens differens, l'un en piroüetant, sur lui même ce qui lui aura fait paroitre les autres vaisseaux tourner autour de lui; & l'autre mouvement de notre vaisseau aura été un mouvement local par lequel il se sera transporté d'un lieu dans un autre; de sorte que le point de vuë de notre vaisseau étant changé; il ne sera pas surprenant, que quoique tous les autres vaisseaux, n'aient pas remués ils paroissent cependant, les uns à l'égard des autres dans une situation differente de celle qu'ils avoient d'abord.

L'on peut se former une idée de ce double mouvement par la comparaison d'une boule qui roule dans une chambre, sur un plancher; cette boule tourne sur elle même en même tems que toute sa masse avance & change de place, à l'egard des parties de la chambre dans laquelle elle se meut.

L'on pourroit faire un troisiéme systeme, composé de deux que je viens de dire; à sçavoir de supposer que tout a été en mouvement; & alors nous pourrions ne donner au vaisseau sur lequel nous sommes que le seul mouvement de revolution sur lui même; & attribuer le changement de situation apparente des autres vaisseaux entr'eux; à quelque petit mouvement réel, qu'ils auroient eus entr'eux sans qu'ils soient obligés de tourner au tour de notre vaisseau; puisque nous les debarrassons de cette revolution, par le mouvement de notre vaisseau sur lui même.

Nous pourrions enfin ne supposer à nôtre vaisseau que le mouvement local, en laissant aux autres le mouvement de revolution autour de nous; & peut etre que ces autres vaisseaux qui nous paroissoient tourner autour de nous avec des vitesses differentes, se sont cependant mûs d'une egale vitesse, qui leur aura fait achever leurs revolutions en même tems; & la difference qui nous y a paru; n'est peut être provenuë que du mouvement local de nôtre vaisseau.

Mais si entre toutes ces differentes manieres on me demande quelle est la vraye? Je repondrai que je n'en sçai rien. Ou au moins que quand j'aurai trouvé differentes hypotheses ou suppositions, qui m'expliqueront Geometriquement parlant toutes les apparences du mouvement, que les observations m'auront fait connoitre; je ne pourrai pas demontrer par ces seules observations, quelle est la veritable hypothese.

A l'application: tout ce que les Astronomes ont pû faire jusqu'ici, n'a été que de bien observer quels etoient les mouvemens apparens des Astres; tant entr'eux; qu'a l'egard de la terre; & ils connoissent effectivement si bien les loix de ces mouvemens, qu'ils sont capables de predire long tems d'avance les Eclypses & autres Phenomenes precisement de la maniére qu'ils arrivent; comme on eprouve assez souvent.

Les Astronomes sçavent outre cela, toutes les hypotheses que l'on peut faire du mouvement réel des corps celestes; lesquelles peuvent rendre raison des apparences qu'ils ont observées; mais pour demontrer, entre toutes ces differentes hypotheses possibles, quelle est la veritable; c'est ce que les Astronomes ne peuvent pas faire par les seules observations.

Pour expliquer cela plus particulierement, considerons l'admirable systeme, qu'a suivi Copernic. Les Astronomes sont persuadés; & peuvent demontrer par les observations à toute personne qui sçait la Geometrie; qu'en supposant à la terre les deux mouvemens que Copernic lui donne; sçavoir le mouvement annuel & le mouvement diurne; il ne sera pas necessaire de donner aucun mouvement aux etoiles fixes, ni au soleil; & que chaque planete principale n'aura qu'un seul mouvement simple autour du Soleil; & qu'enfin la Lune & les Satellites de Jupiter & de Saturne, n'auront chacun que deux mouvemens, l'un autour de leurs planetes principales, & l'autre autour du Soleil.

Au lieu que si l'on suit le Systeme de Tycho Brahé, & que l'on ne donne à la terre que le seul mouvement diurne; chacune des planetes principales devra avoir deux mouvemens: l'un autour du Soleil, & l'autre autour de la terre; & dans ce systeme les Satellites de Jupiter & de Saturne devront avoir trois mouvemens, & la Lune un seul; & enfin les etoiles fixes pourront etre censées immobiles, mais il faudra donner au Soleil un mouvement annuel autour de la terre.

Enfin si l'on ne vouloit accorder aucun mouvement à la terre; ni le mouvement annuel, ni le mouvement diurne, il faudroit, pour faire convenir ce systeme avec les observations; reconnoitre à tous les corps celestes, même aux Etoiles fixes, le mouvement diurne; donner outre cela au soleil un mouvement propre autour de la terre, attribuer trois mouvemens differrens à chaque planete principale; & enfin quatre mouvemens à chacun des satellites de Jupiter & de Saturne; il n'y auroit que la Lune qui n'auroit comme le Soleil que deux mouvemens.

Il en seroit de même de tout autre systeme que l'on voudroit faire: dites à un Astronome lequel des corps celestes vous voulez qu'il suppose immobile; ou quel mouvement vous voulez qu'il lui donne; & alors il vous dira quelle doit etre la composition de tous les autres mouvemens. Mais si vous lui demandez s'il y a quelqu'un de ces corps celestes qui soit veritablement en repos? & quel il est? il vous repondra qu'il n'en sçait rien; ou au moins s'il a quelque opinion la dessus; il ne pourra pas vous la demontrer geometriquement par les seuls faits astronomiques; il sera obligé d'avoir recours à des vraisemblances, & à des hypotheses physiques.

Je reserve à une autre fois l'examen de ces preuves, tirées des principes Physiques, ou prises d'ailleurs. Il me suffit d'avoir fait voir, comment les observations astronomiques seules avec toute la geometrie, que l'on y pouroit emploier, ne suffisent pas pour demontrer, quel est le veritable systeme du monde, & si la terre est mobile, ou non.

Quand je dis que j'ai prouvé cette proposition; je ne pense l'avoir prouvé que de la maniére qu'il convient dans une pareille assemblée en me servant d'une comparaison simple & familiere, à la place d'une demonstration exacte que j'aurois dû y emploier; mais dont le fil auroit pû paroitre trop long & trop ennuieux à suivre, à la pluspart de ceux qui me font l'honneur de m'entendre.

Il me reste, pour achever de parler des faits astronomiques que l'on apporte en faveur du mouvement de la terre, à repondre à ceux qui sont dans l'opinion que le Systeme de Copernic, du mouvement de la terre autour du Soleil, & tout ce qui s'ensuit, seroit demontré; si l'on pouvoit appercevoir quelque parallaxe annuelle aux Etoiles fixes, ou (pour me faire entendre de tout le monde) si l'on pouvoit determiner la distance qu'il y a d'ici aux Etoiles fixes, ce qui demontreroit suivant eux le mouvement local de la terre à l'egard des Etoiles fixes. D'où quelques uns pretendent que le mouvement de la terre est presentement demontré, de puis les observations qui ont été produites par Mrs. Flamsted & Cassini, lequelles determinent cette parallaxe annuelle.

Je reponds à cela que les observations de Mrs. Flamsted & Cassini, ne me paroissent pas prouver le mouvement local de la terre à l'egard des Etoiles fixes, ni ne determinent point sûrement la distance de ces etoiles à la terre, & que quand meme ces observations prouveroient ces deux choses, elles ne demontreroient pas pour cela le mouvement réel de la terre.

On sçait les diverses tentatives que les Astronomes ont fait pour determiner l'éloignement des Etoile fixes à la Terre, en quoi on n'a pû esperer de reussir que de nôtre tems, dans lequel la decouverte des lunettes, & la precision que l'on a pû esperer par leur moyen dans les observations, etoit seule capable de nous conduire sûrement dans une recherche aussi subtile, mais tout ce que l'on a fait jusqu'ici pour cela souffre encore de grandes difficultez, & ne paroit pas suffisant, à tout le monde pour etablir une pareille distance.

La Geometrie pratique enseigne la maniére de mesurer les distances inaccessibles. L'on choisit une base, aux extremitez de laquelle l'on observe les deux angles qui se terminent à l'objet dont on veut sçavoir la distance; mais plus la distance que l'on veut mesurer est grande, & plus faut il que la base le soit. Le diametre entier de la terre, quoi qu'il ait plus de trois milles lieues[1], & qu'il soit la plus grande base que nous puissions mesurer, ce diametre dis-je ne suffit point encore pour determiner l'enorme distance qu'il y a d'ici aux Etoiles fixes: ce qui a obligé les Astronomes, de prendre pour base de cette mesure le diametre même de l'orbe annuel; je veux dire toute l'etendüe du cercle que la terre decrit autour du Soleil, dans le Systeme de Copernic. L'on sçait que le diametre de ce cercle a plus de soixante millions de lieües, cependant par des observations exactes faites aux deux extremitez d'une si grande base, l'on n'a point encore pû decouvrir dans la situation des Etoiles fixes, une variation constante & reglée, qui ne depende que du seul mouvement de la terre, de sorte que cette base quelque grande, qu'elle soit; ne paroit pas encore suffisante pour mesurer une aussi prodigieuse distance qu'il y a de la terre aux Etoiles fixes.

Mrs. Cassini & Flamsted, deux des plus exacts observateurs de France, & d'Angleterre, ont à la verité produit des observations, de l'exactitude desquelles ils se sont rendus garans; & par lesquels ils ont taché de prouver la distance des Etoiles fixes à la terre.

Les observations de Mr. Flamsted, roulent principalement sur l'Etoile polaire qu'il a observé en differens tems de l'année. Il marque les variations qu'il a trouvé dans les hauteurs meridiennes de cette Etoile; ces variations ont cela de particulier, qu'elles reviennent tous les ans les mêmes, aux mêmes mois: d'ou il paroit assez naturel de penser que cela ne provient que du mouvement de la terre; mais Mr. Cassini, accordant à Mr. Flamsted l'exactitude de ses observations à crû prouver quelles n'étoient pas entierement conformes, à ce que demandoit le mouvement de la terre. D'un autre

1 If Delisle puts the diameter of the earth at more than 3000 *lieues*, this means he takes probably the French *lieue commune* as his unit; 1 *lieue commune* = 4.4454 km. His value of «more than» 60 million *lieues* for the diameter of the earth's orbit would then give «more than» 133.3 million km for the distance earth–sun (the exact value is 149.6 million km).

côté, la demonstration de Mr. Cassini a paruë defectueuse à quelques Astronomes Anglois, qui persistent dans l'opinion de Mr. Flamsted.

Comment accorder ce different? Des observateurs exacts conviennent d'une variation annuelle sur une même Etoile, & ils disputent entr'eux de sa veritable cause. On doit regarder cette question, comme un de ces points d'Astronomie qui reste encore à connoitre; & dont on ne peut s'assurer que par de nouvelles observations sur plusieurs autres Etoiles.

Mr. Cassini dixhuit ans aprés ses remarques, sur la pretension de Mr. Flamsted, à rapporté quelques unes de ses propres observations pour établir la distance de Sirius. C'est ainsi que l'on appelle une Etoile qui est dans la constellation du grand chien; & qui est la plus brillante de tout le ciel; mais Mr. Halley celebre Astronome Anglois à fait voir par ses remarques sur les observations de Mr. Cassini, qu'il n'est pas de son sentiment. Et il en donne de fort bonnes raisons. C'est un droit de repressailles que Mrs. les Astronomes Anglois exercent sur les François.

Il y a fort long tems que Mr. Hughens, un des plus exacts, & des plus profonds Astronomes du siecle precedent, pour estimer la distance des Etoiles fixes à la terre, s'avisa de disposer une bonne lunette, de maniére, que le Soleil ne lui parût par cette lunette que de la grandeur dont nous paroit Sirius. L'on sçait qu'une lunette qui le plus souvent est faite pour grossir les objets & pour les faire paroitre plus proches, que cette lunette dis-je etant tournée bout pour bout les diminue autant au dessous de leur grandeur naturelle qu'elle les augmentoit auparavant; & que suivant la proportion que l'on aura donné aux verres de cette lunette; l'on peut sçavoir de combien est l'augmentation en regardant par un bout; ou la diminution en regardant par l'autre bout[2]. De sorte que si deux objets nous paroissent egaux; un à la vuë simple & l'autre par une lunette qui diminüe de cent fois; & si l'on suppose que ces deux objets soient réelement egaux; l'on en conclurra que celui que l'on voit à la vuë simple est cent fois plus eloigné que l'autre.

C'est de cette maniére que Mr. Hughens supposant que l'Etoile fixe appellée Sirius, est un Soleil de la grandeur du nôtre, il en conclut la distance ou il doit etre de la terre pour nous paroitre aussi petit que nous le voyons.

Mais sans discuter les difficultez de cette observation l'on s'aperçoit bien, que l'on ne peut par cette methode conclurre la distance des Etoiles à la terre que dans la supposition que ces Etoiles sont des soleils, de la grandeur du nôtre; ce qui n'est qu'une conjecture qui pourroit aussi bien être fausse que vraye, & en effet, Mr. Cassini dans la recherche qu'il a faite de la distance de la même Etoile à la terre; bien loin de faire la supposition de

2 See §4.4 *The distance to Sirius and its estimate by Huygens*, pp. 15–16 h. v.

Mr. Hughens que Sirius est un Soleil de la grandeur du nôtre, il trouve par les nombres de son observation que le diametre de Sirius, est cent-fois plus grand que celui du Soleil.

Mr. Hughens propose encore une autre maniére de determiner par observation, le mouvement de la terre à l'egard des Etoiles fixes; c'est en observant si deux Etoiles assés proches l'une de l'autre, ne paroissent pas changer tant soit peu de distance apparente en differens tems de l'année; d'ou l'on pouroit conjecturer que cela viendroit du mouvement de la terre par lequel elle se trouveroit tantôt plus prés & tantôt plus loin de ces Etoiles: Je dis conjecturer, car il faudroit toujours supposer que cela ne viendroit pas de quelque mouvement que les Etoiles fixes auroient entr'elles. Mais outre qu'une pareille observation ne demontreroit pas absolument le mouvement de la terre; l'on ne voit pas même, que cette observation ait été faite, ou au moins qu'elle ait parfaitement reussie Mr. Hughens la propose; mais il ne dit pas l'avoir faite.

Le P. Riccioli dans son Astronomie[3] reformée rapporte l'observation d'un nommé Piron, qui dit avoir trouvé une variation de quelques secondes, dans la distance de deux Etoiles fort voisines l'une de l'autre, lesquelles sont dans le front du Scorpion; mais en comparant les mesures qu'il rapporte avec celles que Mr. Maraldi nous a donné de la distance de ces deux mêmes etoiles, dans les memoires de l'Academie des Sciences de Paris, l'on voit qu'il n'y a aucun fonds à faire sur cette observation de Piron. Outre que le P. Riccioli fait voir que la variation que Piron avoit trouvé, dans la distance de ces deux etoiles n'etoit pas conforme à ce que demande le mouvement de la terre.

Je ne connois point d'autres observations, que l'on puisse mettre en parallele avec celles que je viens de rapporter, c'est pourquoi rien ne m'empeche de soutenir ma premiére proposition, que non seulement on n'a point encore demontré par observations le mouvement de la terre; mais qu'il est même impossible de le demontrer par les seules observations Astronomiques, sans le secours des hypotheses physiques, au autres.

Mais demandera t'on, faut il pour cela cesser d'observer? ne doit on pas tenter même de nouvelles observations? & refaire avec plus d'exactitude, celles sur lesquelles on a trouvé quelque chose à redire? ne peut'on pas esperer que si toutes les observations que l'on a deja ne demontrent pas

3 The *Astronomie reformée* is G. B. Riccioli, *Almagestum Novum astronomiam veterem novamque complectens observationibus aliorum, et propriis novisque theorematibus, problematibus, ac tabulis promotam, in tres tomos distributam*, t. I, Bologna: ex typ. hrdi V. Benatii, 1651.

independamment de toute supposition le mouvement de la terre, les nouvelles observations que l'on fera pourront au moins rendre ce mouvement de la terre de plus en plus vraisemblable; de sorte que l'on passeroit ensuite pour ridicule si l'on en disconvenoit.

C'est le meilleur parti; on sçait par exemple que le Soleil, est un million de fois plus gros que la terre[4]; & qu'il est eloigné de nous de plus de 30 millions de lieües; l'on sçait aussi que les deux Planetes superieures Jupiter & Saturne sont encore plus eloignées de la Terre que le Soleil; & qu'enfin les etoiles fixes sont encore dans une beaucoup plus grande distance. Est il concevable? que toute cette entendüe des cieux fasse tous les jours une revolution autour de la terre, plutôt que de donner à la terre, un petit mouvement sur son axe, mouvement d'ailleurs, dont on a des exemples dans presque tous les corps celestes, puisque l'on sçait tres certainement, que la plus part tournent sur eux mêmes; outre le mouvement local que j'ai marqué ci devant que chacun devoit avoir pour expliquer les apparences dans les differens systemes.

Il en est de même de tout le reste du Systeme de Copernic; les mouvemens des corps celestes y sont reduis à une si grande simplicité & harmonie entr'eux que la pluspart des Astronomes & Philosophes charmés de la beauté de ce Systeme ne doutent nullement de sa verité; quoi qu'ils n'en aient pas de demonstration purement Geometrique independante de toute supposition.

Que si l'on ajoute dans la suite, de nouvelles observations, pour s'assurer des faits Astronomiques qui ne sont pas encore entierement constans, l'on ne doit pas douter que l'on n'en tire de nouvelles preuves ou de plus fortes vraisemblances pour le veritable systeme du monde.

C'a eté dans cette vuë, comme aussi pour perfectioner la Geographie & la navigation, que feu l'Empereur PIERRE I. de Glorieuse Memoire a jetté les premiers fondemens de cette tour Astronomique; qui comme un fanal devoit eclairer tout l'Empire.

Feu l'Empereur s'étoit proposé d'orner cet observatoire d'instrumens Astronomiques d'une grandeur & d'une subtilité capable de faire tout ce que l'on n'a pû faire ailleurs faute de pareils secours.

C'est pour conduire cette entreprise, de même que pour cultiver toutes les autres sciences Mathematiques & Physiques, que PIERRE LE GRAND a formé le plan de cette Academie, qui pourroit servir de modele, à toutes les autres Societez de gens de lettres, puisque le merite seul y donne place,

4 The number given, «a million times bigger than the earth», refers obviously to the volumes of both spheres; the radius of the sun, according to recent reference works, is 109 times the radius of the earth.

de quelque nation que l'on soit, & que les recompenses y repondent à la magnificence du fondateur, & à la grandeur de ses desseins.

Mais la mort trop prompte de S. M. aiant suspendu l'execution de ses projets, en faveur des sciences, il etoit reservé au jeune Monarque qui nous commande aujourdhui, (digne Successeur des vertus de PIERRE LE GRAND) d'achever ce qui reste imparfait. La protection singuliere que sa Majesté aujourdhui regnante accorde à son Academie, lui presage des succés, au dessus de l'attente de toutes les nations: même, de celles chez lesquelles les sciences fleurissent depuis si long tems.

Sur tout, le choix que sa Majesté a fait de vos Excellences, comme des personnes singulierement portées par leurs lumiéres & par inclination, à favoriser les Sciences dont l'Academie fait profession, ce choix dis je, ne nous pouvoit être plus avantageux, ni d'une plus grande esperance pour l'avancement des Sciences dans cet Empire. Je pourrois même faire voir ici, ce que les sciences les plus necessaires à l'Etat, ont deja gagnées par les soins & les travaux de vos Excellences, mais je m'aperçois que vous aprehendez, que je n'entre ici dans un detail que vous ne souffririez qu'avec peine. C'est le comble de la vertu, d'aimer plus à faire de grandes actions, qu'a en entendre parler.

Le discours que Mr. de L'Isle vient de faire, contient en substance un examen des deux systemes du monde les plus fameux: savoir de celui de Ptolomée qui suppose la terre entiérement en répos, & de celui de Copernic, qui lui attribue des differens mouvemens. Si nous voulions écouter la pluralité des voix jointe à une espece de droit d'ancienneté, la question dont il s'agit seroit bientôt vuidée: le Systeme de Ptolomée l'emporteroit sur celui de Copernic; mais la chose change de face, si au lieu de conter les voix on les pese: un seul du parti des Coperniciens contrebalance mille, du parti contraire: car on n'y parvient qu'à force de raisonner & de se depouiller des prejugés. Leur sentiment n'est d'ailleurs rien moins que nouveau. Ils pourroient conter jusqu'à vingt siécles; mais il y a eu des tems où l'on ne se le disoit qu'à l'oreille. L'exemple de Galilée un des principaux Geometres de son tems, qui pour avoir vû plus clair que les autres sur ce point, se crût encore trop heureux d'en etre quitte pour une amende honorable, cet exemple, dis-je, à fait une trop grande impression sur l'esprit des savans.

Mais Graces à Dieu, nous vivons dans un Siecle, où la pluspart sont revenus de cette superstition, & sous la protection d'un Monarque, qui loin d'empecher par ces contraintes l'avancement des Sciences, les fait fleurir de plus en plus par ses graces Imperiales. C'est ainsi que S. M. I. non contente de rendre son vaste Empire heureux & florissant par un regne plein de douceur & de sagesse, sait procurer à toute la terre & aux siecles à venir cette source intarissable de felicité, que les Sciences fournissent. Animés par les bienfaits d'un si grand Monarque nous nous sentons tous également engagés à faire nos derniers efforts pour remplir nos devoirs d'une maniére à nous attirer de plus en plus sa puissante protection & à faire des vœux du plus profond de nos cœurs pour la conservation de sa Sacrée Personne. Dieu qui nous l'a donné, & à qui sera consacré desormais le doux anniversaire du Couronnement de S. M. I. ce même Dieu veuille exaucer nos ardentes prieres.

Aprés cette digression, que nôtre zele exigeoit, je reviens au sujet, que j'ai quitté.

Mr. de L'Isle fait voir dans son discours que la seule voye de l'Astronomie ne suffit pas pour decider la question, si la terre a quelque mouvement ou non. Si nous envisageons cette question comme un simple probleme optique, où il faille expliquer les differens changemens qu'on observe dans les Astres, il est certain qu'elle pourroit etre resoluë en differentes maniéres, & c'est en ce sens, que Mr. de L'Isle a dit, que les seules observations d'Astronomie ne suffisent pas pour déterminer la cause du changement des Astres par raport à leur situation; puisque des differentes causes pourroient produire le même effet en apparence. Considerons d'abord les mouvemens journaliers. Posons la terre en repos & tout le systeme se mouvoir autour d'elle dans le tems de 24 heures, & qu'ensuite par un miracle de la nature tous les Astres s'arretant tout d'un coup, la terre commençat à faire ses revolutions d'un sens contraire autour de son axe, pendant 24 heures. C'est une verité incontestable, que nous ne pourrions nous aperçevoir de ce changement. La terre nous paroitroit avoir conservé son repos & les Astres leur mouvement. Ce seroit encor la même chose si les Astres ne perdoient qu'une partie de leur mouvement & que la terre fit ensuite ses revolutions d'autant plus lentement. Ainsi du seul Phenomene du mouvement journalier des Astres, on ne peut encore rien conclure.

Ce même raisonnement peut s'apliquer aussi aux changemens annuels des Astres par raport à la terre; car ils s'expliquent avec la même facilité, soit qu'on dise que le Soleil tourne tous les ans autour de la terre, ou la terre autour du Soleil, ou que tous les deux fassent le tour autour d'un point entre deux, chacun en des tems égaux.

Les Astronomes modernes croïoient d'avoir trouvé une maniére seure pour decider la question; & voici sur quoi ils se fondoient. Un voyageur commençant à voir une ville fort loin, n'en pourra d'abord distinguer les objets. Mais il les developera peu à peu à mesure qu'il s'en aproche; il reconnoit deux clochers qui lui paroissent fort proches l'un de l'autre, & cette distance apparente des deux clochers s'augmente de plus en plus; enfin tous les objets de la Ville semblent se desserrer. Il en seroit de même, si nous pouvions faire quelque voyage vers les Etoiles fixes: elles nous sembleroient aussi se desserer à mesure que nous en approcherions, & fixant nôtre attention particuliérement à deux Etoiles, nous les verrions peu à peu sous un plus grand angle.

Mais nous faisons ce voyage sans nous en apercevoir, s'il est vrai que la terre se tourne autour du Soleil. Il semble donc que toute la question s'est reduite à ce seul point, si on a jamais pû remarquer deux mêmes Etoiles sous des differens angles. Cependant les plus habiles Astronomes n'ont pû encore bien remarquer ces differences, & ce qui paroitra peut-être un peu hardi, c'est que les Coperniciens repondent, sans s'ebranler, que les approchemens que nous pouvons faire vers les Etoiles fixes ne sont pas sensibles & que par consequent ils ne peuvent produire des effets sensibles, non plus que l'approchement de quelques pas, peut avoir quelque effet sensible dans une distance de plusieurs lieües. Un peu de détail developera plus cette reponse des Coperniciens.

Pour faire que les approchemens & réculémens de la terre, par raport à une certaine Etoile fixe, soient les plus sensibles, il faut choisir une des plus grandes Etoiles, les plus grandes etant vraisemblablement les plus proches: ensuite les Etoiles du Zodiaque doivent etre preferées aux autres; la Geometrie enseigne aussi qu'afin que la difference des angles, sous lesquels nous voyons deux Etoiles en differens tems, fut la plus grande, qu'il est possible, il faudroit choisir deux Etoiles, qui sont à peu prés dans leur opposition; mais comme les trop grandes refractions pourroient rendre les observations plus douteuses, on pourra choisir deux Etoiles qu'on voit environ sous un angle droit. Posons ensuite que l'habileté des Astronomes & l'accuratesse des instrumens soient suffisantes pour pouvoir nous asseurer d'une difference de vingt secondes[5]; je demande aprés cela qu'elle devroit etre la distance des Etoiles fixes à la terre, pour ne pouvoir encore s'asseûrer d'aucune difference des angles sous lesquels nous les voyons en differens tems.

5 Bernoulli fixes, quite realistically, the smallest angular variation that could be detected at 20 arcseconds. The smallest detectable parallax α would then be 10 seconds. Without mentioning it he halves this angle and computes how far a star has to be in order to see the distance sun-earth under such an angle of 5 seconds. The ratio of the two distances is then simply

$$\frac{5\pi}{180\,.\,60\,.\,60} = \frac{1}{41236}\,,$$

provided we took the popular (but approximate) value

$$\pi = \frac{22}{7} = 3.142857\ldots$$

Pour repondre à cette question, je dirai que la distance de l'Etoile fixe la plus proche doit etre pour le moins 41 236 fois plus grande que celle du Soleil.

J'ai donc montré, que les Coperniciens n'ont qu'a soutenir que la moindre distance des Etoiles fixes surpasse 40 000 fois celle du Soleil pour éluder la dite objection. Ceux qui ont un peu etudié la nature ne sentent aucune difficulté dans cette enorme distance; ils savent que rien n'est ni grand ni petit en soi même, & qu'il se pourroit qu'il y eut dans un grain de sable, dans un atome, un monde semblable au nôtre, & que tout nôtre monde, que nous voions, ne fut qu'une petite parcelle d'air ou d'ether dans un autre monde. Car pourquoi voudrions nous mettre des bornes à la toute puissance de Dieu. Je ne dois pas oublier ici, que plusieurs Astronomes se sont flattés, d'avoir remarqué une parallaxe dans les Etoiles fixes; & si on avoit pris toutes les mesures indiquées ci dessus, pour remarquer quelque changement, dans la situation de ces Etoiles, elle auroit peut être été tres sensible; mais quelque que soit le resultat de ces observations, j'ai deja fait voir, qu'on n'en peut rien conclurre contre le Systeme des Coperniciens; je dis plus: on n'en peut rien deduire en sa faveur, posé qu'on ne demande ici qu'une solution optique des mouvemens apparens des Astres. Je suis seur qu'on ne s'attendoit pas à cette derniére proposition; tout le monde a crû que le mouvement annuel de la terre autour du Soleil seroit entierement demontré, si on avoit trouvé une paralaxe sensible dans les Etoiles fixes, ou, ce qui revient à peu prés au même, si on pouvoit voir sous des differens angles deux Etoiles fixes.

Quoi! ces raisonables Astronomes oublient donc qu'ils ont à faire à des Ptolomaiciens & ces derniérs assés hardis pour soutenir que les Etoiles fixes tournent tous les jours autour de la terre, n'oseroient-ils pas opiniatrer, qu'un point fixe pris dans le Systeme des Etoiles fixes se meut annuellement dans le Zodiaque autour de la terre, ce qui produiroit les mêmes Phenomenes? quel sujet avons nous de les supposer devenûs tout d'un coup si traitables.

Il est donc vrai, qu'il n'y a point de Phenomenes connus dans l'Astronomie, qui prouvent necessairement[6] & sans l'aide d'aucun autre principe le mouvement de la terre. Mais si nous voulons entrer dans la Physique & Metaphysique, tous les doutes se dissipent & la question cesse d'etre une question.

6 This in the strict mathematical meaning of «necessary», i.e. no known phenomena in astronomy – excluding any other principles from consideration – offer a compelling proof for the motion of the earth.

Je demande d'abord si un Phenomene pouvant etre expliqué de plusieurs façons, on ne doit pas choisir la plus simple, ou au moins rejetter celles qui sont les plus composées? Ne voions nous pas par tout la nature agir par la voye la plus courte? & comment pourroit-elle faire autrement etant dirigée par un Etre souverainement sage.

Ce principe supposé, comme il est incontestable, on n'a qu'a faire un raport entre les mouvemens qu'on suppose dans le systeme de Ptolomée & ceux de l'autre parti, pour voir le ridicule du sentiment des Ptolomaiciens. En verité ils pourroient avec beaucoup plus de raison soutenir, que les corps ne tombent jamais vers la terre, mais que c'est toujours la terre qui s'eléve vers les corps. Ce seroit ici l'endroit à entrer en quelque détail sur les causes finales de tous les mouvemens des Astres; nous en pourrions tirer des preuves convainquantes pour le Systeme de Copernic, mais cela me meneroit trop loin. Je me contenterai d'exposer en peu de mots ces preuves fondées sur la parfaite ressemblance de la terre aux Planetes. Ce sont tous des corps opaques, qui empruntent leur lumière du Soleil. On sçait depuis quelque tems, que deux planetes sont ornées comme notre terre, de lunes; peut-être en decouvrira-t-on aussi avec le tems dans les autres planetes.

Enfin un homme suspendu au milieu de tous ces corps n'y verroit aucune difference essentielle. Doit on douter aprés cela que tous ces corps ne subissent les mêmes loix. Il y a une circonstance qui le prouve absolument.

La voici. On connoit jusqu'a des milliémes parties prés la distance de chaque planete au Soleil; on sçait aussi le tems qu'il faut à chaque planete pour faire sa revolution autour du Soleil; Kepler tres-celebre Astronome a trouvé un certain raport entre la distance de chaque planete & le tems de sa revolution; aprés cétte observation de Kepler[7], que Mr. Newton a ensuite demontrée par des principes de Mechanique, on examina d'abord si

7 If T_1 and R_1 are the time of revolution of one planet and its mean distance from the sun, and similarly for a second planet, Kepler found that

$$\frac{T_1^2}{R_1^3} = \frac{T_2^2}{R_2^3} \, .$$

This is explained by Newton's mechanics: any two masses M and m mutually attracted by a force inversely proportional to their distance squared move on ellipses around their common centre of mass, such that

$$\frac{R^3}{T^2(M+m)} = \frac{G}{4\pi^2} \, ,$$

where G is the constant of universal gravitation. If all planet masses are small vs. the sun's mass, i.e. $m \ll M$, this gives Kepler's law.

la terre observoit aussi cette loix des Planetes, & ce qui est admirable, on lui trouva la même proprieté. C'est donc une consequence necessaire, que la terre est une Planete; elle tourne donc autour du Soleil de même que les autres Planetes. Celui qui niéroit cette consequence pourroit douter avec la même raison, si l'homme a naturellement un cœur, des poumons, un foye &c.

Disons aussi avant que de finir, que dans le Systeme de Copernic, il y a une simplicité une regularité & un ordre, qui ravit. On y rend raison de tout avec une derniére exactitude. Les mêmes loix que nous connoissons sur la terre s'observent dans toutes les Planetes. Enfin un horologer ne sauroit mieux connoitre les artifices & les regles employées dans la construction d'une horloge, qu'un Astronome par rapport au mouvement des Astres. C'est Kepler qui a jetté les premiers fondemens à ces sublimes connoissances, mais nous en devons l'entier developement à Mr. Newton, cet Illustre Anglois, que toutes les nations nomment l'ornement de nôtre siecle. Plût à Dieu que ces connoissances fussent à la portée de tous les hommes: elles rempliroient leurs esprits d'une sainte admiration pour les œuvres du Createur de l'Univers.

Mais quelle obscurité, qu'elle confusion ne regne pas dans le Systeme de Ptolomée? On ne peut pas plus l'éclaircir que le mouvement confus de ces petits corps que les rayons du Soleil agitent. Ce sentiment ne semble-t-il pas favoriser les Epicuréens assés impies & insensés pour soutenir que tout ce que nous voions n'est qu'un assemblage fortuit d'atomes. Quelle injustice d'accuser les Coperniciens d'avoir des sentimens contraires à la Saine Religion, & ceux-ci n'auroient-ils pas plus de droit à faire cette objection à leurs adversaires. Si on avoit lû sans prevention les passages de la Ste Ecriture, bien loin d'y trouver des preuves pour le sentiment de Ptolomée, on en auroit peut être trouvé pour le contraire; c'est ce que Cosme d'Egypte[8], Moine, à pretendu de faire voir dans sa Topographie Chretienne. Pour dire la verité, ces tems, ou l'on ne pouvoit dire sans heresie que la terre est ronde, qu'il y a des Antipodes, & que la terre tourne, ne meritent gueres d'éloges & bien moins ceux, qui etoient les chefs & les auteurs de ce faux zele.

8 See p. 11 note 20 h. v. He founds his vision of the world on the Scriptures and is influenced by oriental legends. He polemises against the false Christians who admit pagan doctrines, a. o. a spherical earth. The work became known in Europe through: B. de Montfaucon, *Cosmae Indicopleustae Topographia Christiana* (texte grec et traduction latine) in *Collectio Nova Patrum et Scriptorum Graecorum Eusebii Caesariensis, Athanasii et Cosmae Aegypti*, t. II, Parisiis: sumptibus C. Rigaud 1706.

AVERTISSEMENT
de Mr. De L'isle

Depuis la lecture de mon discours, j'ai reçû un nouveau traité publié depuis peu en Danemarc[9]*, en faveur du mouvement de la terre; l'Auteur y rapporte des observations astronomiques fort subtiles, qui semblent determiner la distance des Etoiles fixes à la terre, plus sûrement que j'ai encore fait jusqu'a present. Comme je n'avois point eu connoissance de cet ouvrage, ni de ces observations, lors'que j'ay lû à l'Academie le Discours precedent; je n'ai pû y faire mention de ces nouveaux faits astronomiques cités en faveur du Systeme de Copernic. Mais je puis dire (pour me conformer à l'intention de mon discours) que quand même ces nouvelles observations prouveroient & determineroient seurement la distance des Etoiles fixes à la terre, elles ne demontreroient pas pour cela, Geometriquement parlant, le mouvement de la terre, à ceux qui ne voudroient admettre que les faits d'experiences, & les demonstrations d'optique fondées sur ces faits. Mais aussi, on seroit bien ridicule, & extraordinaire d'en agir ainsi, & de refuser d'accepter la nouvelle lumiere, & les nouvelles preuves de vraisemblance que les observations de Danemarc nous fournissent du mouvement de la terre, sur tout y aiant deja en faveur de ce systeme, un grand nombre d'autres preuves, appuiées sur les principes de la Mechanique & de la Physique, aussi certains & aussi convainquans que ceux de la pure Geometrie; laquelle elle même, tire ses premiers principes de la Mechanique & de la Metaphysique.*

9 See the Introduction p. 12 note 21 h. v.

St. 28

Réfléxions sur la meilleure figure a donner aux Ancres, et la meilleure maniére de les essayer

Prix Paris III (pièce VI) 1737 (1737), pp. 49–84

Quelle est la meilleure maniere d'éprouver les Ancres?
Sujet proposé par l'Académie Royale des Sciences, pour le troisiéme Prix de l'année 1737.

Omnia conando docilis solertia vincit.[1]

I. Quoique l'histoire des Ancres nous manque presque entiérement, il ne faut pourtant pas douter que l'invention n'en soit très-ancienne, tant à cause de l'ancienneté du nom, que de leur nécessité indispensable dans les grandes Navigations, telles qu'on a faites depuis des temps immémorials. Il y a donc apparence que les Ancres n'auront guéres manqué d'atteindre à la plus grande perfection dont elles sont capables: toute correction possible & importante aura difficilement échappé à tant de Nations, à tant de recherches & à tant de siécles, mais sur-tout à nos derniers temps, dans lesquels on a poussé les Sciences & les Arts en général, & la Navigation en particulier, à un degré de perfection que nos Ancêtres auroient à peine osé esperer. Je sens donc bien la difficulté de mon entreprise, mais aussi me semble-t-il que dans les choses aussi importantes & aussi perfectionnées, la moindre addition doit être reçûë avec autant de satisfaction qu'on reçoit des inventions toutes nouvelles. C'étoit-là sans doute le motif de l'illustre Académie, de donner pour sujet de ses Prix la perfection des Ancres & de l'ancrage, & de le donner pour la seconde fois[2]. II en est comme des Horloges; l'application des Pendules par M. Huguens n'en est qu'une légere addition, mais qui merite autant d'éloges que tout ce qu'on avoit inventé auparavant sur la Mesure du Temps. Il est vrai qu'on pourroit facilement imaginer des especes d'Ancres toutes nouvelles, ou faire de grands changements à leur structure reçûë: ces nouveautés ou changements pourroient avoir de bonnes apparences: mais lorsqu'on en feroit l'essai, on trouveroit peut-être ces nouvelles

1 Ingenuity, easily taught, masters all by endeavouring.
2 Refers to the question posed in 1735, which received no prizeworthy answers. See the *History of the prizes*, pp. 17–19 h. v.

Ancres bien inférieures aux ordinaires. On verra dans la suite que j'ai examiné avec beaucoup de soin la structure des Ancres, tâchant d'approfondir leur méchanisme avec les fonctions de chaque partie, & il m'a paru d'y voir beaucoup d'invention; j'ai donc crû n'y devoir changer que dans les circonstances qui dépendent absolument de la Géométrie: l'expérience a fait voir l'essentiel, & le raisonnement qu'un homme de Lettres peut faire dans son cabinet, ne doit plus y avoir de prise que dans les choses d'une méchanique bien ménagée, & cela non tant pour décider que pour donner à penser aux personnes intelligentes qui sont à portée de faire de nouvelles expériences.

Voici l'ordre que je me propose dans ce discours: je commencerai par exposer la structure ordinaire des Ancres, après quoi je décrirai la maniére méchanique dont elles agissent; je rechercherai ensuite ce qu'il faut faire pour rendre leur usage le plus sûr & le plus parfait, & pour diminuer quelques inconvénients qui leur restent, en parcourant toutes les parties des Ancres: je donnerai la description d'une nouvelle sorte d'Ancres, & enfin j'adjoûterai quelques réfléxions sur la meilleure maniére d'essayer les Ancres, mais je ne toucherai que légerement ces deux derniers points.

II. Les Vaisseaux portent ordinairement quatre Ancres de différente grandeur; la plus grande est la *maîtresse Ancre*, qu'on réserve pour les cas extraordinaires & les plus périlleux en temps d'orage: la seconde est celle dont on se sert ordinairement: la troisiéme, nommée l'*Ancre d'affourche*, un peu plus petite que la précédente, est celle que l'on mouille opposée à l'Ancre déja jettée, c'est à dire, de sorte que leurs cables forment un angle tantôt plus tantôt moins grand; la quatriéme est appellée l'*Ancre à touer*; elle est beaucoup plus petite que les précédentes, aussi ne s'en sert-on pas pour arrêter les Vaisseaux, mais pour les *touer*. Toutes ces Ancres, quoique de différente grandeur, ne laissent pas d'avoir les mêmes proportions dans leurs parties, de sorte qu'il seroit superflu d'en traiter séparément.

Les Ancres sont composées des parties suivantes. Il y a la *vergue*: c'est une barre de fer d'une épaisseur égale, c'est-à-dire, cylindrique: sa longueur est proportionnée à la largeur du Vaisseau, & cette proportion est dans la maîtresse Ancre comme 2 à 5. Le diametre de la *vergue* fait environ la vingt-sixiéme partie de sa longueur: le bout de la *vergue*, qui joint le *jas*, est quarré, ou plûtôt prismatique sur une section quarrée: il est troué à l'extrémité pour donner passage à l'*arganeau*, qui est un anneau de fer, auquel le *cable* est attaché, & afin que celui-ci ne se pourrisse pas par l'enrouillure de l'*arganeau*, on enveloppe l'*arganeau* avec de vieux cordages. Le bout prismatique de la *vergue* est plus épais que le reste, pour y mieux faire tenir le *jas* par lequel il passe.

Ce *jas* est un assemblage de deux piéces de bois d'une figure égale, empattées fort étroitement ensemble: il fait un angle droit avec la *vergue*, & va en diminuant vers les extrémités: sa longueur ne differe pas beaucoup de celle de la *vergue*, mais il a environ quatre fois plus d'épaisseur dans son milieu que la *vergue* n'en a.

A l'autre bout de la *vergue* est soudée la *croisée* sous un angle droit & perpendiculairement au plan qui passe par le *jas* & la *vergue*. La *croisée* est recourbée vers le *jas*, & sa courbûre forme environ un arc de cercle de soixante degrés, plus ou moins, dont le centre viendroit à peu-près au milieu de la *vergue*: ses épaisseurs vont en diminuant vers les deux bouts: elle a enfin deux branches, dont chacune peut être censée former un arc de cercle d'environ trente degrés.

Les *pattes* sont deux piéces de fer triangulaires, soudées sur le dedans de chaque bout de la *croisée*; leur longueur est un peu plus grande que la base, qu'on appelle les *oreilles*: elles sont recourbées en dedans autant que la *croisée*, pour pouvoir mordre plus facilement dans le sable du fond de la Mer.

Enfin le *cable* peut encore être censé appartenir à l'Ancre: il est attaché, comme j'ai déja dit, par l'*arganeau*: sa longueur est ordinairement de 120 brasses: il passe par les *écubiers* du Vaisseau, qui sont des trous faits aux côtés de l'avant du Vaisseau; les quarrés de leurs épaisseurs, qui en marquent le nombre des fils, ne suivent pas la proportion des poids des Ancres, & on a des Tables là-dessus, qui ne sont fondées absolument que sur l'experience. Voici cependant un exemple. Une Ancre de 500 livres a un cable d'environ 3 pouces de diametre; il sera composé de 375 fils, dont chacun est estimé de 4 livres de poids, de sorte que tout le *cable* pesera 1500 livres, & par conséquent trois fois plus que toute l'Ancre. Les cordages sont tantôt plus tantôt moins pesants suivant qu'ils sont plus ou moins roides, cependant dans l'eau ils sont tous d'une même pesanteur spécifique, & lorsqu'ils sont bien mouillés & imbibés, ils descendent sous l'eau par leur propre poids, étant alors d'une pesanteur spécifique plus grande que l'eau.

J'ai pris ces descriptions & proportions d'un Livre de Navigation, & même quelques-unes des simples figures sans description, sur-tout lorsqu'elles me paroissoient conformes aux observations que j'avois faites autrefois moi-même sur les Ancres; j'établirai ma théorie sur ces proportions, & si on ne leur trouvoit pas assés de précision, il sera facile d'accommoder mon raisonnement aux véritables proportions.

III. Examinons maintenant de quelle maniére les Ancres étant jettées dans la Mer, mordent dans le sable: c'est certainement ici le point principal, & qui demande le plus d'attention: aussi nous faudra-t-il, pour connoître à fond toute la méchanique des Ancres, commencer ces recherches d'un peu loin. Il est clair que l'Ancre ne sçauroit mordre dans le fond à moins de lui présenter sa pointe, c'est-à-dire, d'avoir la *croisée* dans un plan vertical: mais les Ancres peuvent se coucher sur un fond horisontal de deux maniéres différentes: l'une tient la *croisée* couchée sur le fond pendant que le *jas* y est appuyé par un de ses bouts; dans l'autre c'est au contraire le *jas* qui est couché horisontalement, & la *croisée* est soûtenuë par la pointe de l'une de ses *pattes*; ce n'est que dans celle-ci que l'ancrage peut se faire: voyons cependant laquelle de ces deux situations est la plus naturelle aux Ancres: nous examinerons ce point d'abord hors de l'eau; & ensuite dans l'eau.

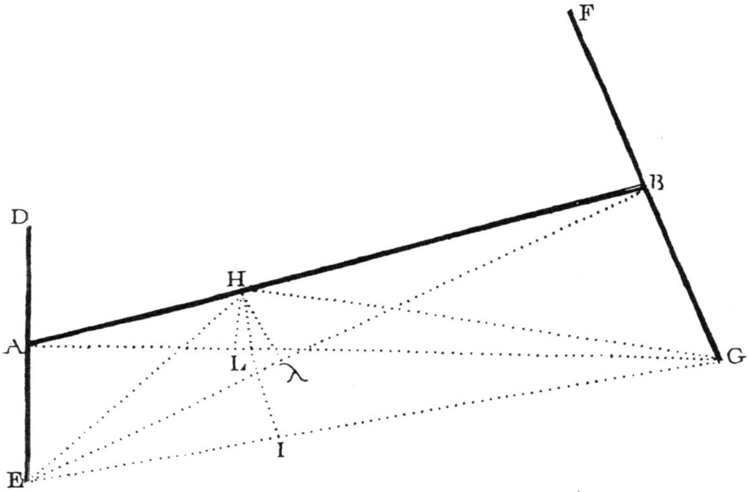

Fig. 1

IV. Soit dans la premiére Figure *DE* la ligne droite qui joint les extrémités des deux *pattes*: on peut la considérer ici à la place de la *croisée*, comme si celle-ci n'avoit point de courbûre: soit *AB* la *vergue*, & *FG* le *jas*: concevons pour la premiére situation la ligne *DE* couchée horisontalement, de sorte que les points *D*, *A*, *E* & *G* soient dans un plan horisontal, & les points *G*, *B*, *F* & *A*, dans un plan vertical[3]. Examinons à présent ce qu'il faut pour changer cette situation de l'Ancre, & lui faire prendre l'autre situation: il faut pour cet effet que l'Ancre tourne autour de la ligne *EG*, qui joint l'une des pointes de la *croisée* avec le point d'appui du *jas*. De cette ma-

3 See the *Review*, pp. 21–28 h. v., for a modern paraphrase of the argument.

niére le centre de gravité de l'Ancre (que je supposerai en H) s'éleve d'abord jusqu'à ce qu'il soit arrivé au point le plus haut, après quoi l'Ancre acheve de se renverser par son propre poids, & prend ainsi la seconde situation qui est requise pour les ancrages. Il s'agit ici de sçavoir combien le centre de gravité H est élevé pendant ce renversement, puisque la force[4] requise pour faire prendre à l'Ancre la situation qu'elle doit avoir, lorsqu'elle ne l'a pas, est proportionnelle à cette même élévation du centre de gravité.

Pour cet effet qu'on tire les droites HE & HG, comme aussi la ligne AG, puis HL perpendiculaire à AG & HI perpendiculaire à EG: on voit que la hauteur initiale du centre de gravité par dessus l'horison est égale à HL, & que sa plus grande hauteur, pendant le renversement de l'Ancre, est égale à HI, de sorte que pour renverser l'Ancre, il faut auparavant donner à son centre de gravité une élévation égale à $HI - HL$: il ne reste donc plus qu'à exprimer analytiquement ces lignes.

Soit $AE = a$, $BG = \alpha$, $AH = b$, $BH = \beta$, on aura

$$AG = \surd(AB^2 + BG^2) = \surd(\alpha\alpha + bb + 2b\beta + \beta\beta):$$

de-là on tire

$$EG = \surd(AE^2 + AG^2) = \surd(aa + \alpha\alpha + bb + 2b\beta + \beta\beta)\ ;$$

pour abbréger, je supposerai $EG = c$: on trouvera HL en prenant la quatriéme proportionnelle à AG, GB & AH, ce qui donne

$$HL = \frac{b\alpha}{\surd(\alpha\alpha + bb + 2b\beta + \beta\beta)}\ .$$

Pour trouver la ligne HI, je considere le triangle EHG, dont la base EG est $= c$; le côté $EH = \surd(EA^2 + AH^2) = \surd(aa + bb)$, & le côté $HG = \surd(GB^2 + BH^2) = \surd(\alpha\alpha + \beta\beta)$: des trois côtés donnés on trouve la perpendiculaire à la base, sçavoir

$$HI = \frac{\surd(\overline{2aa + 2bb} \times \overline{\alpha\alpha + \beta\beta} + 2cc \times \overline{aa + bb + \alpha\alpha + \beta\beta} - \overline{aa + bb}^2 - \overline{\alpha\alpha + \beta\beta}^2 - c^4)}{2c}\ .$$

Connoissant donc les droites HI & HL, leur différence donnera l'élévation cherchée du centre de gravité, pendant qu'on fait prendre à l'Ancre sa juste

4 This «force» is really a living force (*force vive*), i.e. a kinetic energy, because it is proportional to an increase in height which means in potential energy.

position, laquelle élévation doit être estimée proportionnelle à la force requise pour cet effet.

V. Changeons maintenant la proposition, en supposant au contraire que l'Ancre soit dans sa juste position, c'est-à-dire, que la *croisée* soit dans le plan vertical, le *jas* étant couché horisontalement, & voyons quelle est la force qui pourroit faire quitter à l'Ancre cette position nécessaire pour l'ancrage, & la remettre dans l'état qu'elle a été supposée au commencement de l'article précédent. Il est facile de voir qu'on n'a qu'à convertir pour cet effet les lettres a & α, comme aussi b & β; ce que faisant, la valeur de c, qui est $= \surd(aa + \alpha\alpha + bb + 2b\beta + \beta\beta)$, demeure dans les deux cas la même. Si l'on tire donc $H\lambda$ perpendiculaire à EB, ce sera $HI - H\lambda$ qui exprimera la force requise à ce second renversement. La premiére ligne HI est la même que dans le premier cas, & on a

$$H\lambda = \frac{\beta a}{\surd(aa + bb + 2b\beta + \beta\beta)} .$$

VI. Les deux forces respectives étant en vertu des deux articles précédents, comme $HI - HL$ à $HI - H\lambda$, on remarquera ici, qu'il convient que la premiére force marquée par $HI - HL$ soit aussi petite, & l'autre marquée par $HI - H\lambda$ soit aussi grande que les autres circonstances le permettent, par où l'on obtiendra cet avantage, que l'Ancre n'étant pas dans sa juste position, elle s'y mette avec facilité, & qu'y étant, elle ne la quitte que difficilement: il suit de là que plus HL est grande, & $H\lambda$ petite, plus les Ancres prendront facilement leur position requise pour le succès de l'ancrage.

Ce que je viens de dire sert également pour les Ancres submergées, & pour celles qui seroient jettées sur un fond hors de l'eau, & toute la différence qu'il y a, est que le centre de gravité est placé différemment dans ces deux cas, à cause du *jas*; j'appliquerai cependant ces regles à l'un & l'autre cas, tant pour en voir la grande différence dans leur résultat, que pour nous servir de cette application dans la suite.

VII. La *vergue* est d'une épaisseur égale, & il me semble que la *croisée* aura à peu-près le même poids que le *jas* chargé du bout quarré de la *vergue* & de l'*arganeau*. Cela étant, on peut placer le centre de gravité dans les Ancres non-submergées, au milieu de la *vergue*: supposons encore la *vergue* & le *jas* d'une longueur égale, & que la distance des extrémités des *pattes* ou DE soit égale à la moitie du *jas*. Toutes ces positions font $\alpha = \beta = b = 2a$, &

$c = a\sqrt{21}$, ce qui donne $HI - HL = \sqrt{\frac{8}{7}} - \sqrt{\frac{4}{5}} =$ (en fraction décimale) 0.174, & $HI - H\lambda = \sqrt{\frac{8}{7}} - \sqrt{\frac{4}{17}} = 0.584$. Les deux forces dont j'ai parlé aux articles IV & V, sont donc comme 174 à 584, ou à peu-près comme[5] 2 à 7.

VIII. On comprend aisément par-là, qu'une Ancre étant jettée au hasard sur un fond, donnera probablement à sa *croisée* la position verticale & non l'horisontale, si ce fond est horisontal & hors de l'eau: mais on se tromperoit, si on ne faisoit cette probabilité que comme 7 à 2. Il est plûtôt vrai que l'Ancre étant jettée avec force, & que le fond soit bien dur, il est moralement[6] impossible que la *croisée* reste couchée sur le fond, & voici la raison de cette proposition assés paradoxe; c'est que l'Ancre jettée avec une force, que je supposerai plus grande que 7, se roulera d'abord, & qu'à mesure qu'elle se roule, elle perdra de sa force jusqu'à ce que cette force étant moindre que 7, & plus grande que 2, la *croisée* de l'Ancre doit enfin garder nécessairement sa situation verticale: sans cela il faudroit que dans un seul renversement de l'Ancre, elle perdît plus que $\frac{5}{7}$ de sa force, ce qui ne sçauroit arriver sur un fond bien dur[7].

IX. Pour déterminer les mêmes choses dans les Ancres submergées, il faut principalement faire attention que le *jas* est de bois, mais d'un bois fort & pesant, de sorte que pris avec le bout quarré de la *vergue* & avec l'*arganeau*, il ne manquera pas d'avoir à peu-près la même pesanteur spécifique que l'eau, & que par conséquent son poids peut être négligé sous l'eau; & comme tout le reste de l'Ancre est de fer & homogéne, on n'a plus qu'à examiner quelle seroit la place du centre de gravité H dans une Ancre dégarnie de son *jas*. Or toutes choses bien considérées, j'estime que dans ce cas HB sera à peu-près double de HA. Mettant donc $b = \frac{4}{3}a$, & $\beta = \frac{8}{3}a$, en retenant les autres hypotheses & dénominations de ci-dessus, on n'a qu'à suivre les mêmes raisonnements & calculs. De cette maniére, on trouvera[8] $HI - HL = 0.337$, & $HI - H\lambda = 0.287$, ce qui fait voir que les deux forces sont assés égales, & que même la premiére est plus grande que la seconde, marque

5 The result of the argument is $\frac{174}{584} = 0.29795$; the proposed approximation is $\frac{2}{7} = 0.28571$, a 4.7% discrepancy.

6 This word characterizes a near-certainty; it appears for the first time in this sense in the *Ars conjectandi* of Jacob Bernoulli, written in the years up to 1690 and published posthumously in 1713. See Jac. B. *Werke* 3, p. 240. See also Jac. B. *Werke* 2, p. 201 for a German translation.

7 The value of this argument was discussed in the *Review*, pp. 25–27 h. v.

8 The two numbers are, respectively, 0.3352 and 0.2848.

qu'il est plus facile & plus naturel aux Ancres sous l'eau, d'avoir la *croisée* couchée que dressée; cela doit nécessairement rendre l'ancrage mal sûr. On auroit eu de la peine à croire la chose si différente pour les deux cas, sans les calculs que nous venons de faire.

X. L'ordre demande que nous examinions maintenant ce qui arrive aux Ancres jettées au fond de la Mer, lorsqu'elles sont tirées par le *cable*: nous le ferons, après avoir dit deux mots sur la nature du fond propre au *mouillage*.

Il ne doit pas avoir au de-là de 50 brasses de profondeur, puisque la longueur du cable ne surpasse pas 120 brasses, & qu'il doit toûjours faire un angle fort oblique avec l'horison, comme je le démontrerai ci-dessous. D'ailleurs le fond ne doit pas être trop dur, car l'Ancre ne sçauroit y mordre, ni assés enfoncer: si le fond est simplement sablonneux, l'Ancre enfonce facilement, mais elle n'y tient pas assés ferme, & le Vaisseau est sujet à *chasser sur son Ancre*. Suivant les observations de M. le Comte Marsigli[9], le fond de la Mer est le plus souvent d'une conglutination sablonneuse d'argile, de coquillages, & d'autres corps: cette conglutination est formée par la matiére glutineuse qui réside dans les eaux de Mer, & le tout forme une croûte qui n'est pas fort épaisse, mais qui est d'ailleurs d'une consistance fort propre pour le *mouillage*, sçavoir, ni trop dure, ni qui se laisse trop facilement labourer par l'Ancre: au-dessous de cette incrustation, le fond est d'une constitution pierreuse, dans lequel la *patte* ne sçauroit plus mordre. Les corps mêlés avec le sable durci, doivent rendre la surface du fond assés inégale & raboteuse.

XI. On jette l'Ancre ordinairement, pendant que le Vaisseau avance sur sa route, en prenant garde que le *cable* ne se roidisse pas d'abord, & en filant pour cet effet le *cable* autant que la vîtesse du Vaisseau le demande, jusqu'à ce qu'on le voye faire avec la surface de la Mer un angle d'environ 30 degrés; alors l'Ancre étant traînée plus ou moins vîte, il arrivera d'abord qu'elle roule de côté & d'autre, se couchant tantôt sur le *jas*, tantôt sur la *croisée*, ayant pour l'un & pour l'autre une facilité à peu-près égale en vertu du VIIIme article. Mais voici la raison de ce roulement, c'est que la *croisée* se couchant horisontalement, le *jas* sera appuyé sur le fond par un de ses bouts, faisant avec le fond un angle d'environ 63 degrés, si les parties des Ancres suivent les proportions que nous avons supposées dans le second article: or si le fond de la Mer étoit parfaitement uni & poli comme une

9 L.F. Marsili, *Histoire de la mer* (translated by Leclerc), Amsterdam 1725.

glace, il est très-certain que l'Ancre demeureroit constamment dans cette situation, mais comme ce fond est raboteux, & que le *jas* présente sa pointe appuyée en avant, à cause de son inclinaison en arriére, on voit qu'il heurtera continuellement contre les obstacles qu'il trouvera en son chemin, qu'il s'en ébranlera, & que très-aisément il se renversera & se couchera sur le fond; alors c'est la *croisée* qui est dans le plan vertical, & l'une de ses *pattes* présentera au fond sa pointe, quoique sous un angle presque droit, c'est-à-dire, que la tangente de l'extrémité de la patte fait un angle presque droit avec le fond; cette tangente est pourtant un peu inclinée en arriére, mais pas tant que le *jas* l'étoit dans l'autre situation de l'Ancre; ainsi donc la *croisée* heurtera aussi contre les inégalités du fond, de même que le *jas* le faisoit auparavant: mais ces impulsions qui se font contre le bout de la *croisée*, ne seront peut-être pas si sensibles comme celles du *jas*, à cause que la *croisée* n'est pas si longue que le *jas*, ni si obliquement appuyée sur le fond: car on démontre aisément dans la Méchanique, que la longueur & l'obliquité du levier heurtant, tel qu'est ici la branche inférieure de la *croisée*, rendent les impulsions plus sensibles & plus efficaces dans notre cas pour renverser l'Ancre; & quant à l'obliquité en particulier, l'expérience le confirmera, si en s'appuyant sur une canne, on la glisse sur le plancher: car on verra que la canne étant perpendiculaire au plancher, ne s'en trémoussera pas tant, que si elle est inclinée en arriére. Il semble donc, pour les deux raisons apportées, que les impulsions données contre la *croisée* ne la renverseront pas si facilement qu'elles renversent le *jas*, & qu'elles serviront plûtôt à faire mordre la *patte* dans le fond; car après que la *croisée*, par un coup reçû, a été poussée en haut, & qu'elle retombe la pointe contre le fond, cette pointe y entrera par la chûte d'un si grand poids, quelque legere que soit la chûte, & dès-lors l'Ancre n'est plus si sujette à se renverser, & l'action du *cable* l'enfoncera de plus en plus, comme nous ferons voir ci-dessous.

XII. Ce que nous venons de dire sur la méchanique des Ancres jettées & traînées au fond de la Mer, se confirmera par l'expérience à ceux qui la voudront prendre comme je l'ai fait: qu'on fasse un petit modele d'Ancre, mais dont le *jas* ait toute la legereté possible, pour imiter parfaitement la nature des Ancres jettées au fond de la Mer, où le *jas* n'a plus de poids; si on traîne cette petite Ancre sur une table bien polie, on verra qu'elle ne se renversera jamais, que ce foit le *jas* ou la *croisée* qui est couchée: qu'on traîne ensuite cette Ancre sur un plancher moins uni, & on la verra se renverser très-souvent & très-facilement, & cela avec une facilité égale pour l'une & pour l'autre situation: car si d'un côté, en vertu du IXme article, la *croisée* se couche plus facilement que le *jas*, nous avons fait voir au contraire dans le précédent article, que les petits chocs donnés contre la *croisée* ne sont pas si sensibles que ceux du *jas*. Enfin si on traîne la petite Ancre sur un fond sablonneux, dont les inégalités & la dureté soient proportionnées à la grandeur & au poids de l'Ancre, on trouvera qu'après plusieurs roulements de côté & d'autre, la *patte* commencera à s'enfoncer, & qu'elle y entrera après cela si avant, qu'on ne sçauroit plus l'entraîner, sans employer considérablement plus de force, qu'on n'avoit fait au commencement de l'expérience. Tout ceci me paroît prouver exactement la vérité de ce que je viens d'avancer sur la méchanique des Ancres, & nous fait voir en même temps, quelle seroit la meilleure maniére d'essayer les Ancres: c'est sur quoi je m'expliquerai plus clairement à la fin de ce discours. Il nous reste à examiner ce qui arrive aux Ancres, après qu'elles ont déja commencé à mordre dans le fond.

XIII. Dès que l'Ancre a mordu dans le fond, elle laboure d'abord le sable, & par-là même elle résiste avec une plus grande force au *cable*, ou plûtôt au Vaisseau qui est la force mouvante; si on faisoit d'abord trop roidir le *cable*, sa force pourra facilement élever l'*arganeau*, ce qui fera nécessairement renverser la *croisée*, si elle ne tient déja bien ferme dans le fond, & en ce cas c'est à recommencer. On voit par-là que la flottaison du *jas* entre deux eaux ne sçauroit qu'être extrêmement préjudiciable au *mouillage*. Il y en a cependant qui ont cru cette flottaison essentielle à l'ancrage, pensant que c'est pour cette raison, qu'on fait le *jas* de bois. Il est certain que si le *jas* flottoit naturellement entre deux eaux, la *croisée* se coucheroit toûjours sur le fond, auquel cas l'ancrage ne sçauroit se faire: on s'en convaincra, si dans les expériences exposées dans le précédent article, on tiroit l'Ancre assés verticalement pour élever le *jas*, car on verra la *croisée* tomber aussi-tôt. Et si le *jas* étoit élevé après que la *patte* est déja entrée dans le fond, il arrivera, ou que la *patte* tienne déja assés ferme pour empêcher le renversement de l'Ancre, & en ce cas elle tiendra aussi assés ferme pour empêcher que le Vaisseau ne *chasse sur son Ancre*, ou qu'elle ne tienne pas assés ferme pour soutenir l'effort de la *croisée* de se coucher, & en ce cas toute la manœuvre de l'ancrage est renduë inutile. Ainsi donc l'élevation du *jas*, ou bien celle de l'*arganeau*, ne peut jamais avancer le *mouillage*, mais bien le retarder. Il faut donc l'éviter, en tirant le *cable* le plus horisontalement qu'on peut, & en *filant le cable sur les bittes* assés vîte pour qu'il ne soit pas trop roide, ni par conséquent ses efforts trop grands. Nous allons examiner cela de plus près.

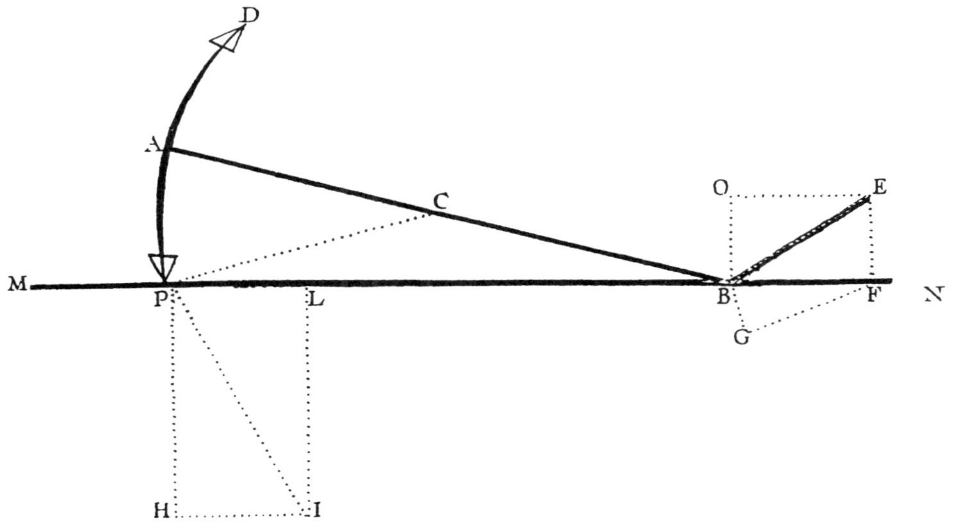

Fig. 2

XIV. Soit dans la seconde Figure, *MN* le chemin que l'Ancre fait sur le fond de la Mer, *PC* la ligne tirée perpendiculairement à l'extrémité de la *patte*; elle coupe, comme j'ai dit au second article, la vergue *AB* au milieu *C*, & par les proportions supposées dans le même article, l'angle *ACP* sera d'environ 30 degrés, & par conséquent l'angle *CPB*, qui en est la moitié, de 15 degrés; c'est cet angle qui fait l'obliquité de la *patte* contre le fond, & sans cette obliquité, elle n'y pourroit mordre qu'autant qu'elle y seroit forcée par son propre poids, qui seul ne suffiroit pas; car outre ledit poids, c'est aussi la force du *cable* qui fait entrer la *patte* plus avant dans le fond, après qu'elle a commencé à y mordre; la direction de cette seconde force est la tangente du *cable* en *B*, celui-ci prenant la figure de la chaînette, qui approche plus ou moins de la ligne droite selon sa longueur & la force avec laquelle il est tiré par le Vaisseau. Soit donc *BE* la direction du *cable* en *B*, & que *BE* exprime en même temps la force[10] qui tire l'Ancre *BDP*: il faut résoudre d'abord cette force en sa verticale *OB*, & horisontale *FB*, en faisant le rectangle *BOEF*. Quant à la force verticale *BO*, elle fait effort pour lever l'*arganeau*, mais de sçavoir si elle l'élevera actuellement ou non, cela dépend de la force absoluë *BE* & de l'angle *EBF*, comparés avec la force que l'Ancre exerce par son poids sur le point *B*, & la résistance que l'enfoncement de la *patte*

10 The force of the currents against the ship is supposed to be strictly horizontal.

peut apporter contre cette élevation. Ce qu'il y a de sûr, est que l'élevation de l'*arganeau* & du *jas* peut très-facilement faire renverser l'Ancre sur sa *croisée*, & retarder par-là le succès de l'ancrage, pendant qu'elle ne sçauroit être d'aucune utilité, & qu'il faut par conséquent diriger la manœuvre de maniére que le *jas* reste couché sur le fond; cela étant, la dite force verticale *OB* reste sans effet. Quant à la force horisontale *FB*, comme sa direction passe par *P*, & que c'est la résistance de la *patte* qui est opposée à cette force, on voit qu'il faut la considérer comme appliquée en *P*, & ensuite la résoudre en deux, l'une parallele à la direction de l'extrémité de la *patte* en *P*, & l'autre perpendiculaire à cette direction: ces directions sont représentées par les lignes *BG* & *GF*, la premiére sert directement à enfoncer la *patte*, l'autre force appliquée en *P*, & parallele à *GF* ne fera que presser fortement la surface de la *patte* contre le sable, qu'elle renversera en labourant le fond, si la force est assés grande; & cela continuëra ainsi jusqu'à ce que la *patte* soit assés enfoncée pour ne plus se laisser entraîner par cette force. Voilà de quelle maniére les Ancres agissent, en arrêtant ainsi peu à peu les Vaisseaux, & les affermissant ensuite contre le vent, contre les courants, & sur-tout contre les coups de Mer: ce sont ceux-ci qui font le plus d'effet, mais ils ne font ordinairement qu'enfoncer de plus en plus les Ancres, qui le sont déja trop pour labourer encore le fond, pendant qu'elles peuvent toûjours y entrer davantage, d'autant que les lames tirent le *cable* brusquement, de sorte que la force résultante approche plûtôt de la nature des chocs, que des simples forces qu'on appelle *mortes*[11]; aussi voit-on que dans les grandes tempêtes, les coups de Mer font plus souvent rompre le *cable* que *déraper* l'Ancre.

XV. Tâchons ici d'avoir quelque idée sur le rapport des forces dont nous venons de parler, pour pouvoir les comparer ensemble, ce qui nous sera d'une grande utilité dans la suite.

11 The *forces mortes* (dead forces) denote forces that are continuously active, such as gravitation or the force due to a steady current against the ship at anchor. They are traditionally distinguished from the *forces vives* (living forces) that manifest themselves during very short times, such as elastic forces during collisions or the forces on a ship due to sudden impacts of water masses. A cable may be strong enough to hold a ship stationary against a strong current, yet when at the upper end of the cable the ship is subject to large accelerations, and hence huge forces pull on the cable, while the inertia of the anchor resists acceleration, the tension in the cable will mount beyond the breaking point.

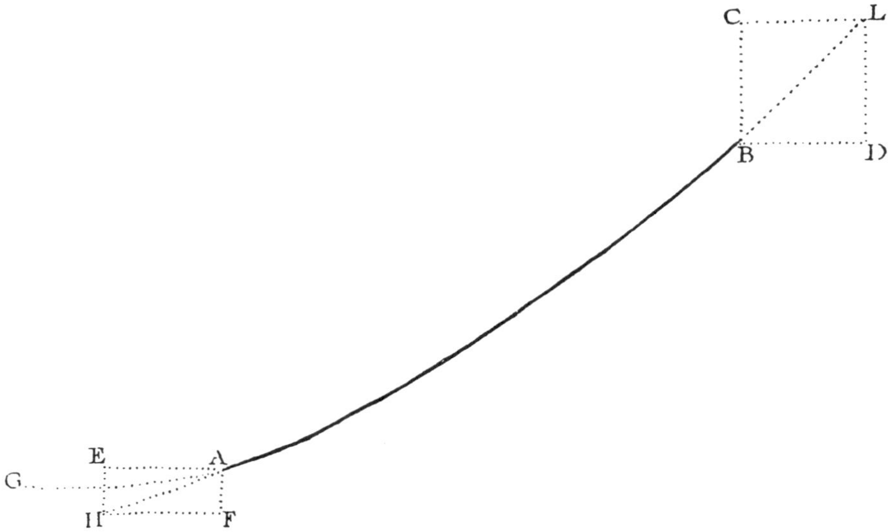

Fig. 3

Soit donc dans la troisieme Figure, *AB* le *cable*, le point *A* représentant l'endroit de l'*arganeau*, & *B* celui de l'*écubier*, en négligeant la longueur du *cable* depuis la surface de l'eau jusqu'à l'*écubier*. J'ai déja dit qu'à cause du poids que le *cable* a sous l'eau, il prendra la figure connuë sous le nom de la *chaînette*. Qu'on conçoive cette courbe *BA* continuée jusqu'en *G*, qui est son point le plus bas, & où la tangente est horisontale: qu'on s'imagine ensuite à chacune des extrémités *A* & *B*, être appliquées deux forces, une horisontale & une verticale, qui toutes quatre soient en équilibre, & tiennent le *cable* suspendu; les grandeurs & les directions de ces forces sont représentées par *AE*, *AF*, *BD* & *BC*. Ceci posé, on sçait qu'en faisant les rectangles *AEHF* & *BDLC*, les diagonales *AH* & *BL*, représentant les deux forces résultantes des quatre forces exposées, seront des tangentes aux points *A* & *B*; outre cette propriété connuë, en voici deux autres.

1.° Je dis que les deux forces horisontales *AE* & *BD* sont toûjours égales entr'elles[12]. J'ai une démonstration analytique de cela tirée de la nature de la chaînette, que j'obmets parce qu'il me semble qu'il suffit de remarquer, que toutes les autres forces, tant celles qui sont représentées par *AF* & *BC*, que

12 This is the same equality, expressed in geometric language, as equation (4) in the *Review*, p. 28 h. v. The curve *AB* is an arc of the catenary (*chaînette*); its apex is the point *G*.

les forces infiniment petites qui tirent chaque point du *cable*, sont verticales, & que par conséquent les deux dites forces horisontales AE & BD doivent se détruire & être égales sous des directions opposées.

2.° Que la force BC est à la force AF, comme la longueur BG est à la longueur AG, car la force BC est précisément égale au poids du *cable* sous l'eau, de la longueur BG, & la force AF égale au poids que le *cable* de la longueur AG auroit sous l'eau.

De ces deux Théoremes, on peut trouver les forces AE & AF, en connoissant les deux forces BD & BC avec le poids du *cable* submergé BA; il suffit même de connoître la force horisontale BD, & l'angle LBD que le cable fait avec la surface de la Mer.

En conséquence de ce que je viens de dire, posons la longueur du cable $BA = l$, son poids sous l'eau $= p$: la force $BD = P$, le sinus total $= 1$, le sinus de l'angle $LBD = s$, son co-sinus $= c$; cela posé, on aura $BC : BD :: s : c$, ou $BC = \frac{s}{c}P$: or la force BC est égale au poids du cable de la longueur BAG, on trouve par conséquent cette longueur BAG par une telle analogie, $p : l :: \frac{s}{c}P : \frac{sP}{cp}l$, qui est la longueur[13] BAG, & de-là on tire la longueur $AG = \frac{sP}{cp}l - l$. De ceci, on trouve la force AF par une telle analogie, $BAG : AG :: BC : AF$, ce qui donne la force $AF = \frac{s}{c}P - p$. Enfin la force AE est, comme nous avons déja dit, égale à la force BD, ou égale à P. Nous tirerons quelques Corollaires de ces valeurs trouvées, après avoir fait remarquer au lecteur, que la force BD est ici celle que le Vaisseau exerce horisontalement: que la force BC marque l'effort que le Vaisseau fait pour s'élever davantage hors de l'eau, car l'action du *cable* fait un peu enfoncer le Vaisseau: que la force AE marque l'effort horisontal soûtenu par l'Ancre (lequel nous avons exprimé par BF dans la seconde Figure): & enfin que la force AF est produite par une partie du poids de l'Ancre; elle est égale & opposée à BO dans la seconde Figure. Voici maintenant quelques Corollaires qu'on peut remarquer pour notre sujet, préférablement à d'autres.

1.° Si le poids du *cable* submergé étoit comme nul par rapport à la force qu'exerce le Vaisseau, on auroit $p = o$, $BAG = \infty$, & la force AF égale à la force BC.

13 The argument can be found in the original publications of Huygens and Jacob Bernoulli; see the *Review*, p. 30 h. v.

2.° Si le poids du *cable* submergé étoit à la force du Vaisseau, comme le sinus de l'angle que le *cable* fait avec la surface de la Mer à son co-sinus, la partie AG deviendroit nulle, de même que la force AF, c'est-à-dire, que la direction du *cable* près l'*arganeau* seroit alors horisontale, & si ledit poids du *cable* submergé avoit une plus grande raison à la force du Vaisseau, une partie du *cable* depuis l'*arganeau* se couchera sur le fond de la Mer; & si enfin cette raison est plus petite que celle de s à c, la force AF sera toûjours positive vers le bas: d'où l'on voit que plus l'angle du *cable* & de la surface de la Mer est petit, & plus la longueur du *cable* submergé est grande, plus la force cherchée AF sera petite.

3.° Notre remarque la plus essentielle regarde ce que j'ai dit dans le XIVme article, *qu'il faut diriger la manœuvre de maniére que le* jas *reste couché sur le fond.* Or j'estime, après avoir bien considéré toutes les proportions des Ancres & de leur poids sous l'eau, qu'une force verticale tirant l'*arganeau* en haut l'élevera, si elle surpasse deux neuvièmes[14], ou la cinquiéme partie du poids absolu de l'Ancre. Il faut donc, pour faire que le *jas* reste couché sur le fond, que cette force soit moindre que ladite cinquiéme partie du poids absolu de l'Ancre, sans quoi l'action du *cable* élevera le *jas*. Posant donc le poids de l'Ancre hors de l'eau $= \pi$, il faut faire que $\frac{sP}{c} - p$ soit toûjours moindre que $\frac{\pi}{5}$. Ce que nous venons d'exprimer par des formules génerales, nous l'expliquerons dans la suite par des exemples particuliers tirés des regles de la Navigation, pour en faire voir l'utilité. Cependant il faudra tâcher d'avoir quelque connoissance, quand même elle seroit fort imparfaite, de la force horisontale que le Vaisseau exerce, désignée par P. Il n'y a que cette force dont la détermination soit difficile & vague; voici cependant quelques réfléxions qui pourront nous donner quelques éclaircissements là-dessus.

14 Assume that $\frac{1}{2}$mass of shank = mass of stock = mass of arms and length of shank = $4a$. For a submersed anchor, where the wooden stock becomes weightless, the centre of weight is at $\frac{4a}{3}$ from the arms and the effective weight is:

$$\frac{3}{4} \times \frac{6.85}{7.85} \times \pi \, .$$

Due to the lever effect a force at the ring equilibrates at $\frac{1}{3}$ of the effective weight, i.e. $\frac{6.85}{7.85}\frac{1}{4} = 0.218$, which Bernoulli approximates as $\frac{2}{9} = 0.222$ and then, more conveniently, as $\frac{1}{5}$.

XVI. Tant que le Vaisseau n'est pas encore arrêté ni le *cable amarré*, on peut modérer la force P, comme on le trouve à propos, en *filant* le *cable* plus ou moins: il est *lové* pour cet effet, c'est-à-dire, disposé en rond pour pouvoir en laisser passer par l'*écubier* autant qu'on veut, sans que le Vaisseau soit retardé beaucoup par l'Ancre, si la chose le demande ainsi. Nous ne dirons donc rien pour ce cas sur ladite force P, puisqu'elle dépend absolument de la discrétion de ceux qui filent le *cable*: il ne tiendra qu'à eux de faire qu'elle ne surpasse jamais un certain degré, & tant qu'elle n'est pas plus grande que[15] $\frac{c}{s} \times \left(p + \frac{\pi}{5}\right)$, elle n'élevera point l'*arganeau*, en vertu du troisiéme Corollaire du précédent article.

Mais le Vaisseau étant arrêté, il soûtient encore l'effort des vents, celui des courants & les coups de Mer. L'effort des vents contre un Vaisseau n'est pas bien grand, quand les voiles sont baissées: celui des courants l'est beaucoup davantage. Voici comment je l'ai déterminé pour une Frégate, sur laquelle j'ai fait autrefois un trajet sur Mer.

Je remarquois un jour qu'avec un vent en poupe d'environ 20 ou 22 pieds par seconde (c'est-à-dire, qui faisoit parcourir à l'air l'espace de quelques 20 pieds par seconde, ce que je connoissois par le moyen d'un certain Instrument que j'avois inventé & préparé à ce dessein) nous faisions 6 pieds par seconde, ce que je connoissois encore par le moyen d'une boule d'yvoire attachée par un fil, laquelle je plongeois dans l'eau, en remarquant l'inclinaison du fil qui en provenoit. (M. Poleni[16] s'est servi ensuite de la même méthode dans sa piéce qui a remporté le Prix de 1733, digne de cette glorieuse récompense, & je l'ai trouvée fort bonne, moyennant quelques regles que l'expérience m'a fait remarquer). Les voiles, qui étoient perpendiculaires à la direction du vent, pouvoient avoir toutes ensemble une surface à peu-près égale à 2000 pieds quarrés. La vîtesse relative du vent contre les voiles étoit donc 14 ou 16 pieds par seconde, qui est telle que la pesanteur naturelle produit dans un corps qui tombe de la hauteur d'environ 4 pieds: d'où il suit, en se servant de la regle de M. Mariotte sur la force[17] des Fluides (que je cite ici, quoiqu'il y ait quelque correction à faire, me réservant de publier un jour ma nouvelle Théorie sur cette matiére, que j'ai confirmée par

15 Misprinted as $\frac{c}{5}$ in the original.

16 See the notes concerning this method in D. Bernoulli St. 44, *Mémoire sur la nature et la cause des courans*: Prix Paris VII (pièce I) 1749–51 (1769), pp. 3–108 – *Werke* 5, pp. 535–611 and in D. Bernoulli St. 47, *Sur la maniere de suppléer à l'action du Vent*: Prix Paris VII (pièce III) 1753 (1769), pp. 3–99 – pp. 35–72 h. v.; for Poleni see: G. Poleni, *De la meilleure maniere de mesurer sur mer le chemin d'un vaisseau, indépendemment des observations astronomiques*: Prix Paris III (pièce VIII) 1733 (1734).

un grand nombre d'expériences très-exactes) que la force du vent contre les voiles étoit égale au poids d'un prisme d'air haut de 4 pieds fait sur une base de 2000 pieds quarrés, c'est-à-dire, au poids de 8000 pieds cubiques d'air; & comme du temps de cette observation l'air étoit assés chaud, je crois ne devoir donner à un pied cubique d'air que le poids d'une once[18], de sorte que toute la force du vent devient égale à 8000 onces ou 500 livres. Or dans un Vaisseau, dont la vîtesse est uniforme, la résistance de l'eau est égale à la force qui le pousse; notre Frégate souffroit donc alors une résistance de 500 livres, & si on l'avoit affermie à l'Ancre contre un courant de 6 pieds par seconde, ce courant auroit fait contre la Frégate un effort horisontal encore de 500 livres (je dirai ici en passant, que le poids de notre *maîtresse Ancre* étoit aussi d'environ 500 livres); c'est cette force que nous avons appellée P ci-dessus. Si la vîtesse des courants est moins grande, leur force contre le Vaisseau en devient aussi moins grande, mais cela ne va pas tout-à-fait, comme on croit communément, en raison quarrée des vîtesses, car un nombre infini d'expériences a fait voir que cette regle, quoique forte exacte dans les mouvements violents, s'écarte beaucoup de la vérité dans ceux qui se font lentement, comme M. Newton[19] a fait voir dans ses *Princ. Mathem. Philos. nat. edit. 3me*. Ce que nous venons de dire suffit pour nous donner

17 The law is obviously that the resistance of a flat plate of area A in a flow of velocity v, ρ being the density of the fluid, is

$$F = \frac{1}{2} A \rho v^2 .$$

The law was mentioned – as a proportionality – by Mariotte i.e. as

$$\frac{F_1}{F_2} = \frac{v_1^2}{v_2^2} .$$

(see E. Mariotte, *Traité de la percussion ou choc des corps*, Paris 1673); Bernoulli discussed it extensively in his articles on the Action of Fluids on solid Bodies (see: D. Bernoulli St. 14a, *Dissertatio de actione fluidorum – Werke* 4). In the D. Bernoulli St. 31, *Hydrodynamica*, Section XIII, § 19, *Werke* 5, pp. 412–413, he refers to those and casts doubts on the related experiments of Newton in Book 2 of the *Principia* (1726).

18 From the data in SI units found in modern tables it follows that one cubic foot of dry air at 25 degrees centigrade weights 1.11 ounces, the foot and the ounce being those of Paris.

19 We find in the *Principia* the following statements (taken from *The Principia: mathematical principles of natural philosophy* a new translation by I. Bernard Cohen and Anne Whitman, assisted by Julia Budenz (University of California Press), Berkeley (California) 1999):

> The resistance that arises from the friction (latin: *ex defectu lubricitatis*) of the parts of a fluid is, other things being equal, proportional to the velocity

une idée de la grandeur de la force absoluë horisontale d'un Vaisseau déja arrêté, puisque cette force provient [pour] la plûpart des courants de la Mer contre le Vaisseau, quoique la cause de ces courants puisse varier. Disons encore deux mots sur l'effort des lames contre les Vaisseaux.

On sçait que les lames ne sont qu'un mouvement réciproque des eaux qui montent & descendent alternativement sans changer de place[20], qu'autant qu'elles sont emportées par les courants: elles n'agissent donc qu'en élevant avec précipitation le Vaisseau, qui est obligé par-là de s'approcher très-vîtement de l'endroit qui répond verticalement à l'Ancre; & comme le Vaisseau ne scauroit obéir assés promptement, le *cable* en reçoit une forte impression, qui fait quelquefois *déraper* l'Ancre, & quelquefois rompre le *cable*: cette impression est d'autant plus grande que le mouvement des lames est prompt & grand, & que le *cable* approche plus de la position verticale. La maniére de connoître en gros cette force, seroit de sçavoir le temps d'une ondulation, la hauteur de laquelle le Vaisseau est élevé, & combien le Vaisseau est obligé par cette élévation de s'approcher de l'endroit qui est *à pic* avec l'Ancre: car si, par exemple, le Vaisseau étoit élevé de 12 pieds en 3 secondes, & qu'il fût obligé par-là de faire un espace de 6 pieds sur la surface de la Mer vers l'Ancre, on pourroit chercher quelle seroit la force qui dans le temps de 3 secondes pût faire parcourir au Vaisseau 6 pieds depuis le repos. Mais ces recherches seroient trop ennuyeuses, & n'appartiennent pas assés à notre sujet principal.

Enfin il arrive aussi que les eaux des lames se roulent près leur surface, & viennent à se briser contre les Vaisseaux: mais cet effort n'est pas fort considérable, parce que ces eaux ne sont pas en grande quantité, & qu'elles ne causent par leur choc qu'un léger trémoussement aux parties du Vaisseau.

> with which the parts of the fluid are separated from one another. (Book 2, Section 9, Hypothesis)

but also:

> The resistance encountered by spherical bodies in fluids arises partly from the tenacity, partly from the friction and partly from the density of the medium. (Book 2, Section 3, Scholium; in 3rd ed. only).

Newton developed a theory showing that the «resistance due to the density» (which is the one Bernoulli is using) should be proportional to $\rho A v^2$; he also experimented extensively with spheres falling or swinging in air and water. From those experiments he concluded that in such phenomena this law is valid to a high degree of confidence (Scholium to Section 7 of Book 2), from which follows that the contributions to the resistance which are constant or proportional to v can be neglected.

20 Bernoulli's model of waves on the ocean is made more explicit in his memoir St. 47, *Sur la maniere de suppléer à l'action du Vent*; see there for comments pp. 35–72 h. v.

Quand on a jetté deux Ancres ou trois, on connoîtra par les regles de la décomposition des forces, quel effort chaque *cable* & chaque Ancre soûtiennent.

XVII. Nous avons examiné jusqu'ici toute la théorie des Ancres, depuis le moment qu'elles ont touché le fond de la Mer jusqu'à celui qu'on veut desancrer. Ce desancrage se fait en tirant le *cable* par le moyen du *cabestan* jusqu'à faire venir l'Ancre à *pic*; quelque fois pour faire plus vîtement, on gouverne le Vaisseau jusqu'au même endroit en le *virant de bord*. Lorsqu'il y a plusieurs Ancres, on desancre par le moyen d'une Chaloupe en *bossant* l'Ancre, c'est-à-dire, en *amarrant la bosse* qui saisit le *cable*, & qui est un bout de corde garni d'un *cul de porc* double à chaque bout.

Je viens maintenant au point principal; c'est de parcourir toutes les parties des Ancres à part, d'examiner leur fonction, & de voir de quels changements & corrections elles sont susceptibles.

XVIII. Commençons par le *jas*. Nous avons déja vû qu'il sert à mettre la *croisée* dans un plan vertical, ce qui est absolument nécessaire pour le *mouillage*. Nous avons démontré encore dans les articles IV, V, VI, VII, VIII & IX, qu'il est beaucoup plus naturel aux Ancres (en donnant à leurs parties les proportions ordinaires) d'avoir hors de l'eau la *croisée* dressée que couchée sur le fond, & même que cette derniére situation ne sçauroit qu'être extrêmement rare à certains égards, mais qu'à cause de la légereté du *jas* dans les Ancres submergées, les deux positions leur sont à peu-près également naturelles, & même que la *croisée* se couche plus facilement qu'elle ne se dresse, sur-tout lorsque le *cable* commence à se roidir, & que la *patte* n'est pas encore entrée bien avant dans le sable, laquelle derniére circonstance nous avons démontrée dans les articles XIV & XV. Or comme la sûreté du *mouillage* demande absolument que la *croisée* présente toûjours au fond l'une de ses *pattes*, il faut sans doute fixer là toute l'attention: mais il suit des articles IV & V, que plus le centre de gravité d'une Ancre submergée est près de l'*arganeau*, & plus le *jas* est long, plus la juste position des Ancres sera sûre. Ne vaudroit-il donc pas mieux de faire le *jas* de fer que de bois, ou du moins de le garnir tout autour d'une grosse plaque de fer? Le succès de cette correction est sûr & infaillible pour donner la position requise aux Ancres. Il semble que ceux qui se sont avisés les premiers de mettre des *jas* aux Ancres, n'ont fait consister leur action que dans la longueur, sans faire attention que leur poids en augmente le plus considérablement l'effet: sans cela je suis sûr qu'ils n'auroient pas manqué de le faire d'abord de fer. Voyons cependant quelle influence ce changement aura sur les autres circonstances, puisqu'une chose est souvent bonne à un certain égard, & mauvaise à un autre. Il ne sera donc pas hors de propos de faire attention ici à ce que j'ai marqué dans le XIVme article, sçavoir que si l'Ancre est dans sa juste position, l'action du *cable* peut facilement élever l'*arganeau*, & par-là renverser l'Ancre: il ne faut pour cela dans les Ancres ordinaires qu'une force verticale qui soit égale à la cinquiéme partie du poids de l'Ancre, comme j'ai dit à la fin du XVme article: mais si on faisoit le *jas* de fer, quoique du même poids qu'on le fait de bois, il faudra une force verticale qui soit environ égale à la moitié du poids de l'Ancre pour élever l'*arganeau*[21], & l'action du *cable* ne peut guére produire une si grande force verticale, à moins que la *patte* ne soit déja entrée bien avant dans le fond, auquel cas l'Ancre ne sçauroit plus se renverser, ni peut-être sa *vergue* être élevée par cette force, quoiqu'assés grande pour l'élever dans une Ancre libre. On me dira peut-être que l'Ancre devient trop lourde ou trop pesante en faisant le *jas* de fer; mais j'ai déja répondu à cela, qu'on peut le faire du même poids qu'on a coûtume de faire les *jas* de bois, ne prétendant pas qu'on lui donne l'épaisseur ordinaire; il suffira de lui donner la moitié de l'épaisseur qu'on donne à la *vergue*, & de cette maniére il ne deviendra pas plus pesant que s'il étoit de bois, & ne laissera pas d'avoir encore autant de force. On pourra diminuer l'épaisseur du *jas* vers les deux bouts, comme on fait aux *jas* de bois, parce que c'est au milieu que le *jas* souffre le plus, & qu'il doit par conséquent être le plus épais. Quant à la longueur du *jas*, il est vrai que plus il est long, mieux il servira pour mettre l'Ancre dans sa juste position; cependant il ne faut pas augmenter son poids sans nécessité, car sa grosseur devant être proportionnée à sa longueur, il deviendroit trop pesant, si on vouloit le faire plus long que de coûtume: sa longueur ordinaire suffira, comme on voit assés par le VIIme article, qui ne doit point être changé pour être appliqué à l'état de submersion, lorsque les *jas* des Ancres sont faits de fer & semblables aux *jas* de bois par rapport à leur poids & à leur longueur.

XIX. Après le *jas*, nous considérerons la *vergue*. Chacun voit un grand nombre d'inconvénients qui proviendroient, si on vouloit faire la *vergue* tout-à-fait courte. Je ne ferai ici attention qu'à trois points, qui sans doute sont les principaux. Le premier regarde l'angle APB dans la seconde Figure, qui en vertu du XIVme article doit nécessairement être obtus: or si on faisoit la *vergue* assés courte pour que le point B tombât entre les points A & C on voit que cet angle deviendroit aigu, & l'Ancre tout à fait impropre pour l'ancrage. Il faut donc que la *vergue* AB soit plus longue que AC, qui

21 Because the centre of weight lies then in the middle of the shank.

marque, pour ainsi dire, la longueur du rayon osculateur de la courbûre *AP* que je suppose déterminée. Si on faisoit *AB* d'une longueur infinie, l'angle *CPB* deviendroit égal à l'angle *ACP* qui est d'environ 30 degrés; mais je démontrerai ci-dessous que cet angle *CPB* doit être d'environ 27 degrés, & il obtient ladite grandeur, en faisant *AB* neuf ou dix fois plus longue que *AC*: mais une telle longueur, quoique la plus avantageuse à cet égard, seroit énorme, étant cinq fois plus grande que la longueur qu'on donne ordinairement à la vergue: il vaut donc mieux aggrandir l'angle *CPB* (qui n'est que d'environ 15 degrés[22]) par un autre changement dans la structure des Ancres, que par celui de la longueur de la *vergue*: mon intention n'a été jusqu'ici que de démontrer qu'on ne sçauroit faire la *vergue* trop longue à l'égard du premier point, dont nous venons de parler.

Le second point, sur lequel la longueur de la *vergue* a quelque influence, regarde la facilité avec laquelle l'Ancre prend l'une des deux positions qui lui sont naturelles, & dont j'ai parlé dans le IIIme article & les suivants. Or les formules des articles IV & V, m'ont fait connoître à cet égard que plus on allonge la *vergue*, plus l'Ancre prendra facilement sa position requise pour l'ancrage, car la raison de $HI - H\lambda$ à $HI - HL$ dans la premiére Figure (dont j'ai parlé au VIme article) en devient toûjours plus grande. Nous avons trouvé dans le VIIme article cette raison comme 7 à 2 pour les Ancres non submergées, & cette même raison convient aussi aux Ancres submergées, dont le *jas* est de fer, & du même poids qu'on donne aux *jas* de bois, tels que j'ai conseillé de faire dans le précédent article; mais si on faisoit la *vergue* d'une longueur infinie, cette raison seroit environ comme 11 à 2, & par conséquent plus grande: cela fait voir qu'on ne sçauroit faire la *vergue* trop longue à l'égard de ce point, non plus qu'à l'égard du premier.

En troisiéme lieu, la longueur de la *vergue* peut faciliter le desancrage: car lorsque la *patte* s'est trop enfoncée dans le fond, le desancrage se fait avec assés de peine, sur-tout après les tempêtes, par la raison exposée à la fin du XIVme article; en ce cas l'Ancre étant *à pic*, la *vergue* sert d'un long levier, moyennant lequel on fait renverser à la *patte* le sable endurci qui la retient.

Il est donc enfin de la *vergue* comme du *jas*: on ne sçauroit dans la théorie les faire trop longs ni l'un ni l'autre: ce qui doit les borner, consiste simplement en ce qu'il ne faut pas augmenter le poids des Ancres sans en tirer une utilité suffisante, d'autant qu'en faisant leurs parties plus longues, il faut aussi les faire à proportion plus épaisses & plus fortes, étant alors plus sujettes à se rompre ou à se plier.

22 These 15 degrees are in the hypothesis that the shank is twice longer than the span of the arms, which is the normal proportion as appears from Table 2, p. 21.

On fait au reste les *vergues* cylindriques, c'est-à-dire, d'une épaisseur égale dans toute leur longueur. Si on n'avoit égard en cela qu'aux risques que la *vergue* court de se courber ou de se rompre par les différents efforts qu'elle souffre, il est certain qu'il faudroit donner une toute autre proportion à ses différentes épaisseurs, & nommément les augmenter vers la *croisée*, & les diminuer vers le *jas*; mais ces changements entraîneroient d'autres inconvénients: car toute Ancre doit avoir un certain poids, & il est indifférent pour le succès de l'ancrage, de quelle maniére ce poids soit distribué, pourvû que le centre de gravité ne soit ni trop près de la *croisée*, ni trop près du *jas*: s'il est trop près de la *croisée*, l'Ancre en prend plus difficilement sa juste position, & c'est-là l'inconvénient d'augmenter les épaisseurs de la *vergue* vers la *croisée*; & si au contraire le centre de gravité étoit trop près du *jas*, la force qui fait entrer la *patte* dans le fond deviendroit trop petite. Je crois donc qu'on peut laisser les *vergues* cylindriques telles qu'on les fait; mais au reste il faut absolument les faire aussi longues qu'il est possible, en conservant la même masse ou le même poids, sans les rendre trop foibles ou trop sujettes à se courber; cette regle est certaine, mais son résultat ne sçauroit être déterminé que par un grand nombre d'expériences. Quant enfin au bout quarré de la *vergue*, on voit bien qu'on le fait quarré, & plus gros que le reste, pour empêcher davantage le *jas* de tourner autour du bout: mais si on fait le *jas* de fer, comme j'ai conseillé de faire par de fortes raisons, on soudera la *vergue* au *jas*, comme on la soude à la *croisée*.

XX. Examinons maintenant la *croisée*. Sa courbure est ce qui se présente d'abord à l'esprit, & qui paroît le plus de conséquence. Il est vrai que c'est une assés petite portion de courbe, qui pourra toûjours passer sans grande erreur pour une portion du cercle osculateur, c'est-à-dire, d'un cercle décrit du rayon PC (*Fig.* 2.) qui est perpendiculaire à l'extrémité de la *croisée*, puisque l'angle ACP, qui est la mesure de toute la courbure de la demi-*croisée*, n'est que de 30 degrés. Mais comme on doit employer une exactitude géométrique dans toutes ses recherches, celle-ci ne sera pas hors de sa place. La principale question sera de sçavoir les conditions auxquelles il faudra satisfaire, c'est sur quoi je suis bien persuadé que chacun aura une idée particuliére, & ce sera à examiner laquelle aura le plus de poids & de vraisemblance. Pour moi je me suis enfin fixé à un simple arc de cercle, car il n'y a que cette courbe dont les parties congruent parfaitement en les appliquant l'une sur l'autre. Cette courbe donne par-là un grand avantage à la *croisée*; car en l'enfonçant davantage, chaque partie postérieure prend la place d'une antérieure, & ainsi le sable du fond n'est déplacé qu'autour du bord de la *patte*. Si l'on prend toute autre courbe, il faudra qu'au moindre enfoncement chaque partie de la *croisée* se fasse jour, & surmonte un nouvel obstacle, ce qui rend les enfoncements plus difficiles, & fait en même temps que l'Ancre se tient moins ferme dans le sable qu'elle aura élargi de tout côté. On pourra faire l'expérience de ce que je viens de dire fort facilement avec un clou courbé de maniére que les deux tangentes tirées aux extrémités fassent un angle donné: car on trouvera que si on donne au clou une courbûre circulaire, il entrera plus facilement, & tiendra ensuite plus ferme que si on lui avoit donné toute autre courbûre, outre que ses parties en souffriront moins. On m'objectera peut-être ici, & une personne d'autorité à qui j'ai communiqué mes pensées sur cette matiére, l'a fait, que ce raisonnement suppose le centre C en repos, & que l'Ancre n'est plus entraînée par le Vaisseau: je réponds à cela, 1.° Que si on vouloit considérer le mouvement progressif de l'Ancre, chaque degré de vîtesse demanderoit une autre courbe, quoique d'une même classe. 2.° Que la courbûre de la *croisée* est indifférente jusqu'à ce que la *patte* soit déja entrée dans le fond, & que dès-lors l'Ancre est ordinairement déja affermie, & qu'il ne s'agit plus que de l'affermir davantage. Voilà la raison qui m'a fait choisir la courbûre circulaire préférablement à une autre, & cela d'autant plus qu'elle est sans doute la plus facile à forger.

La longueur de la *croisée* est relative avec celle du *jas*; plus la *croisée* est courte par rapport à celle du *jas*, plus l'Ancre se couchera facilement sur le *jas*: il ne faut donc pas la faire longue sans nécessité, & cela d'autant moins que la croûte sabloneuse & pénétrable n'est pas fort épaisse, comme j'ai rapporté au Xme article; mais aussi doit-on faire la *croisée* assés longue pour que la *vergue* ne l'empêche pas d'enfoncer davantage, sur-tout lorsque le fond est tel que la *patte* y entre avec beaucoup de facilité, car en ce cas elle doit entrer bien avant pour s'y tenir ferme sans labourer le fond. On peut remarquer encore, qu'en choisissant pour la *croisée* la courbûre circulaire d'un même rayon, l'angle CPB est proportionnel à la longueur de la *croisée*; on auroit donc à cet égard un avantage en la rendant plus longue, puisque les proportions indiquées au second article ne donnent à cet angle que 15 degrés, & qu'il devroit être, comme j'ai déja marqué, de 27 degrés. Mais comme cet angle peut être augmenté d'une autre maniére, qui ne préjudicie pas aux autres points, cette raison ne doit pas nous engager à faire la *croisée* plus longue que de coûtume. A mon avis, il suffira de donner à la *croisée* la longueur d'un arc circulaire de 60 degrés, quoiqu'ordinairement on la fasse un peu plus longue, autant que j'ai pû juger par les Figures.

Les épaisseurs de la *croisée* sont diminuées vers les extrémités pour deux raisons: l'une est que les différentes forces qui agissent sur la *croisée*, & qui pourroient la courber ou la rompre, sont plus sensibles sur le milieu que sur les extrémités: la seconde raison est que l'on donne par-là aux deux *branches* la nature du coin, qui les fait entrer plus facilement dans le fond.

L'angle que l'extrémité de la *croisée*, ou bien de la *patte*, doit faire avec le fond, le *jas* y étant couché horisontalement, est un des points essentiels: si on donne à la *croisée* la figure d'un arc de cercle, si on lui donne 60 degrés d'ouverture, & si on fait le rayon égal à la moitié de la *vergue*, l'angle APB devient égal à 105 degrés, & l'angle CPB à 15 degrés. Mais ils me paroissent trop petits: leur plus avantageuse grandeur dépend du rapport des forces qui font mordre la *patte* dans le fond, dont l'une provient du poids de l'Ancre, & l'autre est la force BF qui tire l'Ancre horisontalement, & dont j'ai expliqué l'action dans le XIV.[me] article.

Soit donc encore, comme dans le XV.[me] article, le poids absolu de l'Ancre $= \pi$, son *jas* étant de bois, j'estime le poids de l'Ancre sous l'eau $= \frac{2}{3}\pi$. Posons d'ailleurs, comme dans le IX.[me] article, le centre de gravité d'une telle Ancre être à la distance d'un tiers de la *vergue* depuis la *croisée*, & on aura la pression que la *patte* exerce sur le fond de la Mer par le poids de l'Ancre, égale à $\frac{2}{3} \times \frac{2}{3}\pi$ ou $\frac{4}{9}\pi$. On trouve à peu-près la même valeur de $\frac{4}{9}\pi$ pour les Ancres garnies d'un *jas* de fer, tel que j'ai conseillé de faire dans le XVIII.[me] article; parce que si d'un côté le *jas* est plus pesant sous l'eau, le centre de gravité est au contraire plus loin de la *croisée*. Voilà ce qui regarde la premiére force.

Quant à la force horisontale BF (que nous avons appellée P dans le XV.[me] article) nous en avons donné un exemple au XVI.[me] article, en faisant voir qu'elle étoit égale au poids de 500 livres, au cas qu'elle fût produite par un courant de 6 pieds par seconde: cette force de 500 livres étoit à peu près égale au poids de la maîtresse Ancre que le Vaisseau en question portoit: mais il s'en faut beaucoup que la force horisontale, dont il s'agit ici, doive être estimée si grande: car outre que les courants sont ordinairement beaucoup moins forts, ne faisant guéres au de-là de 2 ou 3 pieds par seconde, il ne faut pas considérer ici les forces que le Vaisseau soûtient étant déja affermi à l'Ancre, mais celles qui retardent le Vaisseau par la manœuvre de l'ancrage, l'Ancré n'étant pas encore entrée dans le fond. On peut alors considérer la résistance de l'Ancre, qui est égale à la force P, comme produite par le simple frottement que l'Ancre souffre, étant traînée par sa *patte* & sur une surface sabloneuse semblable à celle du fond de la Mer. Les expériences qu'on a faites à cet égard sur différents corps & différentes surfaces, me font estimer ladite résistance égale à un tiers du poids que l'Ancre a sous l'eau, ou égale à deux neuviémes du poids absolu de l'Ancre. Si je me trompe dans cette estime, du moins est-il évident que je n'ai pas manqué dans l'excès: la force horisontale, égale à la même résistance, doit donc être pour le moins posée égale à $\frac{2}{9}\pi$, & ainsi les deux forces en question, dont il s'agit d'estimer le rapport en gros, sont donc comme $\frac{4}{9}\pi$ à $\frac{2}{9}\pi$, c'est-à-dire, comme 2 à 1. Cela signifie que la *patte P* est pressée par deux forces, l'une exprimée par *PH* qui est verticale, produite par le poids de l'Ancre, & l'autre représentée par *PL* qui est horisontale, produite par l'action du *cable*, & que la premiére est tout au plus double de la seconde. Si l'on acheve le rectangle, la diagonale *PI* exprimera la force résultante des deux dites forces, & l'on voit que c'est la direction de cette diagonale que l'extrémité de la *patte* doit avoir, parce que de cette maniére les deux forces combinées enfoncent directement la *patte* dans le fond. On trouve la même chose par la méthode *des plus grands & des moindres*. Puisque donc *PH* est double de *PL*, & que *PI* doit être une tangente en *P*, il suit que l'angle *LPI* est de 63 degrés, & son complément *CPB* (car l'angle *CPI* est droit) de 27 degrés. (Ce calcul doit être un peu changé pour les Ancres qui auroient le *jas* de fer, en ce qu'elles ne perdent pas tant de leur poids par la submersion, de sorte que la force *PL* en devient un peu plus grande; on pourra donc bien faire dans ces Ancres l'angle *CPB* de 30 degrés. La même méthode sert aussi à déterminer l'angle que le coutre d'une charruë doit faire avec la surface des champs qu'on laboure.) Si la force horisontale exprimée par *PL*, est supposée plus grande par rapport à sa compagne *PH*, ledit angle *CPB* devroit être encore plus grand; cependant de la maniére qu'on fait les Ancres, il ne va gueres au de-là de 15 degrés. Il y a plusieurs maniéres de remédier à ce défaut; mais la meilleure, à mon avis, sera de courber davantage la *croisée*, ou, si l'on veut, de garnir le *jas* dans ses extrémités de deux piéces de bois d'une figure sphéroïdique fort applatie, en faisant passer le *jas* par leurs centres. Par-là l'extrémité de la *vergue B* se hausseroit, & l'angle *CPB* en deviendroit plus grand. Ces changements peuvent se faire sans le moindre préjudice pour aucune fonction des parties qui composent l'Ancre.

XXI. Je finirai cet examen par quelques réfléxions sur le *cable*. Il importe beaucoup que le *cable* tire l'Ancre le plus horisontalement qu'il est possible; la force horisontale *BF* (c'est toûjours la seconde Figure) en devient plus grande, & la verticale *BO* plus petite; l'un & l'autre point est avantageux pour l'ancrage en vertu du XIV.me article. Il faut donc, après avoir jetté l'Ancre, filer le *cable* assés vîtement pour empêcher qu'il ne se roidisse avant que le Vaisseau soit à une grande distance, comme par exemple, double de la profondeur de la Mer qui répond à l'Ancre. Si après cela on roidit le *cable* avec une certaine force, il est à remarquer que l'angle *EBN* (qui est celui que le *cable* fait avec le fond de la Mer) dépend encore de cette même force à cause du poids que le *cable* a sous l'eau, & qui lui fait prendre la figure d'une *chaînette* convexe vers le fond, & concave vers la surface de la Mer. On pourroit croire d'abord que la courbûre du *cable* approche si fort de la ligne droite, & par conséquent l'angle EBN si fort de celui que le *cable* fait avec la surface de la Mer, que les différences peuvent être négligées: mais nous verrons le contraire, si nous appliquons les Théoremes du XVme article à la présente question, & ensuite à quelque exemple fondé sur les principes de la Navigation. Car en retenant les dénominations que nous avons faites dans ledit XVme article, nous trouverons le sinus de l'angle EAH (*Fig.* 3.) qui est égal à l'angle que le *cable* fait avec le fond horisontal de la Mer, de la maniére suivante. Achevés le rectangle $EAFH$, & la diagonale AH marquera la force équivalente aux deux forces AE & AF, & ainsi AH sera une tangente en A & égale à $\surd(AE^2 + AF^2)$ c'est-à-dire

$$= \sqrt{\left(PP + \overline{\frac{s}{c}P - p}^{\,2}\right)} = \sqrt{\left(PP + \frac{ss}{cc}PP - \frac{2s}{c}Pp + pp\right)} =$$

(à cause que $cc + ss = 1$) $$\sqrt{\left(\frac{PP}{cc} - \frac{2sPp}{c} + pp\right)} .$$

Or AH est à AF comme le sinus total au sinus de l'angle EAH ou EAG; on a donc le sinus de l'angle cherché EAH

$$= \frac{sP - cp}{\surd(PP - 2scPp + ccpp)} .$$

Cet angle est donc nul, lorsque $sP = cp$, & si sP étoit plus petit que cp, j'ai déja marqué ci-dessus qu'une partie du *cable* se couchera sur le fond, & l'*arganeau* ne laissera pas d'être tiré horisontalement. Appliquons maintenant ladite formule à quelques exemples tels qu'ils sont ordinaires dans la Navigation.

Soit $c = 2s$, c'est-à-dire que le *cable* fasse avec la surface de la Mer un angle de 27 degrés. Supposons P égale à 125 livres, qui font le quart du poids de l'Ancre dont j'ai fait mention dans le précédent article, & par conséquent un peu plus que $\frac{2}{9}\pi$. Si dans ces hypotheses il reste au *cable* submergé seulement un poids de $62\frac{1}{2}$ livres, ce poids suffira pour faire évanouir l'angle en question, & le *cable* aura une direction horisontale près l'*arganeau*. Si ledit poids étoit plus grand, une partie du *cable* sera traînée sur le fond de la Mer; mais s'il n'étoit, par exemple, que de 25 livres, il faut, pour trouver l'angle du *cable* avec le fond de la Mer, poser $P = 125$, $p = 25$, $s = \frac{1}{\surd 5}$ & $c = \frac{2}{\surd 5}$, après quoi on trouve le sinus de l'angle $EAG = 0.22825$, & par conséquent l'angle[23] même de $13°12'$, ce qui fait connoître que s'il ne reste au *cable* baigné que le poids de 25 livres, il ne fera plus, avec le fond de la Mer, qu'un angle de $13°12'$, pendant qu'il forme avec la surface de la Mer un angle de près de 27 degrés; marque que ces deux angles sont toûjours bien différents; & cela est très avantageux au succès du *mouillage*, que j'avois peine à comprendre avant ces Théoremes; car si le premier angle étoit aussi grand que le second, le *cable* ne manqueroit guére, pour peu qu'il se roidît, d'élever l'*arganeau* & de renverser l'Ancre. Au reste il arrivera facilement que le *cable*, qui porte une Ancre de 500 livres (tel est celui dont nous parlons), ait sous l'eau un poids de 20, 30, jusqu'à 100 livres ou plus, selon qu'il a beaucoup de brasses de baignées: car chaque brasse d'un tel *cable* pese ordinairement hors de l'eau depuis 12 jusqu'à 15 livres, & je crois que sous l'eau il pesera pour le moins une livre & demie, de sorte que 42 brasses peseront déja les $62\frac{1}{2}$ livres dont j'ai fait mention dans cet article.

Après ce que nous venons de dire, nous entendrons plus distinctement ce qui a été dit dans le troisiéme corollaire du XV.me article, sçavoir que pour prévenir l'élévation de l'*arganeau* par l'action du *cable*, il faut que $\frac{s}{c}P - p$ soit moindre que $\frac{1}{5}\pi$, ou bien que p soit plus grand que

$$\frac{s}{c}P - \frac{1}{5}\pi \ .$$

Ainsi dans l'exemple que nous venons d'alléguer, il faut mettre $\frac{s}{c} = \frac{1}{2}$, $P = 125$ & $\pi = 500$; d'où il suit, selon la regle, que p doit être plus grand que $62\frac{1}{2} - 100$; par conséquent la valeur de p est en ce cas toûjours assés grande, puisque $62\frac{1}{2} - 100$ est un nombre négatif. Mais si la force P devenoit plus grande, ou si le *cable* faisoit un plus grand angle avec la surface de la Mer, il pourroit arriver que sans le poids du *cable*, l'*arganeau* ne manquât pas d'être élevé, & l'Ancre d'être renversée sur sa *croisée*: car si, par exemple, le *cable* coupoit la surface de la Mer sous un angle de 45 degrés, la force P demeurant la même, on trouveroit

$$\frac{s}{c}P - \frac{1}{5}\pi = 25 \ ;$$

23 There is a computational error here: the angle $EAG = 16°42'$ and $\sin EAG = 0.28735$.

d'où nous pouvons conclurre que l'Ancre se renversera sûrement en ce cas, à moins que le poids du *cable* submergé ne soit plus grand que de 25 livres.

Puisque donc le poids du *cable* sous l'eau est une chose si utile, & quelquefois si nécessaire au *mouillage*, on tâchera de rendre les *cables* d'une pesanteur spécifique plus grande qu'ils ne sont ordinairement, sans pourtant augmenter leur poids absolu. Je m'imagine que cela est facile à faire, & sans déroger à leurs autres qualités. On pourroit aussi charger le *cable* d'un poids de quelques quintaux, plus ou moins, suivant la grandeur de l'Ancre, à la distance de quelques brasses depuis l'*arganeau*. Il naîtra de ceci, outre les avantages déja expliqués, un autre, dont je n'ai pas encore fait mention, qui est que le *cable* s'écartera par-là davantage de la ligne droite, ce que je présume pouvoir être d'une très-grande utilité contre les coups de Mer, lesquels, par une action trop précipitée, font quelquefois *déraper* l'Ancre, & quelquefois rompre le *cable*, comme j'ai dit à la fin du XIV.me article: car le *cable* pouvant ainsi prêter en longueur, amortira peu à peu, par un méchanisme fort simple, l'effet de ces coups de Mer. Il me semble d'avoir lû quelque part une invention de M. Perrault[24] fondée sur un pareil principe, mais beaucoup plus embarrassée, & je n'ai pas entendu qu'on ait mis en execution ses conseils.

On me dira peut-être que lorsque la Mer n'est pas bien profonde dans une rade, le *cable* ne sçauroit s'écarter beaucoup de la ligne droite; mais aussi alors les lames ne sont pas à craindre, n'étant pas fort hautes.

XXII. Voilà mes réfléxions sur la meilleure construction des Ancres; je n'ai pas oublié en les faisant, combien il est facile de changer en pis dans des choses qu'un long examen & une infinité d'expériences ont produites; c'est pourquoi je n'ai rien voulu changer dans la structure ordinaire des Ancres, qui ne fût fondé sur des principes sûrs, tirés de la Méchanique & de la Navigation: j'espere qu'on trouvera ces principes bien établis, & qu'il serviront aux gens de Mer, intelligents en ces matiéres, à suppléer ce qui pourroit manquer dans cette dissertation, & à donner la derniére perfection aux Ancres.

S'il y avoit quelque chose qui pût me faire pencher pour une nouvelle sorte d'Ancre, ou pour plus de corrections à faire aux Ancres, ce seroit le défaut de sûreté dans le succès du *jas*, destiné à dresser la *croisée*, &

24 According to Réaumur he proposed a new model of an anchor in 1700. See: R.-A. Ferchault de Réaumur, *Fabrique des ancres, lue à l'Académie en juillet 1723 par M. de Réaumur. Avec des Notes et des Additions de M. Duhamel*, Paris: Saillant et Nyon, 1761, p. 4 and p. 47; also the Fig. 8 on plate IV of that publication. It is noteworthy that Duhamel du Monceau in his extensive notes does not mention once Johann II nor Daniel Bernoulli.

l'inaction de celle des *branches* qui regarde en haut. Il est facile d'inventer des Ancres qui n'ayent qu'une seule situation naturelle, & pour peu qu'on y veuille penser, on s'en imaginera plusieurs sortes plus ou moins composées: j'en proposerai une pour exemple dans l'article suivant.

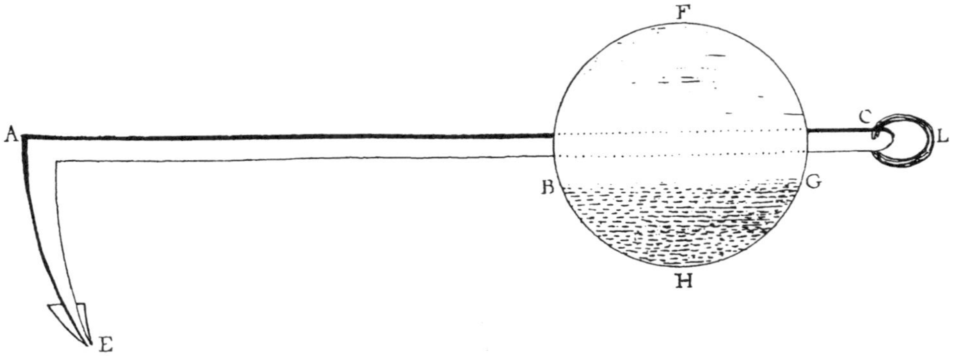

Fig. 4

XXIII. On pourroit faire la *vergue AC* (*Fig.* 4.) avec son *arganeau L*, comme dans les Ancres ordinaires, mais au lieu des deux branches qui composent la *croisée*, il suffira ici d'en faire une représentée par *AE*, & en place du *jas*, on pourroit mettre une sphere *FBHG*, dont la partie supérieure *BFG* soit de bois, & l'inférieure *BHG* de fer. Il est clair qu'on peut donner une telle proportion à ces parties, que *AE* se mette toûjours dans un plan vertical, & présente sa pointe au fond, de quelque maniére que cette espece d'Ancre soit d'abord située, & elle prendra ladite situation, qui seule lui est naturelle, d'autant plus facilement que l'action du *cable* la remuë continuellement, & que le fond de la Mer est, suivant les observations de M. Marsigli, assés dur pour ne pas permettre à la sphere de s'y embourber par son propre poids, lequel enfoncement pourroit peut-être empêcher ou retarder l'Ancre de se mettre dans son état d'équilibre.

Il y auroit beaucoup à dire sur la nature d'une telle Ancre, & sur les proportions qu'il faudroit donner à ses parties: mais comme les choses trop nouvelles, sont rarement écoutées pour l'exécution, à laquelle les Sçavants ont la moindre part, ce seroit un commentaire perdu. Je finirai donc ici mon discours, après avoir dit encore quelques mots sur la meilleure maniére d'essayer les Ancres.

XXIV. J'entends par la *maniére d'essayer les Ancres*, celle de sçavoir tout ce qui leur arrive pendant le *mouillage*, depuis le moment qu'elles ont été jettées dans la Mer jusqu'après le *desancrage*. L'eau & la profondeur de la Mer empêchant de voir ce qui se passe au fond, même dans les rades les moins profondes, je crois que le meilleur expédient seroit de faire l'essai des Ancres sur terre, & d'imiter parfaitement toutes les circonstances qui ont du rapport à l'ancrage: il me semble, après l'examen que nous avons fait du méchanisme des Ancres, que l'exécution de cette idée ne seroit pas tout-à-fait impossible, & je suis confirmé dans ce sentiment par les expériences que j'ai faites, quoique grossiérement, & que j'ai citées dans le XIIme article: voici ce qu'il faudroit observer en suivant ce dessein.

On choisira un terrain propre pour cet effet, sçavoir horisontal, dont la consistance soit d'un sable durci & conglutiné, ayant la surface un peu raboteuse: En un mot, on tâchera d'imiter le fond de la Mer, tel qu'il est ordinairement dans les lieux propres au *mouillage*, le mieux qu'il est possible, en suppléant par l'Art ce qui manquera à cet égard au terrain qu'on aura choisi.

On examinera à part la pesanteur spécifique du *jas*; pour moi je crois qu'elle est à peu-près égale à celle de l'eau: on apprendra par-là combien le *jas* est soûtenu par l'eau de la Mer, & on appliquera au milieu du *jas*, en faisant l'essai des Ancres, une force verticale qui tire le *jas* en haut, & qui soit égale à $\frac{7}{8}$ parties de la force avec laquelle le *jas* est soûtenu par l'eau de la Mer; j'en rabats une huitiéme, à cause que le reste de l'Ancre qui est de fer, perd pareillement une huitiéme de son poids par sa submersion. On pourroit d'une maniére plus simple alleger le *jas*, & rendre par là son effet pareil à celui qu'il a au fond de la Mer, en tirant l'Ancre sous un plus grand angle qu'elle ne l'est ordinairement au fond de la Mer, ce qui feroit presque le même effet. Ces précautions seroient superfluës pour les Ancres qui auroient le *jas* de fer, mais elles sont très-essentielles pour les *jas* de bois: on n'a qu'à confronter ensemble le VIIme & le IXme article pour s'en convaincre, sans parler du reste.

On fera d'abord coucher l'Ancre sur sa *croisée*, & on la tirera ensuite par le moyen d'un cabestan planté à une certaine distance, & sur quelque hauteur. On verra qu'étant tirée avec une certaine vîtesse, elle se dressera d'elle-même, & on examinera avec quelle vîtesse elle doit être ensuite traînée pour la faire mordre plus vîtement dans le sable. On fera l'angle que la direction du *cable* forme avec le fond près l'*arganeau*, depuis 0 jusqu'à 20 ou 25 degrés, & on remarquera que plus cet angle est petit, plus la *patte* s'enfoncera vîtement. On examinera quelle force il faudra d'abord appliquer, & comment ces forces doivent ensuite être variées pour avancer la manœuvre; mais il suffira de l'augmenter jusqu'à ce que la force immédiate qui tire horisontalement l'Ancre soit égale à son poids: car une Ancre qui est assés enfoncée pour pouvoir résister à une force directe égale à son poids, peut être regardée comme tenant assés ferme au fond, puisqu'elle est capable de retenir un Vaisseau contre un courant de 6 pieds par seconde, comme nous avons vû au XVIme article. On remarquera en même temps, par quels degrés les vîtesses de l'Ancre diminuëront après qu'elle a commencé à mordre dans le fond, & qu'elle est tirée avec une vîtesse uniforme.

Quand l'Ancre tiendra assés ferme pour soûtenir la force tantôt nommée, on pourra imiter les coups de Mer, dont j'ai traité à la fin du XIVme article. Cela se fera, en tirant le *cable* brusquement & par intervalle avec beaucoup de force.

Enfin pour connoître comment se fait le desancrage, on n'a qu'à tirer le *cable* sous une direction successivement plus verticale.

J'obmets une infinité d'autres expériences & observations qu'on pourra faire en même temps, si l'on essaye les Ancres de la maniére que je viens d'exposer lesquelles expériences seront également utiles pour découvrir la meilleure figure qu'il faut donner aux Ancres, & pour connoître la meilleure manœuvre qu'on doit observer pour l'ancrage. Pour faire ces expériences & ces essais avec plus d'utilité, on pourra consulter les raisonnements & les calculs que j'ai faits dans le corps de ce discours, & qui en seront, comme j'espere, confirmés.

RECHERCHES

Sur la maniere la plus avantageuſe de ſuppléer à l'action du Vent ſur les grands Vaiſſeaux, ſoit en y appliquant les Rames, ſoit en y employant quelqu'autre moyen que ce puiſſe être.

FONDÉES

Sur une nouvelle Théorie de l'économie des forces & des effets.

Piéce qui a remporté le Prix proposé par l'ACADÉMIE ROYALE DES SCIENCES, pour l'année 1753.

Par M. DANIEL BERNOULLI, *Profeſſeur de Phyſique à Bâle, & Aſſocié Etranger de l'Académie Royale des Sciences.*

St. 47

Recherches sur la maniere la plus avantageuse de suppléer à l'action du Vent sur les grands Vaisseaux, soit en y appliquant les Rames, soit en y employant quelqu'autre moyen que ce puisse être.

Fondées sur une nouvelle Théorie de l'économie des forces & des effets

Prix Paris VII (pièce III) 1753 (1769), pp. 3–99

Quaerendi initium ratio attulit; cum esset ipsa ratio confirmata quaerendo.[1]
Cic. Acad. Quaest. IV. 26.

PREMIERE PARTIE.

I. La mécanique des rames est d'une nature si singuliere que personne n'a encore démontré la proportion qu'il y a entre les forces mouvantes qu'on y emploie & l'effet qui en résulte; c'est cependant un article qui est assez facile dans presque toutes les machines, & c'est en même tems l'article le plus essentiel pour notre sujet. Il s'agit sans doute de connoître, avant toute autre chose, si dans l'usage des rames, les forces mouvantes sont toutes utilement employées pour mettre un navire en mouvement, ou si une partie considérable de ces forces se perd en produisant des effets inutiles & différents de celui qu'on se propose. Dans le premier cas, l'usage des rames est le plus avantageux, & on ne doit pas espérer d'imaginer jamais rien de préférable; mais dans le second cas, il ne faut pas renoncer à l'espérance de pouvoir ou perfectionner l'usage des rames, ou leur substituer d'autres moyens plus avantageux. Mais un tel examen demande premierement une théorie sur les forces de l'homme, & en second lieu une connoissance exacte des forces ab-

1 From Cicero, *Academica II* viii, 26. The full phrase is as follows (words left out are between [ ]):

> «*[Nam] quaerendi initio ratio attulit, [quae perficit virtutem] cum esset ipsa ratio confirmata quaerendo.*» [For] it is reason that initiated research, [reason which has perfected virtue,] since reason herself is strengthened by pursuing research;

Cf. The Loeb classical Library, Cicero XIX, *De natura deorum; Academica*, Warmington: E. H. ed., 1933, pp. 500–501.

solument & indispensablement réquises pour faire singler un navire avec une certaine vîtesse uniforme.

II. Quand je parle des forces de l'homme je n'entens point ici, comme dans le langage ordinaire, ces forces par lesquelles on tient en équilibre de certaines résistances telles que seroit de tenir suspendu un certain poids, ou d'excercer une certaine pression; ce sont-là des forces mortes; il est question ici des forces vives, que l'homme produit pendant son travail; on doit toujours estimer le travail absolu[2] d'un homme par la pression qu'il exerce, par la vîtesse de son point d'appui & par le tems. La seule considération du levier, auquel se réduisent toutes les machines, suffit pour voir cette vérité; il faut cependant avouer que la fatigue de l'homme, qui est la seule chose qu'il faille considérer, n'est pas toujours exactement proportionnelle à son travail pris dans le sens que nous venons de lui donner. Un homme pourra, par exemple, enlever une résistance de vingt livres avec une vîtesse de trois pieds par seconde, & continuer ce travail pendant plusieurs heures de suite; mais il ne s'ensuit pas qu'il puisse enlever quatre livres avec une vîtesse de quinze pieds par seconde, cela seroit tout-à-fait impossible a l'homme; peut-être pourroit-il bien enlever une résistance de dix livres avec une vîtesse de six pieds; mais il est sûr qu'il s'en trouveroit beaucoup plus fatigué au bout d'une heure, que s'il n'avoit employé que la moitié de cette vîtesse avec un effet double. Cette remarque doit apprendre aux Mécaniciens qui s'appliquent à imaginer des machines qui doivent être mises en mouvement par le travail des hommes, à mettre une telle proportion entre leurs parties, que les travaux de ces hommes se fassent avec des vîtesses & des efforts qui leur soient le plus naturels.

III. La remarque que je viens de faire ne doit pas nous empêcher de supposer les fatigues proportionnelles aux travaux tels que nous les avons déterminés, puisque moyennant une juste proportion entre les parties de la machine, on peut faire que la vîtesse & la pression de chaque homme soient constamment les mêmes, & que toute la différence consiste à employer plus ou moins de tems au même travail, & qu'on ne sçauroit douter que dans ces circonstances les fatigues doivent être censées proportionnelles aux tems. Mais je dis bien plus, un grand nombre d'expériences m'ont appris que nonobstant une grande inégalité entre les vîtesses, les fatigues ne laissent pas de suivre la raison composée de la pression, de la vîtesse & du tems, pourvu

2 This *travail absolu* was called *potentia absoluta* in the St. 31 *Hydrodynamica* and is properly our modern concept of «power».

qu'on ne donne pas à ce principe une trop grande étendue, & qu'on ne sorte pas hors de certaines limites. Si un homme peut enlever une résistance de vingt livres avec trois pieds de vîtesse, il pourra aussi y enlever une résistance de soixante livres avec un pied de vîtesse, ou trente livres avec une vîtesse de deux pieds; ou quinze livres avec une vîtesse de quatre pieds, & même douze livres avec une vîtesse de cinq pieds, & tout cela sans se fatiguer ni plus ni moins. Il semble que la nature ait preserit aux animaux une certaine conservation de forces naturelles pareille à celle qu'on connoit aux forces vives produites par la pesanteur naturelle, tant que les animaux ne passent pas les limites de leur économie naturelle. C'est ici une vérité de fait admirable. En voici une autre, que nous devons admirer bien davantage.

IV. Nous n'avons encore considéré que l'effet d'une même espece de travail exercé avec plus ou moins de vîtesse; mais les hommes peuvent subir une infinité d'especes de travail. Dans chaque différente espece il y aura une combinaison différente des muscles que les hommes feront agir: les efforts qu'ils feront seront toujours d'une nature différente; cependant j'ai remarqué qu'avec des fatigues égales, les hommes font constamment des effets à-peu-près égaux; je m'en rapporte aux Mécaniciens qui sçauront faire un juste calcul des effets produits moyennant le travail des hommes employé d'une façon quelconque. Je suis si persuadé de cette vérité, que je n'ai pas hésité d'en rechercher l'explication tirée de l'économie animale, & il m'a paru la trouver, en supposant que la fatigue est causée par la perte qu'on fait d'esprits animaux[3] qui produisent le mouvement des muscles; & que de quelque façon que les esprits animaux agissent sur les muscles, ils doivent être considérés comme de petits ressorts bandés, qui se débandent au moment de leur activité; or, c'est la nature des ressorts de produire constamment la même force vive de quelle maniere qu'on leur permette d'employer leur efficace; & comme on mesure l'effet d'un certain travail par la quantité des forces vives qu'on a produites, soit réelles, soit potentielles, il s'ensuit qu'une fatigue égale, comme causée par la même dépense d'esprits animaux, produit un effet égal dans toutes les especes de travail. L'économie animale ne fournit qu'une certaine quantité d'esprits animaux pendant chaque jour, & c'est cette quantité d'esprits animaux qui fera la mesure de tout son travail jour-

3 The «animal spirits» (*esprits animaux*) are referring to Descartes. In 1726 Bernoulli wrote an article on muscular contraction (D. Bernoulli St. 10, *De motu musculorum*: CP I, 1726 (1728), pp. 297–313 – *Werke* 1, pp. 92–103, in which the *spiritus animales* are the origin of the contraction).

nalier possible, tant en fonctions animales qu'en fonctions vitales. J'estime tout le travail journalier possible d'un homme d'une force & d'une taille moyennes à pouvoir élever 1 728 000 livres à la hauteur d'un pied chaque jour[4], sans apporter du préjudice à sa santé; & je fonde cette estime sur un grand nombre d'observations. Quant aux fonctions vitales, il seroit bien difficile de l'évaluer avec autant de justesse; il n'y a que le travail du cœur, qu'on peut déterminer assez exactement, puis qu'on sçait qu'il fait environ 115 200 battemens dans un jour, qu'il pousse environ deux onces de sang dans chaque systole, & que les observations & expériences de Mr. Hales[5] semblent prouver que le sang soit jetté hors du cœur avec une vîtesse à pouvoir s'élever à la hauteur d'environ huit pieds; ce n'est-là que l'effet du ventricule gauche, celui du ventricule droit en fera environ le quart, & on pourra estimer le travail journalier du cœur égal à celui d'élever 144 000 livres à la hauteur d'un pied, ce qui fait la douziéme partie de ce que l'on peut appeller le travail journalier d'un homme. J'estime le travail des muscles qui servent à la respiration plus grand; & quand on considére qu'il ne se fait, selon toutes les apparences, aucune fonction animale sans le concours & la coopération des nerfs & des esprits animaux, il est à présumer que la nature a destiné les esprits animaux dans une proportion à-peu-près égale aux mouvemens vitaux nécessaires & aux mouvemens volontaires.

V. Nous pourrons donc supposer, que tous les hommes d'une constitution égale seront également fatigués après avoir fait des effets égaux, de quelle maniére que ces différens hommes ayant été employés; mais il semble encore que la constitution des hommes puisse être extrêmement différente, sans que leurs travaux journaliers, dont ils sont capables pendant un grand nombre de jours de suite, soient considérablement différens. Tel homme charnu & vigoureux pourra peut-être faire trois ou quatre fois plus de travail pendant quelques heures de tems, qu'un autre décharné & d'une constitution beaucoup plus foible ne pourra faire dans un tems égal; mais si chacun de ces deux hommes si différens en vigueur, étoit appliqué pendant un grand nombre de jours de suite à une même sorte de travail jusqu'à se fatiguer

4 Bernoulli admits that an able-bodied man can, as an average over 24 hours, produce a power of 20 pounds feet/second, i.e. 31.2 Watts. This number relies on experiments he did in Geneva, in the presence of the Genevese scientists, as he recalls in the St. 31 *Hydrodynamica*, in the appendix to Section 9, *Werke* 5, pp. 311–312.

5 Bernoulli refers to the discoveries of Stephen Hales, an English clergyman, who became famous by his investigation of blood pressure and the circulation of sap in plants and trees. His *Haemostatics* was published in London in 1733 and a French translation appeared in Geneva in 1744.

également, je doute si leurs effets seroient fort inégaux. Cette vérité se manifeste assez clairement dans les bêtes. C'est sans doute parce que l'économie animale ne sçauroit permettre de faire une plus grande dépense journaliere en esprits animaux, que ce qu'elle lui fournit de nouveau chaque jour, & qui vraisemblablement n'est pas fort différente dans les hommes ou dans les animaux d'une même espece.

VI. Ces remarques paroîtront peut-être bien étrangeres pour notre sujet, mais la suite fera voir qu'elles sont très-essentielles, & telles qu'elles nous feront voir quel est le plus grand effet possible qu'on puisse tirer du travail des hommes pour la navigation, & jusqu'où l'usage des rames s'en approche. Mais il faudra auparavant nous engager dans une autre discussion; c'est sur les effets que les hommes peuvent produire par leurs travaux.

VII. L'effet de travail peut & doit toujours être réduit à une certaine quantité de forces vives, quoique ces forces vives puissent avoir des apparences tout-à-fait différentes: elles peuvent pourtant toutes être réduites à une certaine masse élevée à une certaine hauteur verticale. Ces deux articles multipliés ensemble feront constamment la mesure de la force vive provenue d'un certain travail; si ce travail étoit employé à donner continuellement à de nouveaux corps un certain degré de mouvement horizontal, il n'y auroit qu'à voir quelle est la hauteur verticale à la quelle ces corps pourroient s'élever avec leur vîtesse imprimée, & on aura auffi-tôt leur force vive sous la forme désirée: si le travail consistitoit à bander des ressorts, il faudroit examiner à quelle hauteur ces ressorts pourroient jetter une certaine masse en se debandant. Quand on ne fait que tirer horizontalement un corps qui souffre un certain frottement, ce frottement fait le même effet que s'il s'agissoit de bander continuellement de nouveaux petits ressorts; enfin l'effet du travail sera toujours équivalent à celui de lever une certaine masse à une certaine hauteur, & cet effet se montrera presque toujours sous les quatre formes que je viens d'indiquer.

VIII. Examinons à présent quelle est la plus grande quantité de force vive que l'homme puisse produire dans un certain tems sans s'épuiser. Il n'y a que l'expérience qui puisse décider cet article; mais il faut bien distinguer plusieurs différens cas. Un homme appliqué tous les jours à un certain travail, & chaque jour pendant huit heures de tems, pourra, à mon avis, élever vingt livres à la hauteur de trois pieds à chaque seconde, ou bien soixante livres à la hauteur d'un pied; cela fera 1 728 000 livres à la hauteur d'un pied pendant huit heures de tems. J'ai adopté ce résultat sur un grand nombre d'observations, & avec toute la circonspection requise; j'ai vu des cas, où l'homme faisoit trois fois plus d'effet pendant chaque seconde, mais il n'auroit pû soutenir ce travail que pendant quelques minutes de suite. Si on ne vouloit imposer aux hommes que quatre heures de travail par jour, je crois qu'on pourroit leur donner la tâche d'élever chacun 120 livres à un pied de hauteur à chaque seconde de travail. Cependant le partage le plus conforme à la constitution de l'homme est, à mon avis, celui de huit heures de travail par jour, & je supposerai que les rameurs soient destinés à cette fatigue.

IX. On remarquera donc que tout homme appliqué à un tel travail qu'il puisse soutenir pendant huit heures chaque jour, fera un effet équivalent à celui d'élever soixant livres à la hauteur d'un pied par seconde; du moins ce sera là l'effet moyen quand ce travail sera fait par plusieurs hommes. Mais le plus souvent le travail est employé en grande partie à des effets inutiles, & c'est-là le seul article essentiel à éviter; on n'a qu'à satisfaire à cette seule condition pour être assuré d'avoir employé le meilleur moyen qu'il étoit possible pour exécuter l'ouvrage qu'on se proposoit; & dès-lors il ne faut plus entreprendre d'y ajoûter le moindre degré de perfection. Cette réflexion nous conduit d'abord à cette grande & principale maxime: *Que dans tout ouvrage qu'on se propose il faut commencer par examiner quel est l'effet essentiellement & nécessairement attaché à cet ouvrage, effet qui soit inévitable par la nature même de l'ouvrage, & éviter ensuite autant qu'il est possible tout autre effet.*

X. Pour nous conformer à cette regle, nous rechercherons quel est l'effet essentiellement & nécessairement requis, quand on se propose d'entretenir un navire dans un cinglage uniforme; or un navire ne sauroit faire chemin sans donner aux eaux un mouvement qu'elles n'avoient point; nous considérerons ce seul effet comme essentiel & inévitable, en faisant abstraction de la résistance de l'air, & en supposant les eaux comme n'ayant aucune ténacité sensible, ce que les expériences physiques confirment assez bien; encore cet effet n'est-il essentiel que dans un certain sens, puisqu'il n'est pas sûr qu'il soit impossible de mettre à profit le mouvement qu'on avoit imprimé aux eaux: ainsi, par exemple, si on avoit un tuyau d'une largeur uniforme réplié en cercle & rentrant en soi-même, tout rempli d'eau, on voit qu'un globe du même diametre, que celui du tuyau, pourroit se mouvoir dans le tuyau, sans imprimer jamais le moindre mouvement nouveau aux eaux; je nomme donc ledit effet essentiel, que parce que je n'entreprendrai pas de l'éviter ni en tout ni en partie; & je ne crois pas que personne s'avise de l'entreprendre.

C'est donc le mouvement qu'on est obligé d'imprimer continuellement aux eaux, qui cause ce qu'on appelle la résistance des fluides & c'est uniquement dans la production de ce mouvement, que doit être employé le travail des hommes, tout autre effet étant en pure perte.

XI. Pour bien traiter notre sujet, il nous faudroit ici une théorie parfaite sur la résistance des fluides, théorie que je n'espere pas qu'on parvienne jamais à découvrir, parce qu'il sera toujours impossible de déterminer exactement le nouveau mouvement qu'on produit à chaque instant dans chaque goutte d'eau, & d'en tirer la nouvelle petite force vive; c'est cette nouvelle force vive dans les eaux produites à chaque moment par le mouvement du vaisseau, qui donne la pression du vaisseau contre les eaux, ou la résistance des eaux contre le vaisseau, & elle est en même tems l'effet essentiel du travail des rameurs. Ne pouvant donc partir de cette source, nous nous contenterons de la théorie connue sur la résistance des fluides, d'autant plus qu'elle est assez conforme à une infinité d'expériences qu'on a faites sur ce sujet. Nous supposerons en particulier que la résistance directe des eaux contre une surface plane est égale au poids d'un prisme d'eau, dont la base seroit ladite surface plane, & la hauteur celle qui répond à la vîtesse du plan; c'est-à-dire, de laquelle un corps tombant librement acquiere ladite vîtesse.

XII. Sur ces hypothèses, qui me sont communes avec tous les Physiciens & Géometres, j'examinerai ce qui arrive à un certain navire cinglant avec une certaine vîtesse. Après avoir substitué à la surface de la proue une surface plane d'une égale résistance, je supposerai que cette surface plane contienne autant de pieds quarrés qu'il y a d'unités en n, & je nommerai a la hauteur verticale génératrice de la vîtesse du navire; je supposerai qu'un pied cube d'eau pese 70 livres, & j'exprimerai ladite hauteur a en pieds. Ces dénominations & suppositions donnent le poids du prisme d'eau qui marque la résistance contre le navire $= 70na$ livres; il suit de-là qu'on peut substituer au travail requis pour ce cinglage un autre qui sera tout-à-fait le même, & dont on connoît immédiatement l'effet; c'est celui qu'on auroit si un poids de $70na$ livres étoit attaché à un cordeau qui passât sur une poulie beaucoup plus haute que le poids, & qu'on tirât horizontalement l'autre bout du cordeau avec la même vîtesse que celle du navire; or l'effet d'une telle action est d'élever un poids de $70na$ livres avec une vîtesse qu'un corps acquiert en tombant de la hauteur a. Il n'est donc plus question que de sçavoir combien de pieds un mobile peut parcourir dans une seconde de tems avec ladite vîtesse; on sçait par les élemens de la mécanique que ce nombre de pieds est $= 2\sqrt{15a}$, en supposant qu'un corps tombant librement fait 15 pieds dans la première seconde, quoique cette hauteur soit plus grande d'environ un pouce. Ainsi tout l'effet du cinglage en question est $= 140na\sqrt{15a}$; c'est-là aussi le travail essentiellement requis. Si on veut exprimer cet effet immédiatement par la vîtesse du vaisseau, on pourra supposer que le vaisseau fait dans une seconde de tems autant de pieds qu'il y a d'unités en c & puis faire $2\sqrt{15a} = c$ ou bien $a = \frac{cc}{60}$, après quoi le travail essentiel requis pour le cinglage en question, sera $= \frac{7}{6}nc^3$, ce qui veut dire qu'il y faut autant de travail qu'il y en auroit à élever à chaque seconde à la hauteur d'un pied autant de livres qu'il y a d'unités en $\frac{7}{6}nc^3$.

XIII. Cette dernière formule nous apprend que les travaux essentiels pour faire aller un navire plus ou moins vîte, sont en raison cubique des vîtesses. Ainsi donc quand on seroit parvenu à employer utilement toutes les forces, il foudroit un travail 8 ou 27 fois plus grand pour donner au navire une vîtesse double ou triple, & réciproquement les vîtesses observent la raison des racines cubiques des travaux utilement employés. Cela fait qu'on gagne beaucoup sur le travail en perdant peu sur la vîtesse. Il n'étoit pas difficile de prévoir ce théoreme, puis qu'une vîtesse double donne une résistance quadruple qu'il faut enlever avec une vîtesse double; ce qui donne l'idée d'un travail 8 fois plus grand.

XIV. La formule $\frac{7}{6}nc^3$, qui marque le travail essentiel & indispensablement réquis, nous donne d'abord à connoître une chose qu'il eût été bien difficile de déterminer en partant d'autres principes; c'est de sçavoir, en conséquence des loix que la nature a prescrites aux forces de l'homme & de leurs limites, quel est le plus petit nombre d'hommes possible pour fournir audit travail pendant 8 heures par jour; ou bien quel seroit le nombre d'hommes réquis pour un tel travail, en supposant qu'ils ne fissent aucune perte dans la maniere d'employer leurs forces; c'est-à-dire, que tout leur travail fût utilement employé sans faire aucun autre effet que celui qu'il n'est pas possible d'éviter, & qui consiste à donner continuellement une nouvelle force vive aux eaux que le navire parcourt. Pour répondre à cette question, on n'a qu'a diviser la formule $\frac{7}{6}nc^3$, par 60 (§. 8.). Si nous dénotons donc ledit nombre d'hommes par N nous aurons cette équation

$$N = \frac{7}{360}nc^3 .$$

En convertissant cette équation, on obtient

$$c = \sqrt[3]{\frac{360N}{7n}} .$$

Laquelle derniere équation marque quelle est la plus grande vîtesse possible qu'un certain nombre d'hommes puissent donner à un certain navire, & qu'ils lui donneroient réellement, s'ils étoient employés d'une maniere à ne faire aucune perte de leurs forces, ou à excercer leurs forces toutes utilement; c'est-à-dire, s'ils ne faisoient aucun autre effet que celui qui est inséparable d'avec le cinglage.

XV. Nous voici donc en état de décider si dans l'usage des rames on employe utilement toutes les forces que les rameurs exercent, ou bien si une partie considérable en est détournée par des effets, que le cinglage ne renferme pas essentiellement, & qui par conséquent doivent être censés inutilement produits. Dans le premier cas l'usage ordinaire des rames n'admet aucune correction, tout changement seroit dangereux, il ne pourroit qu'être nuissible, sans pouvoir jamais être profitable; mais dans le second cas il reste quelque espérance soit de perfectionner l'usage des rames, soit d'imaginer quelqu'autre moyen plus profitable ou moins défectueux. Pour décider cette question, il faut connoître trois choses; 1° le nombre des rameurs, que nous avons nommé N; 2° le nombre des pieds que le bâtiment fait dans une seconde de tems indiqué par c, & facile à déterminer par expérience; & 3° le nombre des pieds quarrés composant la surface plane d'une résistance égale à celle de la proue du bâtiment, nombre que nous avons désigné par n; c'est ce dernier nombre qui me fait encore quelque peine, faute d'observations ou d'expériences suffisamment détaillées; je crois cependant pouvoir le déterminer en quelque façon pour les galeres, auxquelles je me propose d'appliquer cette théorie.

XVI. M. Chazelles, dans les Mémoires de l'Académie pour l'année 1702, nous en fournit une description & des remarques qui nous mettent assez en état de nous satisfaire là-dessus: voici les circonstances qui nous intéressent[6].

6 This author published in 1702 (see p. 43, note 23 of the Introduction) various observations on the performance of a royal galley. They were used subsequently by Bouguer (see p. 48, note 26 of the Introduction), by Bernoulli in his *Hydrodynamica* and by Euler in E413, *De promotione navium sine vi venti (Mémoire sur la manière de suppléer à l'action du vent)*: Prix Paris VI (pièce I) 1753 (1771) – Euler, *Op. Omnia* II/20, pp. 190–228. It is a pleasant thing to read how Bernoulli, as the (anonymous)

La chiourme étoit de 260 hommes, qui ramoient tous à la fois; M. Chazelles a remarqué que la galere voguant de la plus grande force à pouvoir durer longtems en calme (ce qui fait précisément le cas que nous considérons), a fait 72 toises par minute, ou $7\frac{1}{5}$ pieds par seconde; ces circonstances rendent $N = 260$ & $c = 7\frac{1}{5}$. Voici à présent les circonstances rapportées par M. Chazelles, que je considére pour en tirer en quelque façon la valeur n. La surface réunie de toutes les pales étoit de 130 pieds quarrés. La chiourme faisoit 24 palades dans une minute de tems, & ainsi chaque palade duroit $2\frac{1}{2}$ seconde; mais une palade étant composée de trois mouvemens, desquels il n'y a qu'un seul qui pousse les pales contre les eaux, il n'y a que ce seul mouvement à considérer; & j'ai remarqué que si on connoissoit exactement le petit tems de ce mouvement, on en pourroit conclure assez au juste la résistance de la galere, & par conséquent la valeur n; mais j'ai remarqué aussi qu'une très-petite faute dans l'estime de ce petit tems devient très-considérable pour le résultat que nous cherchons. Nous embrasserons d'abord notre question dans toute son étendue, c'est pourquoi je nommerai T, le tems entier de chaque palade, que nous avons vu être de $2\frac{1}{2}$ secondes, & t tout le tems qu'on employe pendant l'action de pousser les pales contre les eaux; en ce cas il est évident qu'on peut substituer à l'action interrompue des rameurs qui poussent une surface de 130 pieds quarrés, celle de pousser continuellement & sans aucune interruption un plan de $\frac{t}{T} \times 130$ pieds quarrés, mais avec la même vîtesse. Je crois cependant que la surface de 130 pieds quarrés demande quelque diminution; car outre que les rameurs ne ployent peut-être pas entiérement les pales, ils poussent les pales dans une direction qui n'est ni parfaitement parallele à la quille, puisqu'ils font un arc de cercle, ni parfaitement horizontale; ce qui paroît assez par leur manœuvre, ni parfaitement perpendiculaire au plan des pales. Toutes ces raisons, & quelques autres, m'engagent à substituer 100 pieds quarrés aux 130. M. Chazelles dit ensuite que la partie intérieure de la rame avoit douze pieds, la partie extérieure 24 pieds, la pale 5 pieds de longeur, & que le vogue-avant faisoit un mouvement de 6 pieds dans le tems t qu'on doit exprimer en seconde; ainsi la vîtesse de ce mouvement répondoit à $\frac{6}{7}$ pieds par seconde.

Il s'agit a présent de sçavoir dans quel endroit il faut placer le centre d'effort des pales contre les eaux; on pourroit se contenter de le placer au

author of this essay gently criticizes himself as the author of the book; he corrects an error he made in the *Hydrodynamica*, viz. the length of an oar is $12+24$ feet and not $6+12$. In addition he finds – for various reasons, some of them are not mentioned – the resistance should be smaller (going from 130 to 100 square feet) and he improves on the way to fix an effective working length for the oars.

milieu des pales, c'est-à-dire à $21\frac{1}{2}$ pieds de distance depuis l'apostis: mais comme c'est ici une question purement géométrique & facile à résoudre, en prenant la somme des efforts de chaque élément de la pale, il vaudra sans doute mieux de déterminer géométriquement ce centre d'effort pour des pales d'une largeur uniforme telles qu'on à coutume de les faire; voici le résultat de cette solution.

Soit la vîtesse de l'extrêmité intérieure de la rame $= c$ la vîtesse de la galere $= v$, la longueur de la partie intérieure de la rame $= a$; la distance entre l'apostis, & le commencement de la pale $= b$; la longeur de la pale $= d$, la distance cherchée du centre des efforts depuis l'apostis $= \lambda$, je dis qu'on aura[7]

$$\lambda = \frac{v}{c}a + a\sqrt{\frac{a}{3d}} \times \left[\left(\frac{b+d}{a} - \frac{v}{c}\right)^3 - \left(\frac{b}{a} - \frac{v}{c}\right)^3\right].$$

Dans cette équation il faudra faire $v = 7\frac{1}{5}$, $a = 12$, $b = 19$, $d = 5$.

$$\lambda = \frac{432}{5c} + 12\sqrt{\frac{4}{5}}\left[\left(2 - \frac{36}{5c}\right)^3 - \left(\frac{19}{12} - \frac{36}{5c}\right)^3\right]$$

Sur cette équation je me propose d'examiner deux cas; le premier sera celui de supposer chaque palade composée de trois tems égaux, ce qui fera $t = \frac{1}{3}T = \frac{5}{6}$ secondes, pendant lequel petit tems le vogue-avant appliqué, l'extrêmité intérieure de la rame fait six pieds; de maniere qu'il faudra faire $c = 7\frac{1}{5}$, d'où l'on tire pour ce cas

$$\lambda = 12 + 12\sqrt{\frac{1108}{1728}} = 21.608 .$$

Le second cas consistera à supposer $t = \frac{2}{5}T$ car comme le mouvement de pousser les pales contre les eaux est le seul qui souffre une grande résistance, il est à présumer qu'il se fait un peu plus lentement que les deux autres employés simplement à soulever & puis à porter les pales en avant, dans cet exemple nous aurons $t = 1$ sec. & par conséquent $c = 6$ pieds, ce qui fait[8]

$$\lambda = 21.645 .$$

7 In the following equation the term $3d$ does not appear.

8 The original gives erroneously 21.537.

Ce n'est donc pas la valeur de λ, quoique la formule en soit fort compliquée, qui puisse jetter quelque doute considérable sur notre question. Reprenons à présent notre question dans toute sa généralité, nous aurons la vîtesse du centre des efforts relative à la galere pendant que les rameurs poussent les pales contre les eaux $= \frac{\lambda}{a}c$ & cette vîtesse relative aux eaux devient $= \frac{\lambda}{a}c - v$; or le quarré de cette vîtesse multiplié par la surface de toutes les pales diminué autant que l'interruption du plongement des pales, & quelques autres raisons exposées le demandent, est égal au quarré de la vîtesse de la galere, multipliée par le plan que nous cherchons, & que nous nommerons s; cette considération nous donne

$$s = \left(\frac{\frac{\lambda}{a}c - v}{v}\right)^2 \times \frac{t}{T} \times 100 ,$$

dans laquelle équation la quantité $\frac{t}{T} \times 100$ marque la surface de toutes les pales reduites au cas d'agir sans interruption & diminuées en raison 130, à 100, à cause de plusieurs petites obliquités dans le mouvement des pales & de quelques autres raisons. Cette équation un peu mieux rangée donne

$$s = \left(\frac{\lambda c}{av} - 1\right)^2 \times \frac{t}{T} \times 100 .$$

Sur cette équation nous chercherons la valeur de s pour les deux cas mentionnés ci-dessus. Dans le *premier* cas nous avons supposé $\frac{t}{T} = \frac{1}{3}$, ce qui fait $c = 7\frac{1}{5}$ & $\lambda = 21.608$; dans le *second* cas nous avons supposé $\frac{t}{T} = \frac{2}{5}$, ce qui fait $c = 6$ & $\lambda = 21.537$; pendant que dans l'un & l'autre cas il faut faire $a = 12$ & $v = 7\frac{1}{5}$: toutes ces substitutions nous donnent pour le cas $s = 21.37$ & pour le second[9] $s = 10.126$. Cette grande diversité entre les valeurs de s est uniquement provenue de ce que dans le *premier* cas nous avons supposé $\frac{t}{T} = \frac{1}{3}$ & dans le *second* $\frac{t}{T} = \frac{2}{5}$.

M. Bernoulli, dans son Hydrodynamica a pareillement examiné cette même question, & il trouve à la fin pag. 301 que la résistance de la galere étoit égale au poids de $13\frac{1}{3}$ pieds cubiques d'eau, ce qui donneroit $s = 15.74$.

9 The original gives erroneously $s = 9.82$.

Mais quoique ce nombre ait assez de vraisemblance, & que j'approuve sa méthode, il me paroît qu'il n'a pas assez pesé toutes les circonstances, on voit même que sa mémoire l'a trompé à l'égard de la longeur des rames, qu'il ne fait que de 18 pieds. Erreur cependant qui influe très-peu sur sa méthode. M. Euler[10], dans les Mémoires de l'Académie de Berlin, pour l'année 1747 pag. 204, a encore considéré le même exemple, mais il n'a fait dans cet endroit que suivre le sentiment de son compatriote, prenant sans doute la question pour trop-peu déterminée, & ne voulant pas l'examiner par lui-même.

Enfin, M. Bouguer, dans son Traité du Navire, pag. 116, suppose pour les galeres pareilles à celles de M. Chazelle $s = 10$ pieds quarrés; mais il faut avouer que ce n'étoit que sur une simple estime qu'il s'est déterminé sur ce nombre, ne faisant attention qu'au peu de profondeur, & à la grande saillie de la carene; & nous voyons assez par ce que cet illustre auteur nous dit lui-même, pag. 419, combien ces sortes d'estimes peuvent être trompeuses, puisqu'il dit qu'il pensoit que pour un grand vaisseau de guerre du premier rang, ce plan pouvoit être réduit à 77 pieds quarrés, & que l'ayant examiné par sa méthode de résoudre la surface de la partie antérieure de la caréne en plusieurs triangles, il avoit trouvé ce plan le double de ce qu'il avoit pensé. Si M. Bouguer, s'étoit servi de la même méthode pour les galeres, je n'écouterois plus absolument que le suffrage d'un aussi grand poids.

Après tout je supposerai $s = 16$ pieds quarrés en attendant qu'on détermine ce plan avec plus de précision par des expériences.

XVII. Si nous supposons donc[11] $c = 7\frac{1}{2}$ (je reprends ici les dénominations que j'ai faites avant le 16me §) & $n = 16$ (on remarquera que dans la digression du § 16, j'ai substitué aux lettres c & n, les lettres v & s, qui ont dans cette digression la même signification que c & n dans les articles précédens) nous aurons $N = \frac{7}{360}nc^3 = 116$, d'où il suit que 116 hommes suffiroient pour entretenir une telle galere dans un singlage à faire $7\frac{1}{5}$ pieds par seconde, si ces hommes pouvoient être employés pour cet effet d'une maniere à ne faire aucune perte de leur forces; or il y a eu 260 rameurs, nous devons donc conclure, que ces rameurs n'ont employé que $\frac{116}{260}$ parties de leur travail utilement, & que les $\frac{144}{260}$ parties du travail ont été employées inutilement en produisant des effets qui ne sont pas essentiellement renfermés dans le cin-

10 L. Euler E 116, *Memoire sur la force des rames*: Mém. Berlin, 1747 (1749), pp. 180–213 – Euler, *Op. Omnia* II/20, pp. 101–129.

11 Misprinted as $c = 7\frac{t}{2}$ in the original.

glage, c'est environ $\frac{5}{9}$ de perte, & ce résultat nous laisse espérer de pouvoir tirer un plus grand parti du travail des hommes pour le cinglage des navires.

XVIII. Examinons aussi quelle est la plus grande vîtesse possible que 260 hommes puissent donner à la galere en question, en supposant qu'ils ne fassent aucune perte de leurs forces; c'est la seconde équation du § 14 savoir $c = \sqrt[3]{\frac{360N}{7n}}$ qui nous apprend cette plus grande vîtesse possible, & qui nous la donne de 9.42. Il n'est donc pas possible que 260 hommes donnent à la galere une plus grande vîtesse que celle de faire $9\frac{2}{5}$ pieds par seconde, de quelle maniere qu'ils emploient leur travail, il est cependant possible d'y atteindre; ainsi la derniere vîtesse possible ne surpasse la vîtesse actuelle observée par M. Chazelles, que d'environ $2\frac{1}{5}$ pieds par seconde, & la vîtesse actuelle faisoit plus que les trois quarts de toute la vîtesse possible, d'où nous voyons qu'une grande perte de forces n'entraine qu'une perte très médiocre du côté de la vîtesse, c'est que les forces augmentent en raison cubique des vîtesses.

XIX. Si au lieu de $n = 16$ pieds quarrés, on vouloit suivre l'hypothese de M. Bouguer, en supposant $n = 10$, la quantité du travail utilement employée en deviendroit plus petite en raison de 16 à 10; on trouvera $N = 72\frac{1}{2}$ & on en conclurra une perte[12] de $\frac{187\frac{1}{2}}{260}$ du travail de la chiourme; une telle perte me paroît moins vraisemblable, & c'est une raison qui m'a fait préférer la supposition de $n = 16$. Je suis surpris qu'on ne se soit pas appliqué avec plus de soin à connoître ces plans pour toutes sortes de bâtimens, qui influent si fort sur un grand nombre de questions-très utiles pour la navigation. Voici la maniere que je crois la plus commode, & la plus sûre de connoître ce plan pour tel bâtiment que l'on veut. On amarrera ce bâtiment à un autre, qui allant à voiles, tirera le premier après soi avec la même vîtesse; la corde par laquelle on attache les deux bâtimens sera longue, afin que le bâtiment entraîné ne souffre rien du sillage de l'autre, ou du mouvement des eaux causé par le bâtiment antérieur: un bout de la corde passera sur une poulie verticale & parallele à la quille, & on attachera à ce bout de la corde un bassin pour y mettre des poids, pendant que l'autre bout est attaché fixement à l'autre bâtiment; on mettra d'abord assez de poids sur le bassin pour l'empecher de s'élever, jusqu'à ce que le tout soit réduit à un état de

12 Misprinted as $\frac{187\frac{1}{2}}{269}$ in the original.

permanence; après quoi on diminuera peu-à-peu le poids, jusqu'à ce que le bassin commence à s'élever un peu, pendant qu'un autre observe exactement la vîtesse des bâtimens. Si le frottement de la poulie étoit assez grand pour mériter quelque attention, il sera facile de corriger le poids qu'on aura observé. De cette maniere on connoîtra exactement la force qui tire le bâtiment avec une certaine vîtesse. Si ledit poids augmenté de celui du bassin, est exprimé en livres, dont le nombre soit P, on aura $n = \frac{6P}{7cc}$ ainsi supposant la vîtesse c de 4 pieds par seconde, & P de 250 livres, le plan cherché sera de $13\frac{11}{28}$ pieds[13] quarrés. De cette façon on apprendra en même tems, s'il est vrai que les résistances soient parfaitement en raison quarrée des vîtesses, ou si cette regle a besoin de quelque correction, principalement pour les corps d'une aussi grande surface, tels que sont les grands vaisseaux.

XX. Il me paroît tellement nécessaire de partir de nos principes, pour bien traiter la question proposée par l'Académie, que n'étant pas entiérement sûr des résultats que nous à fourni l'exemple des galéres, j'ai voulu m'en éclaircir par l'examen de quelques nouvelles experiences & observations faites sur des petits bateaux auxquels deux rameurs donnent ordinairement une vîtesse a faire $4\frac{1}{2}$ pieds par seconde.

Nous avons choisi deux de ces bateaux assez égaux, & nous les avons attachés ensemble de la façon décrite dans le précédent article; nous avons placé à la pouppe du second bateau entraîné par le premier, un quart de cercle, pour observer continuellement l'inclinaison d'un fil, auquel il y avoit attaché une boule plongée dans l'eau, & nous nous étions bien assurés que cette inclinaison nous donnoit assez au juste les différentes vîtesses[14].

13 Misprinted as $13\frac{1}{28}$ in the original.

14 The device seems to have been first mentioned by Guglielmini in the *Aquarum fluentium mensura nova methodo inquisita* (1691) (see C. S. Maffioli, *Out of Galileo, The Science of Waters 1628–1718*, Rotterdam: Erasmus Publishing, 1994, p. 202).

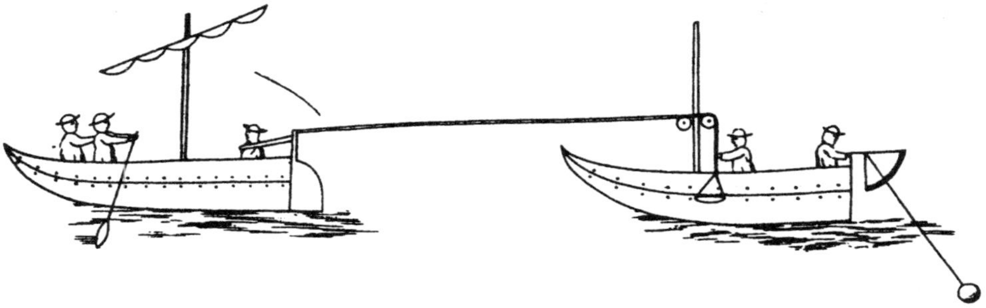

Fig. I

La figure premiere fera assez voir tout le procédé de nos expériences; on remarquera d'abord ici que les rameurs placés dans le premier bateau étoient obligés de surmonter la résistance des deux bateaux que nous supposons égaux, & que le poids mis dans le bassin du second bateau, joint a celui du bassin, marque la résistance du seul second bateau; ensorte que le double de ce poids total exprime la résistance des deux bateaux considérés comme un seul. Quand on poussoit les pales contre les eaux, le bassin chargé montoit un peu & redescendoit dans les intervalles, il n'étoit pas difficile d'augmenter ou de diminuer les poids, de maniere que les élévations du bassin devinssent égales aux descentes, & dans cette disposition nous avons pris le poids du bassin chargé, que nous nommerons poids total, pour la mesure de la résistance d'un seul bateau. Il est vrai que pendant toutes ces expériences & observations, nous remarquâmes presque continuellement plusieurs petites inégalités, qui provenoient de l'inégalité dans l'action des rameurs, & il étoit facile de commettre quelque petite faute dans le poids total du bassin, mais surtout dans l'observation de l'inclinaison du fil qui devoit marquer les vîtesses des bateaux. C'est ce qui m'a engagé a considérer attentivement toutes les circonstances, & à corriger un peu, suivant les circonstances, les résultats des vîtesses, quand j'ai vu clairement qu'il faloit le faire; mais ces changemens n'ont jamais été que très-légers. Je dois me louer ici des offices d'un ami[15] pour l'exécution de ces expériences, mais j'espére qu'on me dispensera de le nommer dans une occasion comme celle-ci.

15 The friend is Gabriel Cramer. The method with the towing ship is apparently his invention. See Introduction, pp. 52–55 h.v. for a review and a discussion of the experiments.

XXI. Voici à présent de toutes les expériences celles qui pourront nous servir pour notre sujet.

I. Deux rameurs mis dans le premier bateau ont donné aux deux bateaux, en employant leur forces ordinaires, une vîtesse de 3.50 pieds par seconde, & le poids total du bassin chargé étoit alors de 90 onces ou de $5\frac{10}{16}$ livres.

Je prendrai, pour plusieurs raisons, cette premiere expérience, pour celle qui ait été faite avec le plus d'attention & de justesse, & je corrigerai dans les suivantes les vîtesses observées, suivant la regle que les résistances sont en raison quarrée des vîtesses.

II. Les deux mêmes rameurs avertis d'employer un peu plus de force, ont donné au bateau la vîtesse de 3.80, & le poids total étoit de 100 onces; la vîtesse corrigée suivant ladite regle devient = 3.69.

III. Les deux rameurs exhortés à ramer plus vigoureusement jusqu'à donner au bateau la vîtesse 4.00, ont soutenu un poids de 122 onces, la vîtesse corrigée devient = 4.07.

IV. Les deux rameurs faisant force de rames jusqu'à convenir qu'ils ne pourroient pas continuer une telle fatigue pendant une demie-heure, ont donné au bateau la vîtesse 4.30, soutenant un poids total de 138 onces, la vîtesse corrigée devient 4.33.

La correction étant ici insensible, & y ayant ainsi un parfait accord entre les deux expériences extrêmes, cela justifie les 2 petites corrections que nous avons employées pour les deux expériences moyennes.

V. Mettant enfin quatre rameurs dans le premier bateaux, & ceux-ci avertis encore d'employer leur forces ordinaires, ont donné aux bateaux la vîtesse 4.50, c'est précisement la vîtesse que deux rameurs donnent ordinairement à un seul bateau en ramant plusieurs heures de suite.

XXII. Avant que de tirer de ces expériences, la conclusion principale sur la quantité des forces utiles dans l'action de ramer, je ferai quelques autres remarques essentielles sur ce sujet.

(a) Pour trouver le plan d'une résistance égale à celle de la proue du second bateau, on se servira de l'équation $n = \frac{6P}{7cc}$ (§. 19.) & cette équation donnera la même valeur dans chacune des 4 premieres expériences, sçavoir $n = 0.393$, ou à-peu-près $\frac{2}{5}$ d'un pied quarré; mais on se souviendra d'exprimer le poids P par livres, ou bien de diviser le nombre d'onces par 16, & de mettre pour c les vîtesses corrigées: sans ces corrections les valeurs de n deviendroient un peu inégales; cependant la valeur moyenne, pour les quatre premieres expériences non corrigées, se trouve encore telle que nous venons de le dire; ce qui confirme nos petits corrections.

(*b*) L'effet des rameurs devant être censé proportionnel à la quantité[16] $P \times c$, ou à c^3, il s'ensuit que ces effets, pour la premiere & la quatrieme experience ont été entr'eux comme 92 à 175, & les travaux ont eu la même proportion; mais il est vraisemblable, qu'un homme ne pourroit soutenir que pendant quatre heures par jour un travail tel que celui de la quatrieme expérience, pendant qu'il pourroit soutenir durant huit heures par jour celui de la premiere expérience; & si cela est, les travaux journaliers reviendroient au même, conformement à notre idée, sur le travail & la fatigue des hommes.

(*c*) Comme dans la premiere & la cinquieme experience, les rameurs faisoient des efforts égaux, & tels que les hommes peuvent faire pendant huit heures par jour, ces deux expériences confirment admirablement bien le Théoréme du §. 13, sçavoir, que les travaux essentiels requis pour faire aller un certain bâtiment plus ou moins vîte, sont en raison cubique des vîtesses. Dans ces deux expériences les travaux étoient, sans doute, comme 2 à 4, puisqu'il y avoit d'un côté deux rameurs, & de l'autre quatre rameurs, qui s'efforçoient également, & les vîtesses étoient comme[17] 3.50 à 4.50; $1 : \sqrt[3]{2} = 3.50 : 4.41$. Ainsi la vîtesse que les quatre rameurs donnoient aux bateaux étoit à fort peu-près celle que notre Théorême donne; elle étoit plus grande d'environ un pouce par seconde, & je donnerai même ci-dessous la raison de ce petit excès.

(*d*) En supposant les deux bateaux parfaitement égaux, on a, pour le premier bateau, pareillement $n = 0.393$; de sorte que les deux bateaux ensemble souffriroient une résistance égale à celle de 0.786 pieds quarrés: mais nous reconnûmes bientôt que le premier bateau étoit un peu-plus pesant que le second; cela nous engagea à changer l'ordre des bateaux pour déterminer pareillement la résistance de l'autre bateau. Je ne rapporterai que la seule expérience qui suit.

VI. *Expérience*. Après avoir changé l'ordre des bateaux & leur vîtesse, comme étant 4.22 pieds par seconde, le poids total fut remarqué de 160 onces.

XXIII. Il suit de cette sixieme expérience que si la vîtesse avoit été de 3.50, comme dans la premiere expérience, le poids total eût été de 110 onces, les résistances des deux bateaux étoient donc comme 90 à 100, & la résistance entiere des deux bateaux ensemble étoit de 200 onces ou de $12\frac{1}{2}$ livres, cette résistance étoit surmontée par deux hommes avec une vîtesse de $3\frac{1}{2}$ pieds;

16 Misprinted as P, c in the original.
17 4.41 was misprinted as 441 in the original.

l'effet utile de leur travail étoit donc égal à celui d'élever $12\frac{1}{2}$ livres à la hauteur de $3\frac{1}{2}$ pieds par seconde, multipliant ces 2 nombres & divisant le produit par 120 qui marque l'effet entier du travail ordinaire de deux hommes, nous obtenons $\frac{43\frac{1}{2}}{120}$ ou $\frac{175}{480}$ ou $\frac{35}{96}$ ou 0.365. De là nous concluons, qu'exprimant le travail entier par 1000, les rameurs en ont employé[18] 365 utilement, & qu'ils ont fait une perte de 635, en produisant des effets inutiles, qui n'étoient pas renfermés essentiellement dans le cinglage des deux bateaux.

XXIV. Comparons ces résultats avec ceux que nous avons trouvés au §. 17 pour les galeres; en leur supposant $n = 16$ pieds quarrés, nous y avons vu que le travail utile de la chiourme faisoit $\frac{116}{260}$ du travail entier, c'est en fractions décimales 0.446. Ces résultats sont donc à-peu-près comme 9 à 11, & par consequent presque les mêmes pour les bateaux & pour les galeres: & si on n'avoit donné aux galeres qu'une résistance égale à celle d'un plan de 13 pieds quarrés, ces pertes des forces seroient devenues entiérement égales de part & d'autre. En tous cas je m'assure que cette conformité, soit entière, soit approchante, paroîtra merveilleuse à ceux qui considéreront la diversité infinie qu'il y avoit des deux côtés. Les rames étoient très-petites sur le bateau, & très-grandes sur la galere; la vîtesse de la galere étoit plus que double de celle du bateau, ce qui rendoit la maniere de voguer tout-à-fait différente: sur la galere on donnoit 24 palades par minute, & sur le bateau, on en faisoit ordinairement plus que 30. Les palades si faisoient à-peu-près en trois tems égaux sur la galere, & en deux sur le bateau; il y avoit 260 rameurs sur la galere, & deux rameurs pour les deux bateaux; enfin il n'y avoit sur le bateau qu'un rameur à chaque rame, & il y en avoit cinq sur la galere, & ces cinq rameurs travailloient d'une maniere extrêmement différente, celui qui est le plus près de l'apostis ne faisant presque point de mouvement & le vogu'avant faisant un mouvement énorme qui en peu de tems le met tout en sueur &c. Cependant, il se trouve à la fin que les rameurs ont fait à-peu-près chacun le même effet sur les bateaux & sur la galere. Celà marque une conservation merveilleuse des forces humaines pareille à la conservation des forces vives, qui a fait tant de bruit; & nous en devons conclure que les hommes, avec des fatigues égales, font toujours des effets à-peu-près égaux; de sorte que dans toutes les questions méchaniques, dans lesquelles il s'agit d'obtenir un certain effet par le travail des hommes, il n'y a qu'à éviter autant qu'il est possible tout autre effet pour être assuré qu'on aura embrassé le meilleur moyen qui étoit possible.

18 Misprinted as 356 in the original.

XXV. Je me crois donc suffisamment fondé à dire, que dans l'action de voguer sur les bâtimens de bas-bord, les rameurs employent 0.446 de leur travail utilement, c'est-à-dire, uniquement à surmonter la résistance des eaux, qui seule fait tout l'effet qu'on se propose, & que 0.554 de leur travail se perdent, étant employés à des effets, qui ne sont pas essentiellement renfermés dans le cinglage. Mais comme tout travail a son effet, qu'on peut toujours réduire à une certaine quantité de forces vives, il est nécessaire de rechercher, en quoi peuvent consister les effets des 0.554 parties du travail inutilement employées; ce n'est qu'après ces recherches, qu'on sera en état d'examiner, si ces effets sont d'une nature à pouvoir être évités en conservant l'usage des rames, ou non. Dans le premier cas il sera possible de perfectioner la méchanique des rames, & dans le second cas il faudra tâcher de substituer aux rames quelqu'autre moyen qui entraîne une moindre perte dans les forces que l'on veut employer.

XXVI. Voici d'abord une source de pertes, qui est bien certaine & qu'on peut déterminer assez au juste; rémarquons que la surface réunie de toutes les pales, aura toujours une certaine grandeur déterminée qu'on doit pousser avec une certaine vîtesse contre les eaux, que par cette action on donne à une certaine quantité d'eau un certain mouvement, & que par-là on produit continuellement une certaine quantité de forces vives, qu'on doit censer n'être pas essentiellement renfermées dans l'effet qu'on se propose, puisque si on étoit à même de se servir de perches, qu'on pût chaque fois appuyer directement contre un point fixe, ou qu'on pût présenter à chaque palade un point fixe pour y appuyer l'extrémité de la pale, on éviteroit entiérement de produire une semblable force vive inutile. Tâchons donc d'évaluer cette perte des forces pour la galere que nous avons considérée.

XXVII. Pour cet effet il faut commencer par réduire la surface réunie de toutes les pales, qui étoit de 130 pieds quarrés, mais qui n'est poussée contre les eaux que par intervalles à un autre plan beaucoup plus petit, qui étant poussé constamment & uniformement contre les eaux, donne à la galere la vîtesse de $7\frac{1}{5}$ pieds par seconde; soit donc ce plan $= s$, la vîtesse uniforme avec laquelle ce plan est constamment poussé contre les eaux $= v$, la surface plane qu'on peut substituer à la proue $= n$, & la vîtesse de la galere $= c$; là dessus nous aurons d'abord $svv = ncc$, parce que la force motrice doit être égale à la résistance, & par consequent $v = c\sqrt{\frac{n}{s}}$. Mais dans cette manœuvre l'effet n'est plus simple. Celui qui résulte du mouvement qu'on donne aux eaux par la surface n, & que nous avons démontré au §. 12 devoir être exprimé par $\frac{7}{6}nc^3$, il survient un autre effet qui consiste dans le mouvement qu'on donne aux eaux par la surface s, & qui est pareillement $= \frac{7}{6}sv^3$. Ce second effet n'est pas essentiellement compris dans le cinglage; mais il devient essentiel aussitôt qu'on se propose de produire ce cinglage moyennant une surface plane s, qu'on pousse directement & sans interruption uniformement contre les eaux avec une vîtesse relative v: si on nomme donc le premier effet E, le second effet accessoire sera $= \frac{sv^3}{nc^3}E$, ou $= \frac{v}{c}E$, ou bien $= E\sqrt{\frac{n}{s}}$. Si donc par exemple, on supposoit $s = n$, cet effet accessoire seroit égal à l'autre effet essentiel; on voit en effet que ce seroit comme faire aller deux galeres égales avec des vîtesses égales, l'une vers l'orient, l'autre vers l'occident.

XXVIII. Ce que nous venons de démontrer sur le rapport entre les deux dits effets, nous pouvons aussi le déduire du rapport qu'il y a manifestement entre les travaux; imaginons-nous encore qu'on trouve en tous lieux des points fixes d'appui, contre lesquels on puisse appuyer directement des perches, & donner par-là à la galere la vîtesse c; un tel travail seroit entiérement utile, puisque ce seroit enlever immédiatement & directement une certaine résistance sans qu'on perde la moindre partie de son travail. J'appellerai ce travail T, qui est essentiellement réquis; qu'on conçoive après cela d'autres points d'appui mobiles, (au lieu de points fixes) dont la vîtesse soit v en sens contraire, ou des points d'appui qui cedent avec la vîtesse v, on voit manifestement que c'est la vraie maniere d'envisager notre question. Or il faudroit tant pour les points d'appui fixes que pour les mobiles, pousser les perches avec la même force pour donner à la galere la même vîtesse, mais dans le premier cas, on les pousseroit avec la vîtesse c, & dans le second cas avec la vîtesse $c + v$; & les travaux étant en ce cas proportionels aux vîtesses (§. 2.), il s'ensuit que le travail dans le second cas, sera $= \frac{c+v}{c}T$; ainsi le travail accessoire est $= \frac{v}{c}T$, ou bien $= \frac{v}{c}E$, puisque T, exprime un travail, qui produit son effet E tout entier. C'est précisement la même valeur que nous avons trouvée dans le précédent article; nos principes sont trop fondés dans sa nature pour se dementir.

XXIX. Pour appliquer notre Théorême à la question dont il s'agit d'évaluer l'effet accessoire que la chiourme produit sur la galère, il faut tâcher de substituer aux lettres E, n & s leur juste valeur; considérons pour cet effet que la surface reunie de toutes les pales est 130 pieds quarrés sur 52 rames; mais comme les rames ne sont plongées dans les eaux que pendant environ le $\frac{1}{3}$ du tems, il faudra envisager la chose comme s'il n'y avoit que 17 à 18 rames poussées sans interruption contre les eaux; donc la surface des pales ne sera plus que d'environ 43 pieds quarrés; on voit ensuite qu'il y a naturellement encore plusieurs petites diminutions à supposer pour ces 43 pieds, toutes m'engagent à supposer finalement $s = 36$ pieds quarrés. Je suis persuadé que c'est la plus grande valeur qu'on doive admettre, mais une si petite correction n'est pas assez importante pour nous engager là-dessus dans de grandes discussions; nous ferons donc $s = 36$ pieds quarrés, $n = 16$ pieds quarrés, & $E = 0.446$ (§. 25.) & nous aurons $E\sqrt{\frac{n}{s}} = \frac{2}{3}E = 0.297$, c'est-là l'effet accessoire produit par le travail des rameurs, qu'on exprime par l'unité, lequel effet accessoire prend sa source dans le mouvement qu'on donne aux eaux par le corps de pale; & pour être persuadé, que ce n'est-là qu'un effet accessoire, on n'a qu'à supposer la surface de toutes les pales, comme infinie, ou extrêmement grande, ce qui n'implique aucune contradiction, & cet effet s'évanouira entiérement ou se réduira à très-peu de chose.

XXX. Si nous ajoutons ensemble les deux effets provenans du travail de la chiourme, & déja indiqués, la somme sera $= 0.743$ de l'effet entier que 260 hommes peuvent produire lorsqu'ils employent toutes leurs forces utilement, & ces 0.743 parties consistent uniquement dans le mouvement qu'on imprime, & communique continuellement aux eaux, tant par la proue de la galere que par la surface des pales.

XXXI. Il nous reste à découvrir en quoi peut consister le reste de l'effet entier qui se réduit à 0.257; car je le repete, tout travail doit avoir son effet, & nous n'avons encore indiqué que les trois quarts de l'effet entier. Je trouve que le quart restant doit être attribué pour la plus grande partie à l'agitation irréguliére des corps des rames; je dis: qu'il faut un travail considérable pour remuer les rames de la maniere que les rameurs le font sur une galere, & cela independamment de la résistance des eaux, considérant ici la simple inertie des rames; ce travail réquis pour remuer les corps des rames est sans doute encore accessoire puisqu'il s'évanouiroit, si on pouvoit faire des rames suffisamment roides & fortes avec très-peu de matiere. C'est-ici une nouvelle matiere, que je tâcherai déclaircir assez pour pouvoir évaluer en quelque façon cette perte, que les rameurs sur une galere font de leur force, & de leur travail.

XXXII. Les rames sur les galeres ayant 36 pieds de longueur, & étant en même tems assez fortes pour soutenir des efforts de plus de soixante livres, à la distance de $21\frac{1}{2}$ pieds (§. 16.) depuis l'apostis, ne fauroient manquer d'être des corps fort lourds & pefants, ce sont cependant ces corps qu'il s'agit de remuer fortement par des mouvemens réciproques & répétés vingt-quatre fois dans une minute de tems; si ces agitations des rames se faisoient d'une maniere que les forces vives une fois imprimées aux rames se conservassent d'elles-mêmes, comme cela arrive à tous les corps qu'on abandonne à eux mêmes, nous trouverions très-peu de perte dans l'agitation des rames, & il n'y auroit guere à considérer, que le petit frottement de chaque rame contre l'apostis. Mais qu'on fasse bien attention aux mouvemens que les rameurs donnent aux rames, & on verra aussitôt qu'il s'agit alternativement de leur imprimer une certaine vîtesse, & ensuite de les arrêter; chacun de ces changemens réciproques demande un certain travail, que nous devons considérer comme accessoire, & qui est en pure perte. Il est difficile d'évaluer au juste cette perte, parce que nous ne connoissons pas exactement les changemens en direction & en vîtesse des rames, en tant que ces changemens sont produits par les efforts des rameurs. C'est pourquoi je me contenterai d'envisager le mouvement des rames sous la forme la plus simple & la plus susceptible de calcul, mais qui ne laisse pas d'imiter assez bien le vrai mouvement tel qu'il est.

XXXIII. Je considérerai dans les palades deux mouvemens, celui qui se fait hors de l'eau, & celui qui se fait sous l'eau; dans le premier mouvement il faudra faire attention à là force vive imprimée à la rame, lorsque cette force vive est la plus grande, & je ferai consister dans cette seule force vive tout l'effet inutilement produit par les rameurs; ensuite le mouvement de la rame sera retardé, & enfin entierement arrêté; je n'attribue aucun travail à cela, parce que les rameurs n'ont qu'à laisser entraîner leur corps pour retarder, & pour arrêter les rames vers la fin de ce premier mouvement; quant au second mouvement, il est vrai que les rameurs impriment d'abord un nouveau mouvement aux rames; mais ce mouvement est ensuite employé utilement à pousser les eaux vers la pouppe; & pourvu qu'on suppose que les rames n'ayent plus aucun mouvement sensible dans le moment qu'on retire les pales hors de l'eau, on n'aura fait aucune perte de ce côté-là: on voit bien qu'en envisageant de cette maniere notre sujet, je suis bien éloigné de surcharger la perte des forces causée par l'inertie des rames. Voici à présent un Théorême, qui nous servira à évaluer cette perte.

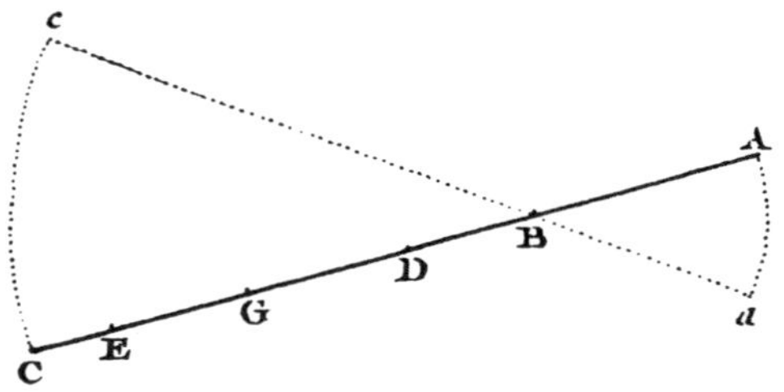

Fig. II

Soit AC, ou ac la longeur de la rame agitée réciproquement de AC en ac, & ac en AC, autour du point B, qui est la place de l'apostis; qu'on marque par M, le poids, ou la masse de toute la rame, qu'on pourra supposer inégalement chargée de matière d'une façon quelconque; soit ensuite le centre de gravité de la rame entiere en D; il faut aussi s'imaginer la rame être suspendue par le point B, en forme d'un pendule simple, & en chercher le centre d'oscillation, que nous supposerons en E; qu'on fasse enfin BG égale à la moyenne proportionelle entre BD & BE: après cette construction, je dis qu'on pourra considérer toute la matiere de la rame, comme concentrée au point G, qu'on peut appeller le centre des forces vives; connoissant le point G, il faudra examiner quelle est la vîtesse de ce point G, pendant que la rame est poussée de AC en ac, soit la hauteur verticale génératrice de cette vîtesse $= h$, & on a démontré que la force vive de la rame est égale à Mh.

XXXIV. On voit donc que pour chaque rame les rameurs produisent sur chaque palade inutilement une force vive exprimée par Mh, & par conséquent $24Mh$ pendant chaque minute, ou $\frac{2}{5}Mh$ pendant chaque seconde; & comme on applique cinq rameurs à chaque rame, on aura $\frac{2}{25}Mh$ pour chaque homme par seconde. Si on veut exprimer le poids de la rame M, par livres, & la hauteur verticale h, par pieds, il faudra comparer la quantité trouvée $\frac{2}{25}Mh$ avec 60, qui est la mesure du travail entier d'un homme pour chaque seconde de tems. Ainsi si le travail absolu de la chiourme est exprimé par l'unité, nous dirons que le travail inutilement employé causé par l'inertie des rames est $\frac{1}{750}Mh$.

XXXV. Je suis bien sûr, que ladite perte est pour le moins $\frac{1}{750}Mh$; mais je ne suis pas sûr, qu'elle ne soit pas plus grande. Or après avoir fait attention a toutes les circonstances décrites par M. Chazelles, de même qu'à celles qui déterminent le point G, conformement au Théorême du §. 33, j'estime h de $\frac{5}{4}$ de pieds, & sur cette estime la perte en question sera $= \frac{1}{600}M$; mais j'avoue que je n'ai aucune idée sur le poids d'une rame de galere; si ce poids est supposé de 100 livres, la perte en question sera 0.167; en tous cas nous voyons qu'elle est considérable. Supposons-là en attendant d'en être instruit = 0.167, & ajoutons cette valeur à celle de 0.743, que nous avons adoptée à la fin du §. 30, & nous aurons 0.910. De cette maniere nous aurons déja trouvé les 0.910 parties de l'effort total du travail de la chiourme.

XXXVI. Nous ne nous mettrons pas beaucoup en peine pour découvrir à quoi peuvent avoir été employés les 0.090 dernieres parties, il est facile de voir qu'il y aura encore plusieurs autres petites pertes; la principale, à mon avis, de toutes ces petites pertes consiste en ce que sans doute la partie extérieure de la rame sera un peu plus pesante que la partie intérieure, ce qui fera qu'à chaque palade, on est obligé d'élever un certain poids, & ce travail doit encore être considéré comme entierement inutile. Nous connoissons donc à présent à-peu-près tout l'emploi que les rameurs font de leur force. Le principal & essentiel consiste à déplacer les eaux qui s'opposent à la proue, c'est ici le seul utile, & il doit être égal à 0.446. Le second consiste à deplacer les eaux, que rencontrent les surfaces des pales; je traite celui-ci d'accessoire, & inutile, & je le fais = 0.297. Le troisieme sert à remuer le corps des rames, & à surmonter leur inertie; il est de même accessoire & inutile, & je l'estime = 0.167. Enfin le dernier emploi des efforts des rameurs pareillement inutile consistera à lever quelques autres obstacles accidentels & sur tout celui qui provient du défaut d'équilibre dans les deux bras de la rame. Je suis aussi bien persuadé, que si on pouvoit éviter tous ces effets inutiles sans tomber dans d'autres pertes, on atteindroit par-la à la meilleure maniere qu'il soit possible de naviger par les efforts des hommes, car nul travail ne reste sans effets, & si tout l'effet est utile, on tire le meilleur parti qui soit possible du travail des hommes. Mettons ces principes à profit dans notre Seconde Partie.

XXXVII. Avant que de finir cette Premiere Partie, nous ferons encore quelques réflexions sur le moyen fondamental, qui est le seul dont les hommes puissent profiter pour faire cingler un navire par leur travail en pleine mer. Ce moyen unique est l'inertie des eaux, quand on se trouve en pleine mer, mais on peut mettre à profit cette inertie de plusieurs manieres, qui sont fort différentes entre elles. Dans l'usage ordinaire des rames on pousse les pales contre les eaux, & ce mouvement se fait avec une vîtesse finie & comme uniforme, dans cette action, on donne pendant chaque instant une vîtesse *finie* à une nouvelle masse d'eau *infiniment petite*, mais on peut changer la chose de maniere que pendant chaque instant on donne un incrément de vîtesse infiniment petit à une masse *finie*; cette considération pourroit bien nous induire à faire quelque changement à la structure des rames, ou même à leur substituer quelqu'autre moyen, en tout cas ce nouveau sujet appartient de trop près à notre question principale pour le passer sous silence.

XXXVIII. Imaginons-nous d'abord toute la surface de la mer parsemée de corps flottans, mais qui n'empêchent pas le sillage, de maniere qu'au lieu d'enfoncer les pales dans l'eau, on puisse simplement appuyer l'extrêmité de la rame contre un de ces corps, & le pousser vers la pouppe, & supposons que ces corps flottans ne resistent à la rame que par leur inertie; sur ces suppositions, nous nous trouverons dans le cas, duquel j'ai fait mention dans le précédent article, qui est de donner à chaque instant une accélération vers la pouppe infiniment petite au corps flottant qui fait une masse finie. Si chacun de ces corps flottans avoit une masse comme infinie, il ne céderoit pas sensiblement à l'effort de la rame, & le rameur ne perdroit rien à cet égard de sa force; mais si ces corps sont considérés, comme ayant une certaine masse, ils céderont par un mouvement accéléré, & la perte du travail des rameurs est la force vive qu'auront acquis ces corps dans le moment qu'on les quitte pour appuyer la rame contre un nouveau corps flottant. Nous substituerons ensuite à ces corps flottans des masses d'eau assujetties à ne pouvoir être déplacées que toutes entieres à la fois.

XXXIX. Maintenant je me propose de chercher la proportion qu'il y auroit entre le travail, si le poids p de chaque corps flottant étoit infini, & qui seroit essentiel, à celui qu'on est obligé d'employer lorsque le poids p est d'une certaine valeur déterminée, ce calcul se fera à-peu-près comme au §. 28. Il n'y a que cette différence, qu'ici la vîtesse avec laquelle le corps flottant est poussé vers la pouppe est uniformément accéléré, pendant que cette vîtesse étoit considérée, comme uniforme dans le §. 28. Soit la vîtesse du corps flottant relative aux eaux calmes, dans un instant quelconque $= v$, que ce corps soit pressé avec une force quelconque π pendant un tems infiniment petit dt, & il faudra toujours estimer le travail proportionel à $\int \pi v\, dt$. Cette considération nous conduit d'abord à cette propriété, que le travail essentiel est au travail accessoire provenant du mouvement qu'on donne aux corps flottans, comme Pct à $\int Pv\, dt$; mais pour avoir une idée plus nette de cette proportion, considérons les espaces parcourus; supposons que depuis le moment qu'on a commencé à toucher le corps flottant, ce corps ait parcouru l'espace s, pendant que le navire parcourt l'espace S, & alors on pourra mettre ds à la place de $v\,dt$, & S à la place de ct; après quoi ladite proportion sera comme PS à $\int P\,ds$, ou P étant constante comme S à s. Nous voyons donc qu'à présent le travail essentiel & entiérement utile est au travail accessoire duquel il est ici question, comme le chemin que fait le navire pendant tout le temps qu'on employe à pousser en arriere chaque corps flottant, est au chemin parcouru par le corps flottant pendant le même tems. Ce Théorême seroit le même que celui du §. 28, si nous avions pu supposer ici la vîtesse v constante, parce que en ce cas les espaces s & S, seroient comme les vîtesses v & c.

XL. Connoissant la perte qu'on souffriroit par une telle sorte de travail, je crois que cette connoissance pourroit bien nous fournir quelques nouveaux moyens d'épargner les forces des rameurs, & d'en tirer plus de fruit. C'est dans cette vue que j'ajouterai quelques réflexions sur notre Théorême.

I. Soit la force, avec laquelle chaque corps flottant est poussé en arrière $= \pi$, le poids de ce corps $= p$, le petit tems, pendant lequel ces corps sont pressés à chaque fois $= t$ exprimé par secondes, & alors on aura

$$s = \frac{\pi}{p} \times tt \times 15 \text{ pieds},$$

pendant qu'on a $S = tc$ pieds, en entendant par c le nombre de pieds que le navire fait dans une seconde de tems; donc le rapport de s à S devient égal au rapport de $15\pi t$ à pc.

II. Comme il y a sur une galere 52 rames, qui agissent par intervalles en 3 tems, nous pourrons leur substituer $\frac{52}{3}$ rames qui agissent sans interruption, & la pression continuelle que chaque rame exerce contre les eaux, doit être censée d'environ cinquante-six livres, qui font $\frac{3}{52}$ de la résistance d'un plan de 16 pieds quarrés mus contre les eaux, avec une vîtesse de $7\frac{1}{5}$ pieds par seconde. Outre cela le tems employé à pousser les pales contre les eaux sur les galeres est le tiers d'une palade ou $\frac{5}{6}$ de seconde; supposons donc qu'on pousse pareillement chaque corps flottant avec une force de cinquante-six livres, en y employant chaque fois un tems de $\frac{5}{6}$ de secondes, & on trouvera[19] $s = \frac{\pi}{p} \times tt \times 15 = \frac{1}{p} 583\frac{1}{3}$ pieds, pendant qu'on auroit S ou $tc = 6$ pieds: si on veut donc faire ensorte que le travail inutile ne fasse

19 Misprinted as $5.83\frac{1}{3}$ in the original.

que les deux tiers du travail utile, qui est la proportion que nous avons trouvée au §. 29 pour la maniere ordinaire de ramer sur la galere, il faudra faire $\frac{s}{S} = \frac{2}{3}$, & par conséquent $p = 145\frac{5}{6}$ livres[20], ce qui fait environ le poids de chaque corps flottant de deux pieds cubes d'eau. Cette supposition rendroit le travail entiérement égal pour la maniere ordinaire de ramer, & pour celle d'appuyer l'extrêmité des rames contre des corps flottans.

Qu'on fasse à présent attention à la maniere dont nous avons trouvé tous les résultats de cet article, & des deux précédentes rémarques, si différentes en apparence de nos méthodes antérieures, & on sera surpris de l'admirable accord qu'il y a de part & d'autre pour les résultats absolus. J'ai dit que le travail inutile est toujours la force vive inutile qu'on en tire; dans notre cas c'est la force vive que l'on donne à chaque corps flottant, cette force vive est pour chaque corps flottant $= \pi s = \pi\left(\frac{\pi}{p} \times tt \times 15\right)$, faisons $\pi = 56$, $p = 145\frac{5}{6}$, $t = \frac{5}{6}$, lesquelles positions font le travail de pousser en arriere les corps flottans, le même que celui de ramer sur la galere, & nous trouverons

$$\pi s = \pi\left(\frac{\pi}{p} \times tt \times 15\right) = 224 .$$

Ainsi la force vive de chaque corps flottant devient = 224, mais comme il y a cinq hommes à chaque rame, nous aurons pour chaque homme $\frac{224}{5} = 44\frac{4}{5}$, qui font la force vive que chaque homme produit inutilement à chaque palade qui dure $2\frac{1}{2}$ seconde, ce qui fait par seconde pour chaque homme $17\frac{23}{25}$ de perte; cette perte étant comparée avec le travail entier que nous estimons toujours 60 par seconde, on trouvera la perte $= \frac{448}{1500}$ parties du travail entier, ou en fraction décimale 0.298, & nous avons trouvé à la fin du §. 29. Cette perte en tant quelle est produite par l'action ordinaire des rameurs = 0.297. Un aussi parfait accord prouve incontestablement la vérité de tous les principes, dont nous nous sommes servis.

IV. Rémarquons enfin que l'effet inutile consistant dans la force vive imprimée aux corps flottans, étant à l'effet essentiel & inevitable comme s à S, ou comme $\frac{\pi}{p} \times tt \times 15$ à tc, ou enfin comme $15\pi t$ à pc, on peut diminuer ce rapport de deux manieres sans rien changer à la valeur π nécessaire pour entretenir la vîtesse uniforme du navire ni à cette vîtesse même exprimée par c, la *premiere* maniere consiste à augmenter le poids des corps flottans

20 I.e. $p = \frac{3}{2}\frac{583\frac{1}{3}}{6} = 145\frac{5}{6}$ livres.

exprimé par p, & la *seconde* maniere est de diminuer le tems t; cette seconde maniere a besoin de quelque explication.

Supposons, comme dans la seconde remarque[21] $\pi = 56$, $p = 145\frac{5}{6}$, $t = \frac{5}{6}$, & on aura $s = 4$ pieds, pendant que S sera $= 6$ pieds; mais si on faisoit $t = \frac{5}{12}$, on trouveroit $s = 1$ & $S = 3$, & l'effet inutile ne feroit plus que le tiers de l'effet utile & essentiel. Mais on dira que de cette façon, les palades *se* succédéroient avec trop de rapidité, & qu'au lieu de 24 il en faudroit 48 dans une minute; à cette objection je réponds, que cela n'est point nécessaire; car, comme nous entendons par t le petit tems, pendant lequel l'extrêmité de la rame reste appuyée contre chaque corps flottant, il n'y a qu'à supposer les corps flottans plus serrés, de maniere à pouvoir appuyer à chaque palade successivement contre deux corps flottans. Ainsi notre seconde maniere de diminuer l'effet inutile consiste proprement à changer dans un même coup de rame le plus souvent qu'il est possible de corps flottans. Si par un seul coup de rame on poussoit successivement dix corps flottans; l'effet inutile deviendroit par-là dix fois plus petit, que si on employe tout le coup de rame à pousser un seul corps, parce que chacun des dix corps ne recevra que la centieme partie de force vive.

XLI. M. Bernoulli expose à la fin de son Hydrodynamique une idée singuliere de voguer sans rames[22], elle consiste à pomper des eaux pour les laisser écouler ensuite vers la pouppe, si M. Bernouilli avoit examiné la chose suivant nos principes, il auroit jugé comme moi, qu'il faut attendre moins d'effet du travail des hommes qui pompent, que de ceux qui rament; car en adoptant toutes les proportions qu'il trouve les plus avantageuses, je remarque que tous les effets inutiles sont au moins trois fois aussi grands que l'effet utile, de sorte que l'effet utile n'est que le $\frac{1}{4}$ de l'effet entier, il y auroit donc $\frac{3}{4}$ de perte, pendant que nous n'avons trouvé que $\frac{144}{260}$ de perte sur la galere (§. 17.), en supposant $n = 16$, & $\frac{187\frac{1}{2}}{260}$ (§. 19), en supposant avec M. Bouguer $n = 10$. Il faut cependant avouer, qu'on donneroit à la galere presque autant de vîtesse en pompant, qu'en ramant; & je suis surpris que M. Bernouilli, après plusieurs positions, qu'il ne pouvoit faire qu'avec une

21 Misprinted as $p = 145\frac{5}{61}$ in the original.

22 The remark on jet propulsion refers to the *Hydrodynamica*, Section XIII, §20 to §29, *Werke* 5, pp. 413–422. The conclusion there was that the highest velocity v of the ship is attained – for a given input of power – if the velocity c at which the water leaves the nozzle is double of v. This conclusion is flawed, as we have shown in the Introduction, pp. 65–72 h. v.

certaine estime, ait pu si bien former la même conclusion finale. Je ne ferai donc aucun usage dans ma seconde partie de cette nouvelle idée.

XLII. On sera peut-être surpris de voir, que dans tout le corps de cette premiere partie, je n'aie rien dit sur la longueur & les proportions les plus avantageuses des rames; c'est que suivant nos principes, la longueur absolue, & la proportion de la partie extérieure à l'intérieure sont par elles-mêmes indifférentes, puisque faisant abstraction de l'inertie des rames, & de la trop grande obliquité dans le mouvement des rames trop courtes, il y a toujours la même proportion entre l'effet utile & inutile, quelles que soyent les longueurs des parties extérieures & intérieures des rames. Il n'y a dans ce choix aucune économie essentielle à observer, & toutes les considérations qu'on doit faire sur cette question ne sauroit tendre qu'à éviter quelques inconveniens accidentels qui surviennent, les uns aux rames trop courtes, & les autres aux rames trop longues; je ne m'étonne donc pas, que ceux qui ont traité cette question l'ayent décidée tout différemment les uns des autres. J'aurai occasion, dans la seconde Partie, de faire sur ce sujet quelques réflexions passageres.

SECONDE PARTIE.

I. L'Académie propose de trouver la maniere la plus avantageuse de suppléer à l'action du vent sur les grands vaisseaux, soit en y appliquant des rames, soit en y employant quelquautre moyen que ce puisse être. J'espere bien qu'un bon usage de nos principes nous apprendra ce que cette illustre Compagnie demande; mais je crois devoir prévenir le lecteur, que le plus grand effet possible qu'on puisse se promettre du travail des hommes étant fort borné, la maniere la plus avantageuse de suppléer à l'action du vent sur les grands vaisseaux par le travail des hommes pourroit facilement paroître assez defectueuse à en juger autrement que par nos principes. Je me propose donc avant toutes choses d'examiner, quelle est la plus grande vîtesse possible qu'un certain nombre d'hommes puissent donner à un grand vaisseau donné.

II. Pour me conformer davantage à l'intention de l'Académie, je m'attacherai aux vaisseaux du premier rang, & je supposerai, qu'un tel vaisseau a une résistance égale à celle d'un plan de 150 pieds quarrés. C'est ainsi que M. Bouguer détermine ce plan dans son Traité du Navire, pag. 419; non sur une simple estime, comme il a fait par rapport aux galeres, mais par une méthode qui ne pouvoit guere tromper ce grand homme aussi parfaitement instruit de tout ce qui pouvoit contribuer à rendre cette détermination plus exacte. Il n'y a donc qu'à supposer $n = 150$ dans l'équation générale que nous avons donnée au §. 14. de la premiere partie sçavoir:

$$c = \sqrt[3]{\frac{360N}{7n}}$$

& nous aurons $c = \sqrt[3]{\frac{12}{35}N}$, qui nous marque quel est le nombre de pieds c, que le vaisseau peut faire dans une seconde de tems par le travail ordinaire d'autant d'hommes qu'il y a d'unités en N, en supposant tout leur travail utilement employé. Cette vîtesse n'est donc possible que dans la théorie pure, & plus on s'en approchera, plus la maniere de voguer sera parfaite.

III. Si les hommes, au lieu d'un travail ordinaire, soutenable pendant huit heures par jour, vouloient faire un travail forcé, tel qu'ils puissent soutenir tout au plus une demi-heure de suite, ils pourroient alors faire un effet double à-peu-près, & ils augmenteroient la vîtesse du vaisseau presque en raison de 1 à $\sqrt[3]{2}$, ou plus exactement en raison de 1 à $\sqrt[3]{\frac{175}{92}}$; voyez la premiere & la quatrieme expérience du §. 21. de la premiere partie, & la note (b)

du §. 22; on aura donc en ce cas $c = \sqrt[3]{\frac{105}{161}}N$. Sur ces deux équations, qui expriment les plus grandes vîtesses possibles, tant pour le travail ordinaire, que pour le travail forcé, j'ai construit la Table suivante.

Table[23] *fondamentale des vîtesses possibles d'un vaisseau du premier rang.*

Nombre d'hommes employés à la navigation.	Vîtesses possibles entieres pour un travail ordinaire & durable.	Vîtesses possibles entieres pour un travail extrêmement forcé
10	1.50 pieds par seconde	1.86 pieds par seconde
20	1.90	2.35
30	2.17	2.69
40	2.39	2.96
50	2.58	3.19
60	2.74	3.39
70	2.88	3.57
80	3.01	3.74 †
90	3.13	3.89 †
100	3.25	4.03
120	3.45	4.27
140	3.63	4.49
160	3.80	4.70
180	3.95	4.89
200	4.09	5.06
220	4.22	5.22
240	4.34	5.39 †
260	4.46	5.53 †
280	4.57	5.65
300	4.68	5.79
350	4.93	6.10
400	5.16	6.38
450	5.36	6.63
500	5.55	6.86
550	5.73	7.09
600	5.90	7.31
650	6.06	7.50
700	6.21	7.69
800	6.50	8.04
900	6.76	8.37
1000	7.00	8.68

23 In the original Table, the numbers marked † were wrongly printed as: 3.72 for 3.74; 3.87 for 3.89; 5.37 for 5.39 and 5.51 for 5.53.

Cette Table nous apprend quelle vîtesse un certain nombre d'hommes pourroient donner à un vaisseau du premier rang, soit par un travail ordinaire & durable, soit par un travail forcé, en supposant qu'il fût possible de mettre à profit tous leurs travaux sans la moindre perte. Je m'assure qu'on trouvera sans difficulté, dans ces vîtesses de l'une & de l'autre espece, tout cet air de vérité qu'on peut attendre sur ces sortes de matieres, quand on ne considere la Table, que depuis environ 200 hommes jusqu'à 1000; mais cette apparence de vérité se perd peu-à-peu, à mesure qu'on diminue le nombre d'hommes, elle s'évanouit enfin entiérement, & prend tout l'air d'une fausseté manifeste. En effet, il sera bien difficile à ceux qui n'auront pas compris toute la force de nos principes, de se persuader qu'il soit possible, suivant notre Théorie, à dix hommes de donner à un vaisseau du premier rang une vîtesse à faire un pied & demi par seconde par un travail ordinaire, & une vîtesse à faire $1\frac{6}{7}$ pieds par un travail forcé; ils se récrieront contre ces énormes vîtesses pour un si petit nombre d'hommes, & laisseront-là toute cette théorie; j'aurois peut-être été tenté de donner une autre tournure à mes pensées, si j'avois pû quitter pour un moment cette haute opinion que j'ai de mes Juges. Rassuré par leurs lumieres, je ne crains point d'avouer, que ce sont précisement ces réflexions qui me paroissent les plus favorables à nos idées. J'espére qu'on en portera la même opinion après qu'on aura vu les changemens que l'action de ramer demande par la Table des vîtesses que nous venons de donner.

V. L'objet de cette Table n'est qu'un être de raison, auquel on n'atteindra jamais, mais qui doit nous guider dans nos recherches, & duquel on approchera d'autant plus, qu'on employera plus utilement le travail des hommes. La Table suppose un travail entiérement utile, mais elle prendra une toute autre face aussitôt que nous considérerons la chose autrement qu'*in abstracto*, & que nous appliquerons notre théorie à un travail déterminé, duquel nous connoissons déja les effets pour de certains cas. Choisissons ici l'action de ramer ordinaire, envisageons donc notre grand vaisseau de premier rang comme une grande galere, sur laquelle les hommes puissent ramer aussi commodement qu'ils le font sur les galeres ordinaires, on conviendra que c'est ici une supposition bien libérale & avantageuse, & on seroit sans doute assez content, si quelqu'un indiquoit une maniere de faire aller les grands vaisseau sur mer si peu défectueuse. Voyons donc quelles vîtesses réelles on donneroit par ce travail au grand vaisseau en question.

VI. J'imiterai sur ce grand vaisseau toutes les circonstances que nous avons remarquées sur la galere. La surface réunie de toutes les pales sur la galere étoit de 130 pieds quarrés, ce qui fait la moitié d'un pied quarré pour chaque homme; c'est pourquoi nous donnerons pareillement la moitié d'un pied quarré [par] pale à chaque homme sur le vaisseau. Nous avons ensuite réduit sur la galere les 130 pieds quarrés à 36 pieds quarrés (Part. 1. §. 29.), qui seroient poussés contre les eaux, sans aucune interruption, & tout à fait directement. Ce n'est donc plus que $\frac{36}{260}$ pieds quarrés de pale pour chaque homme, mais considérée comme agissant sans interruption, & nous supposerons la même chose pour les vaisseaux du premier rang. Si le nombre des rameurs est $= N$, on aura $\frac{36}{260}N$, pieds quarrés, qui poussent continuellement les eaux en arriere pour avancer le vaisseau. C'est ce plan que nous avons nommé S, au §. 27 de la premiere Partie: il est d'autant plus petit qu'il y a moins de rameurs, & nous avons démontré aux §§. 26 & 27, que plus ledit plan est petit, plus la perte du travail des rameurs est grande. Voilà pourquoi les vîtesses de la Table du §. 3, excédent de beaucoup les vîtesses réelles qu'un petit nombre de rameurs peuvent donner au grand vaisseau; cependant on comprendra assez, qu'il n'est pas absurde de dire que dix hommes puissent donner à un vaisseau du premier rang une vîtesse à faire un pied & demi par seconde; on n'a qu'à s'imaginer ce vaisseau attaché à un poteau par une longue corde, que dix hommes tireroient à soi; le vaisseau mu avec ladite vîtesse souffriroit une résistance d'environ 400 livres suivant la théorie commune sur la résistance des fluides, ainsi le travail de chaque homme ne consisteroit qu'à surmonter une résistance de quarante livres, avec la vîtesse d'un pied & demi par seconde, ce qui fait précisément le travail naturel d'un homme. Si par une telle manœuvre les dix hommes ne pouvoient donner au vaisseau ladite vîtesse, ce seroit une marque infaillible, que la théorie sur la résistance des fluides, s'écarte sensiblement de la vérité pour les grands corps mus avec peu de vîtesse, cela changeroit bien les calculs à faire, mais non pas nos principes, ni les maximes que nous en tirerons pour employer le travail des hommes à la navigation avec plus de profit. Il est donc visible, que si les vîtesses que la Table marque pour un petit nombre d'hommes paroissent d'abord excessives & même absurdes, ce n'est que parce qu'on considére ces hommes sur le pied d'autant de rameurs, & nous allons voir qu'il s'en faut de beaucoup qu'un pareil nombre de rameurs puisse donner au vaisseau toute la vîtesse possible.

VII. Pour connoître à présent la vîtesse réelle de notre grand vaisseau produite par un nombre de rameurs quelconque N, nous nommerons E, l'effet utile de ces rameurs, & nous aurons, en vertu des §§. 27 & 28, Partie premiere, $E\sqrt{\frac{n}{s}}$ pour cet effet inutile, qui consiste dans le mouvement qu'on donne aux eaux par les coups de rame; dans cette expression $E\sqrt{\frac{n}{s}}$ il faut pour un vaisseau du premier rang faire $n = 150$ pieds quarrés, & $S = \frac{36}{260}N$ (§. 6. Part. 2.), & de cette maniere ledit effet inutile devient[24] $E = \sqrt{\frac{150 \times 260}{35N}}$, ou à-peu-près $= \frac{33}{\sqrt{N}}E$; ainsi la somme des deux effets est $= E + \frac{33}{\sqrt{N}}E$. Outre cette perte exprimée par $\frac{33}{\sqrt{N}}E$, nous avons remarque au §. 31. Part. prem. que l'action de remuer les corps des rames, & de vaincre leur inertie fait encore une perte, laquelle jointe à quelques autres petits défauts peut faire environ le quart de l'effet total, ce qui fait un tiers des deuxdits effets, ou un tiers de $E + \frac{33}{\sqrt{N}}E$; donc la somme entiere de tous les effets est $= \frac{4}{3}E + \frac{44}{\sqrt{N}}E$. Ainsi l'effet utile est à la somme de tous les effets comme E est à $\frac{4}{3}E + \frac{44}{\sqrt{N}}E$, ou comme[25] 1 est à $\frac{4}{3} + \frac{44}{\sqrt{N}}$. Il faut donc diminuer chaque vîtesse de la Table en raison de $\sqrt[3]{\left(\frac{4}{3} + \frac{44}{\sqrt{N}}\right)}$ à[26] 1, parce que les vîtesses sont en raison des racines cubiques de travaux utilement employés (§. 13. Part. prem.) ou essentiellement requis, & que la Table suppose le travail entier utilement employé, ce qui donne cette Analogie; comme la racine cubique de l'effet total, ou $\sqrt[3]{\left(\frac{4}{3}E + \frac{44}{\sqrt{N}}E\right)}$ est à la racine cubique de l'effet utile, ou $\sqrt[3]{E}$, ainsi chaque vîtesse de la Table fondamentale à la vîtesse réelle, que les rameurs donneront au vaisseau; ainsi on déterminera les vîtesses réelles, en divisant chaque vîtesse de la Table fondamentale, par le nombre $\sqrt[3]{\left(\frac{4}{3} + \frac{44}{\sqrt{N}}\right)}$. Comme je ne doute pas que ces nouvelles vîtesses ne repondent assez bien aux expériences qu'on pourra faire, je ne me suis point fait de peine de construire sur cette regle la Table ci-jointe; mais on se souviendra que cette Table n'est faite que pour les vaisseaux du premier

24 Misprinted as $= \sqrt{\frac{150 \cdot 260}{36N}}$ in the original.
25 The 1 was misprinted $\int$ in the original.
26 The 1 was misprinted $\int$ in the original.

Table[27] *des vîtesses réelles qu'on peut donner à un vaisseau du premier rang, par le moyen des rames.*

Nombre de rameurs.	Vîtesses pour un travail ordinaire & durable.	Vîtesses pour un travail extrêmement forcé.
10	0.60 pieds par seconde.	0.75 pieds par seconde.
20	0.85	1.05
30	1.03	1.27
40	1.18	1.46
50	1.31	1.62
60	1.43	1.77
70	1.54	1.91
80	1.64	2.03
90	1.73	2.14
100	1.81	2.24
120	1.96	2.43
140	2.10	2.60
160	2.25 †	2.79 †
180	2.36	2.92
200	2.48	3.07
220	2.59	3.21
240	2.69	3.34
260	2.79	3.46
280	2.89	3.5
300	2.98	3.69
350	3.19	3.95
400	3.39	4.20
450	3.57	4.43
500	3.73	4.63
550	3.88	4.82
600	4.03	5.00
650	4.17	5.17
700	4.30	5.34
800	4.55	5.65
900	4.79	5.94
1000	5.01	6.21

27 In the original Table, the numbers marked † were wrongly printed as: 2.23 for 2.25 and 2.76 for 2.79.

rang, dont les résistances répondent à un plan de 150 pieds quarrés, qu'on suppose pour chaque rameur la moitié d'un pied quarré de pale, & que ces rameurs travaillent de la même maniere, que sur les galeres, quoique cette derniere supposition ne soit pas fort essentielle; voici cette nouvelle Table.

VIII. Cette Table corrigée sur la précédente, conformément à nos principes, & à nos calculs sur la perte, que les rameurs font de leur travail en ramant sur les galeres, ne me paroît plus rien renfermer qui puisse blesser l'estime naturelle qu'on peut se former là-dessus. On voit à présent par cette Table, que dix rameurs ne pourront plus donner au vaisseau qu'une vîtesse à faire $\frac{3}{5}$ de pieds par seconde, ou environ sept pouces par un travail naturel, & neuf pouces par un travail forcé. Ces deux vîtesses font $\frac{2}{5}$ des vîtesses totales correspondantes, mais les effets utiles ne font, que la quinzieme partie des effets totaux, d'où l'on voit la grande perte que les dix rameurs font de leurs travaux. Mais lorsqu'il y a mille rameurs sur le vaisseau, ils lui donnent les $\frac{5}{7}$ de la vîtesse totale, & ils employent plus que le $\frac{1}{3}$ de leurs travaux utilement; ces grandes différences proviennent uniquement de ce qu'en augmentant le nombre des rameurs, on augmente en même tems la surface réunie de toutes les pales, c'est aussi-là la raison pourquoi les quatre rameurs dans la cinquieme expérience du §. 21. Part. prem. ont donné aux bateaux rélativement aux deux rameurs dans la premiere expérience du même §, une vîtesse un peu plus grande, que ne demande le Théorême du §. 13. Part. prem. Voyez la note (*c*) du §. 22. Part. prem.

On voit enfin par chacune des deux Tables, combien peu de profit on fait en augmentant de beaucoup le nombre des rameurs, puisqu'une différence de neuf-cent à mille rameurs ne fait qu'environ trois pouces de différence dans les vîtesses.

IX. Je m'en rapporte à présent à ceux qui pourront juger de nos résultats par leur propre expérience: ils n'auront pas manqué surtout de remarquer à-peu-près les vîtesses des grands vaisseaux qu'on remorque sur une chalouppe, ou sur une galere, si nous faisons abstraction de la petite résistance, qui survient en attachant la chalouppe au vaisseau, & même de celle qui survient en y attachant la galere; notre Table nous enseigne que dix hommes sur une chalouppe pourroient remorquer le vaisseau par un travail naturel, avec une vîtesse de sept pouces par seconde, & vingt hommes avec une vîtesse de dix pouces. J'aurois souhaité sur tout, de trouver là-dessus quelques expériences ou observation dans quelque auteur; je présume, que ces vîtesses sont encore un peu trop grandes à cause d'une autre résistance, à la quelle nous n'avons pas fait attention, & qui ne sauroit presque se manifester que dans les très-petites vîtesses; c'est l'adhésion des eaux contre la carene du vaisseau, qui vraisemblablement ne croît qu'en raison simple des vîtesses, pendant que la résistance essentielle des fluides croît en raison du quarré des vîtesses; mais comme je ne fais ces remarques, que pour faire voir tout l'usage de nos principes, & que dans l'application il ne s'agit pas des petites vîtesses qu'un petit nombre de rameurs peuvent donner aux grands vaisseaux, nos conclusions ne seront pas beaucoup changées par l'adhérence des fluides, quand même l'expériences démontreroit, qu'elle est assez considérable dans les mouvemens très-lents. La derniére Table nous apprend encore que 260 hommes, sur une galere pourroient remorquer un vaisseau du premier rang avec une vîtesse d'environ deux pieds neuf pouces par seconde, ou même avec une vîtesse de presque trois pieds $\frac{1}{2}$ avec un travail extrêmement forcé; ces deux vîtesses doivent pourtant être diminuées en raison de $\sqrt[3]{11}$ à $\sqrt[3]{10}$, parce que la galere attachée au vaisseau augmente la résistance à-peu-près en raison de 11 a 10, après lesquelles diminutions le vaisseau ne seroit plus remorqué, qu'avec une vîtesse d'environ deux pieds huit pouces, ou en faisant force de rames, avec une vîtesse d'environ trois pieds quatre pouces. Je ne doute pas que ces vîtesses ne se confirment assez bien par l'expérience, ce qu'on trouvera peut-être de moins doit encore être attribué à l'adhérence des eaux, laquelle cependant, je présume ne pouvoir plus être guerre considérable en ce cas.

X. Je fais avec plaisir toute les remarques, qui pourront servir de pierre de touche à notre théorie, & comme il est faute d'essayer les vîtesses d'une galere semblable à celle dont parle M. Chazelles poussée successivement par 10, 20, 30, jusqu'à 260 rameurs, en observant simplement les tems employés à parcourir un même espace, je mettrai ici encore la formule, qui marque généralement combien de pieds la galere parcourra dans une seconde de tems. Pour trouver cette formule, on n'a qu'à faire tout ce que nous avons fait dans cette seconde partie avec cette seule différence qu'au lieu d'avoir supposé $n = 150$ pieds quarrés, il faut ici faire $n = 16$ pieds quarrés, là-dessus on trouvera pour le travail ordinaire,

$$\sqrt[3]{\frac{405N}{168 + 112\sqrt{\frac{260}{N}}}}$$

& pour le travail forcé il faudra multiplier cette formule par 1.24. Si l'Académie, aux heureux auspices de laquelle la République des Lettres est redevable de toutes les grandes découvertes de notre siécle, trouvoit mes réflexions assez dignes de son attention pour engager quelque Capitaine de galere entendu à faire ces expériences, j'oserois me flatter, qu'elles s'accorderoient assez, avec ladite formule, pour constater entiérement toute notre théorie; je trouve, par exemple, que dix rameurs en travaillant avec la même fatigue, qu'ils le font lorsque toute la chiourme fait parcourir à la galere $7\frac{1}{5}$ pieds par seconde, pourront lui donner la vîtesse 1.76; il n'y a que la force de l'adhérence des eaux qui pourra un peu diminuer cette vîtesse. Mais il est tems que nous nous approchions de notre sujet principal.

XI. Supposons un certain équipage destiné au travail, comme par exemple 500 hommes. La Table fondamentale nous fait voir, que toute la vîtesse possible du vaisseau du premier rang sera alors de 5.55 pieds, pour un travail ordinaire & durable, ou de 6.86, pour un travail extrêmement forcé. C'est donc-là une vîtesse de laquelle on peut bien tâcher d'approcher, mais qu'il ne faut pas esperer d'atteindre entiérement. On conviendra même que ce sera bien assez, que d'obtenir le même effet sur les grands vaisseaux, qu'on obtient par l'usage des rames sur les vaisseaux de basbord. Cependant les 500 hommes ne pourront donner alors même au vaisseau qu'une vîtesse de 3.73, 4.63; & nous avons vu, que les deux sources principales de ce défaut consistent dans l'insuffisante grandeur des pales, & dans l'inertie des rames: ainsi toute la perfection qu'on peut donner à l'usage commun des rames consiste uniquement à augmenter la surface des pales autant qu'il est possible sans tomber dans de nouveaux inconveniens, & à diminuer l'inertie des rames; on donne aux pales des grandes rames cinq pieds de longueur, ce seroit peut-être outrer la chose que de les faire plus longues, mais j'avoue que je ne vois pas assez, pourquoi on ne leur donne qu'un demi-pied de largeur, il me semble plutôt, qu'il n'y auroit aucun inconvenient considérable à leur donner un pied de largeur; cela supposé, car j'avoue que c'est avec peu d'assurance que je touche aux choses qu'un si long & universel usage à établies; voici les avantages qu'on en rétireroit. Sans rien changer aux pressions que les rameurs exercent contre les rames, ni aux tems qu'ils employent, tant aux palades entieres qu'aux coups des pales contre les eaux, ou aux saccades, les agitations des rames en deviendroient plus petites, & en même tems l'inertie des rames en auroit moins d'effet, on ménageroit doublement le travail, & la perte en deviendroit plus petite: quand à l'inertie des rames, on en diminuera les effets en choisissant le bois le plus fort, & le plus leger, & en ne les chargeant dans aucun endroit inutilement de matiere, il ne faut pas lui donner près de l'apostis plus d'épaisseur qu'il en faut pour pouvoir soutenir sans se casser un poids d'environ 100 livres appliqué au milieu de la pale; depuis l'apostis il faut diminuer les épaisseurs, de maniere que les cubes[28] des diametres soient proportionels aux distances depuis le milieu de la pale, car les forces des cylindres d'une même matiere sont à-peu-près en raison triplée des diamétres. C'est aussi suivant cette regle, qu'il faudroit diminuer les épaisseurs depuis l'apostis vers l'autre bout de la rame; mais comme de cette façon le défaut de l'équilibre seroit trop grand je voudrois qu'on remédiât à ce défaut en chargeant de plomb la partie interieure de la rame, ou son manche tout près de l'apostis.

XII. C'est peut-être la seule raison de ne point surcharger les rames de matiere, sur tout aux pales, le mouvement desquels est le plus rapide qui a empêché les mariniers de donner plus de largeur aux pales; si c'étoit-là la raison, je voudrois qu'on tentât tout avant que de renoncer à un avantage si essentiel, peut-être vaudroit-il mieux de substituer aux pales de bois des quadres, ou bordures de fer avec de la grosse toile étendue dessus, on diminueroit par-là en même tems un autre inconvenient, qui est que l'équilibre dans les deux parties de la rame est trop dérangé par les immersions & émersions alternatives des pales, ce qui emporte certainement une force vive, qu'on perd entiérement, je voudrois même que la toile fût assez ample pour former un espece de sac, pendant la saccade, qui pût contenir trois ou quatre pieds cubes d'eau. Ces trois ou quatre pieds cubes d'eau tiendroient lieu de ces corps flottans, dont j'ai parlé dans notre premiere Partie aux §§38, 39 & 40. L'inertie de cette masse d'eau qu'il faut mettre en mouvement augmente considérablement la résistance contre les pales, ce qui doit toujours faire l'objet principal de ces recherches. En ce cas il faudroit faire ensorte que ces sacs se vuidassent parfaitement sur la fin de la saccade, & il seroit bon d'imaginer quelqu'obstacle pour chaque rame, qui l'arretât brusquement à la fin de la saccade, & jettât en même tems la pale hors de l'eau; un tel choc jetteroit encore les eaux vers la pouppe, & ces eaux par leur réaction donneroient, en ce moment un coup vigoureux au navire.

XIII. J'ai n'ai fait ces remarques passageres, que pour suivre le fil de nos principes. Je n'ignore pas qu'on ne doit pas esperer de pouvoir appliquer

28 The breaking resistance of a cylinder fixed at one end and subject to a force at right angles is proportional to the cube of its diameter. This comes about because of the combination of the resistance to longitudinal forces, proportional to the area of the section, and the leverarm proportional to the diameter. It was discussed by G. Galilei in the *Discorsi e dimostrazioni matematiche intorno a due nuove scienze attenenti alla mecanica ed i movimenti locali*, Leiden 1638 – Galilei, *Opere* (Ed. Naz.), t. VIII, pp. 160–162, prop. IV. It is a key property in the memoir of Johann II Bernoulli on anchors (*Discours sur les Ancres*: Prix Paris III (pièce IV) 1737 (1752)).

avec succès les rames ordinaires sur les grands vaisseaux. Voici comme en parle M. Bouguer dans son excellent Traité du Navire, pag. 118. «Jusques à présent (dit-il) on n'a appliqué les rames avec succès qu'aux seuls navires de basbord, quoiqu'on ait senti combien, il seroit important de pouvoir les appliquer aussi dans certains; cas aux vaisseaux proprement dits, la hauteur de ces derniers a rendu inutiles les différentes tentatives qu'on a faites de tems en tems pour tâcher de leur procurer ce secours. On a principalement insisté sur ce que les rames fussent tournantes, comme les rames des moulins à eau, mais comme on n'a pas pu leur donner assez de vîtesse, elle n'ont point eu d'effet, ou n'en ont eu que très-peu.»

L'idée des rames tournantes, comme les aîles des moulins à eau, est trop naturelle, pour ne pas se présenter d'elle-même. M. Chazelles en parle aussi dans les Mémoires de l'Académie de l'année 1702. Cela m'engage à examiner tout le succès qu'on peut s'en promettre; ce sera ensuite aux mariniers de voir s'ils voudront se contenter d'un tel succès.

XIV. Il faut d'abord examiner qu'elle est la surface entiére de toutes les ailes qui poussent sans interruption les eaux directement en arriére, si les parties de chaque aile ont une obliquité, & une inégalité de vîtesse sensible, on pourra toujours par un juste calcul substituer aux ailes une surface plane équivalente poussée perpendiculairement contre les eaux. Je nommerai cette surface s, comme nous avons fait ci-dessus, & la lettre n, exprimera de même le plan, dont la résistance soit égale à celle de la proue du vaisseau. Là-dessus, je dis que sans faire attention à la maniere de mettre en mouvement les rames tournantes, qu'on pourra varier à l'infini, on aura toujours par cette manœuvre deux effets, l'un utile & essentiellement requis, l'autre inutile & simplement accessoire par la nature de la manœuvre, & cela entiérement dans le même sens que nous l'avons démontré au sujet des rames ordinaires. Ainsi l'effet utile étant encore E, l'autre sera nécessairement $E\sqrt{\frac{n}{s}}$ (§§ 27 & 28. Part. prem.). C'est-là un principe constant & infaillible; il faudroit donc donner aux ailes une surface infinie, si on vouloit éviter entiérement cette perte.

La plupart de ceux qui ont écrit sur cette matiere, n'ont pas assez approfondi ce quelle renferme en quelque façon de Méthaphysique. Ils ont bien vu que la pression totale des ailes contre les eaux, doit être égale à la résistance du vaisseau, & qu'on peut produire telle pression que l'on se propose, moyenant une surface quelconque; que plus cette surface est petite, plus il faut lui donner de vîtesse; ils ont donc voulu reparer par la vîtesse, ce qu'ils ne pouvoient obtenir commodement par la grandeur de la surface; mais il semble qu'ils n'ont pas fait assez attention, qu'il coûte plus de travail pour obtenir une même pression moyennant une petite surface que moyennant une grande. Il est même à remarquer que ce surplus de travail devient essentiel & inévitable aussitôt qu'on s'est déterminé à une certaine surface indiquée par s, & que ce seroit tenter l'impossible d'y vouloir rémédier autrement qu'en augmentant cette surface.

XV. Nous avons vu au §29. Part. prem., que cette perte faisoit les $\frac{2}{3}$ du travail utile pour la galere en y supposant $\frac{n}{s} = \frac{16}{30}$. Outre cette perte on en fait quelques autres sur la galere par l'usage ordinaire des rames, leur somme peut aller à un quart du travail total; il est vrai qu'à examiner les sources que nous en avons indiquées, nous ne voyons rien de semblable dans les rames tournantes. Leur inertie n'est plus d'aucun empêchement aussitôt que ces rames, ont acquis leur plein mouvement il n'y a ici non plus aucun défaut d'équilibre qui puisse causer quelque travail inutile, mais il y aura en échange des frottemens qui ne se trouvent pas dans une machine aussi simple que celle d'une rame ordinaire, outre cela les impulsions des ailes contre les eaux se feront naturellement avec plus d'obliquité vers le commencement & la fin de leurs immersions, que celles des pales. Nous supposerons donc ces dernieres petites pertes de part & d'autre égales, ou plutôt nous en ferons abstraction en ne considérant que la perte essentiellement attachée à une surface déterminée s.

XVI. Après ces remarques nous verrons aussitôt quel effet on peut se promettre tout au plus sur les suppositions que fait M. Chazelles pag. 100, des Mémoires de l'Académie 1702. «On ne doit pas douter (dit-il) que la force de cent hommes par exemple, poussant continuellement un volume d'eau de 18 pieds quarrés de chaque côté, ne mette bientôt en mouvement le plus gros vaisseau, puisqu'une simple chalouppe se fait sentir &c.» Si nous faisons donc, conformément aux suppositions de M. Chazelles $s = 36$ & $n = 150$, comme il convient pour les plus gros vaisseaux, dont il parle, il est impossible que le travail inutile soit moindre que $E\sqrt{\frac{150}{36}}$, ou que $2.04E$. Ainsi sur 304 hommes il n'y auroit que 100, d'utilement employés, & le travail de 204 hommes seroit entiérement perdu, pendant que sur une galere poussée par 260 hommes l'effet inutile n'étant que $E\sqrt{\frac{16}{36}}$, ou $\frac{2}{3}E$, on peut dire dans le même sens, que du travail de cinq hommes, on ne perd que celui de deux. Cette grande disproportion fait voir combien cette navigation proposée par M. Chazelles seroit désavantageuse & déffectueuse. Je trouve effectivement, que moyennant les deux roues, qui portent les rames tournantes, cent hommes ne pourroient donner au grand vaisseaux dont il est question que la vîtesse marquée dans la Table fondamentale du § 3, qui repond au nombre de cent hommes multiplié par $\sqrt[3]{\frac{100}{304}}$, c'est-à-dire, pour le travail ordinaire $3.25 \times \sqrt[3]{\frac{100}{304}}$, ou 2.24 pieds par seconde, & pour le travail extrêmement forcé $4.03 \times \sqrt[3]{\frac{100}{304}}$, ou 2.77. Voilà pour les deux cas toute la vîtesse qu'il est possible de donner au vaisseau sur les hypothèses de M. Chazelles. Il est vrai que ces vîtesses sont sensibles, qu'elles vont même jusqu'aux $\frac{2}{3}$ des vîtesses totales, ou des vîtesses que cent hommes pourroient donner au vaisseau, s'ils ne faisoient aucune perte de leur travail, mais il est vrai aussi que dans cette hypothèse, 100 hommes pourroient fournir le même effet, que 304 hommes fournissent dans le cas proposé par M. Chazelles. C'est que les pertes des travaux sont beaucoup plus grandes, que les pertes des vîtesses, puisque les travaux utiles sont en raison triplée des vîtesses réelles. Encore doit-on remarquer, que dans les évaluations que nous venons de faire, nous avons fait abstraction de plusieurs autres petites pertes qu'on fera nécessairement, outre celle que nous avons indiquée qui est la principale, & qui est si essentiellement attachée aux hypothèses, qu'une plus petite perte impliqueroit une contradiction contre les loix de la nature. Je repete souvent ces sortes d'expressions de peur qu'on ne confonde les défauts essentiellement attachés aux différentes manœuvres, avec ceux qui ne sont qu'accessoires, on ne doit pas espérer de diminuer les premiers défauts sans changer les hypothèses, mais on peut bien l'esperer par rapport aux défauts accessoires ou accidentels. Je ne crois pas, après ces réflexions & ces évaluations, qu'on s'avise d'employer les rames tournantes avec les proportions que M. Chazelies leur donne.

XVII. Après l'examen que nous venons de faire des suppositions de M. Chazelles, je vais examiner ce qui seroit essentiellement requis pour tirer des rames tournantes sur un vaisseau du premier rang autant de profit qu'on obtient par le secours ordinaire des rames sur une galere. Or dans une galere, allant de toutes rames nous avons eu[29] $\frac{S}{n} = \frac{36}{16}$. Il faut donc, pour faire travailler l'équipage du grand vaisseau avec un avantage égal, rendre la valeur de $\frac{S}{n}$ aussi $= \frac{36}{16}$, comme nous supposons pour un vaisseau du pre-

29 Bernoulli changes notation: the area of the blades is now called S; formerly it was s.

mier rang $n = 150$, nous aurons $S = 337\frac{1}{2}$ pieds quarrés. Il faudroit donc de nécessité absolue faire ensorte, que les rames tournantes présentassent continuellement aux eaux une surface de $337\frac{1}{2}$ pieds quarrés. Une telle surface seroit requise sans aucune interruption, quand même elle seroit poussée contre les eaux dans une direction parfaitement parallele à la quille, ce qui cependant n'est pas, & la nature du mouvement circulaire dans un plan vertical demanderoit cette surface encore plus grande. De-là nous voyons qu'il faudroit sur notre grand vaisseau dix-neuf ou vingt moulins, que M. Chazelles demande. Si on faisoit ces moulins deux fois plus grands, il en faudroit encore dix, & quel seroit enfin le succès de ces énormes tentatives; ce seroit de se procurer autant d'avantage, qu'en ont les rameurs sur une galere, qui va à toutes rames; supposons encore qu'on se soit procuré tout cet avantage, & voyons qu'elle seroit alors la vîtesse que 500 hommes pourroient donner au vaisseau. Je dis que pour trouver cette vîtesse, il faut dans la Table fondamentale du §3. multiplier les deux vîtesses qui répondent au travail de 500 hommes par $\sqrt[3]{\frac{3}{5}}$, parce que sur cinq parties de travail il y en auroit trois d'utilement & deux d'inutilement employées §29. Part. prem., & cette regle donne la vîtesse du vaisseau par le travail ordinaire des 500 hommes = 4.68 pieds par seconde, & pour le travail forcé = 5.78 pieds. Nous avons cependant supposé dans ces résultats, qu'il ne se fait aucune autre perte, que celle qui provient de l'insuffisante étendue de la surface totale S; ainsi ces vîtesses seront encore trop grandes, & on ne pourra jamais y atteindre entiérement, si les frottemens & quelques autres empêchemens emportoient encore le quart de travail entier; il faudroit multiplier lesdites vîtesses par $\sqrt[3]{\frac{3}{4}}$, après quoi on auroit pour le travail ordinaire la vîtesse 4.25, & pour le travail forcé 5.253 voilà les vîtesses finales qu'on est fondé à attendre de tout cet appareil; il est vrai que ces vîtesses sont considérables, & qu'elles font plus que les $\frac{3}{4}$ des vîtesses possibles totales, je laisse juger les gens de marine, si on doit se donner sur un grand vaisseau tout l'embarras que ces moulins causeroient, & ces embarras pourroient bien être beaucoup plus grands, qu'il ne paroissent d'abord. Si cependant on ne vouloit pas rénoncer à ces rames tournantes, je serois d'avis qu'on se relâchât un peu sur la grandeur de la surface S, quand même la perte du travail en deviendroit plus grande, & qu'on se procureroit l'avantage que nous fournit l'idée exposée au §38. Part. prem., en faisant que les moulins, au lieu de pousser simplement les eaux par les ailes, puisent en même tems de grandes masses d'eau qu'ils emportent à la fois. Il y auroit ensuite un grand nombre de regles à observer pour occuper les hommes par un espece de travail qui leur soit naturel, tel qu'est celui d'exercer une force d'environ vingt livres avec une vîtesse d'environ trois pieds par seconde, mais j'avoue que je ne fais pas assez grand cas des rames tournantes pour entrer dans toutes ces discussions.

XVIII. Tâchons plutôt de trouver quelques nouveaux expédiens moins embarrassants & plus profitables, mais sans perdre jamais de vue les principes que nous avons si bien établis. S'il ne faut pas espérer de pouvoir se servir des rames ordinaires sur les vaisseaux de haut-bord, il ne faut pas pour cela renoncer entiérement à ce principe de mouvement; il n'est pas difficile d'imaginer plusieurs manieres de se servir commodement de rames sur les grands vaisseaux, en leur donnant une autre façon. En voici une qui me pa-

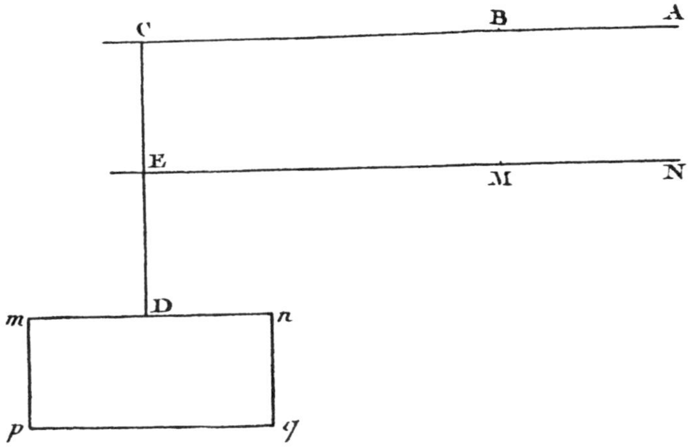

Fig. III

roît assez commode & utile, je n'en ferai qu'une ébauche fort légere, *ABC* est une piece de bois perpendiculaire à la quille, & d'abord horizontale, qui repose au point *B* sur l'apostis: *CD* représente une barre de fer mobile autour du point *C*, moyennant une charniere dans un plan vertical; à cette barre on affermiroit la pale *mnpq*; au point *E* on attacheroit un ou plusieurs bouts de corde, par lesquels plusieurs hommes tireroient la barre de fer vers la pouppe & pousseroient la pale *mnpq* contre les eaux. Voici à présent comme il faudroit manier cette espece de rames; au moment qu'on auroit fini de tirer la pale avec force contre les eaux, & même un peu plutôt, un homme peseroit au point *AC*, & rétireroit la pale entiérement hors de l'eau, sur quoi la barre & la pale par leur propre poids se remettroient dans la position verticale, & la passeroient même: au moment que ce balancement se finiroit, l'homme en *A*, cesseroit de peser sur *A*, & laisseroit par-là replonger la pale dans les eaux, & au même instant les hommes recommenceroient à tirer fortement la pale contre les eaux.

XIX. Il s'en faut sans doute de beaucoup que cette description soit suffisante pour voir en detail tout ce qu'il faudroit faire pour l'exécution de ces nouvelles rames, aussi n'ai-je prétendu, qu'exposer simplement leur façon d'agir; il y auroit bien de la matiere, si l'on vouloit entrer dans tout le détail que cette idée demande; mais je suis fermement persuadé, que si on observe tout ce que je vois qu'il faut, & que l'on peut faire pour tirer le meilleur parti de l'usage de ces nouvelles rames, le travail des hommes pourra être plus utilement employé, que n'est celui de la chiourme sur une galere. On pourra placer facilement sur un grand vaisseau cinquante de ces rames, appliquer dix hommes à chacune, & donner vingt pieds quarrés à chaque pale, ce qui feroit en tout 1000 pieds quarrés; & si toute la manœuvre se faisoit en deux tems égaux, nous aurions 500 pieds quarrés qui agiroient sans interruption, pendant qu'il n'en faudroit que $337\frac{1}{2}$ pour se procurer autant d'avantage qu'on a en se servant de toutes les rames sur une galere. Il est vrai qu'ici le travail qu'il faut employer au point A, pour retirer chaque fois la pale hors de l'eau est en pure perte; mais quand il faudroit employer deux hommes pour cet effet, cette perte ne feroit que la cinquième partie du travail entier; & nous avons vu par raport aux galeres qu'on y perd jusqu'au $\frac{1}{4}$ du travail entier. La plus grande difficulté sera de placer les hommes, & de les mettre à même de tirer le point E, dans une direction horizontale, & perpendiculaire au plan du système; il me semble qu'on pourra remédier à cet inconvenient, s'il est nécessaire, moyenant des poulies, qu'on trouvera toujours où placer, & qui serviront à conduire les cordes là où l'on voudra placer les hommes destinés au travail. Cet expédient me paroîtroit assez convenable, s'il ne falloit pas craindre les frottemens; ceux qui penseront là-dessus comme moi, préféreront peut-être cet autre moyen: au lieu d'attacher au point E une corde, par le moyen de laquelle on tire fortement la pale contre les eaux, on pourroit se servir d'un autre levier E, M, N, dont le point d'appui seroit en M, & qui entreroit dans l'entre-deux des ponts. Les hommes agiroient sur le manche MN, à-peu-près de la même maniere que les rameurs le font sur les rames, & seroient mis par-là en état de pousser fortement les pales contre les eaux par le moyen d'un plat ME. Dans cette maniere de ramer, le mouvement du levier AC se feroit constamment dans un plan vertical, & celui du levier EN, dans un plan horizontal, & ces mouvemens seroient beaucoup plus commodes, que ne sont ceux d'une rame simple sur une galere. Il y auroit sur tout cela un grand nombre d'explications à donner, si d'abord les lumieres de mes juges, & puis celle des articles, au cas que mes idées eussent merité quelqu'attention favorable, ne m'en dispensoient.

XX. Voilà ce qui peut suffire sur la perfection de l'usage des rames, & sur la meilleure maniere de les appliquer aux grands vaisseaux; mais remarquons que nous n'avons encore considéré que l'action des rames en tant que les pales sont poussées directement contre les eaux, c'est-à-dire, dans une direction perpendiculaire à leur plan. Il y a cependant une infinité d'autres mouvemens, qu'on peut donner aux pales pour pousser les navires, & je suis surpris que personne ne se soit encore avisé d'en examiner le mécanisme; le mécanisme infiniment diversifié des différentes sortes de poissons dans leur action de nager, & même l'exemple des oiseaux, dans leur façon de battre des ailes pour voler, étoient bien propres pour exciter cette idée. J'ai donc cru, pour satisfaire entiérement à la question de l'Académie, devoir entrer dans cette nouvelle discussion. Elle nous fournira une maniere toute nouvelle de voguer, préférable, à mon avis, à tout autre, les rames pourront être appliquées très-commodement aux plus grands vaisseaux en aussi grand nombre qu'on voudra, sans que les rames s'embarrassent les unes les autres, comme cela arrive même sur les galeres, à moins que la mer ne soit calme.

Mais comme c'est ici un sujet tout nouveau, il nous faut nécessairement remonter à nos principes, sans lesquels il me paroît absolument impossible de traiter ces matieres avec aucune précision. Rien n'est communement plus vague que l'examen des machines qu'on propose pour de certains effets, on les réduit à rien quand on a formé l'intention de les réjetter, & on en grossit les effets à l'infini quand on veut faire valoir ses nouvelles idées. Mais suivant nos méthodes, on n'est plus maître de rien, on consulte la nature, & on est obligé de s'en tenir simplement & purement à sa décision. Je ne me propose point de forcer la nature à nous donner ce qui n'est pas en son pouvoir; voyons donc ce que nous pouvons nous promettre de l'action oblique des rames.

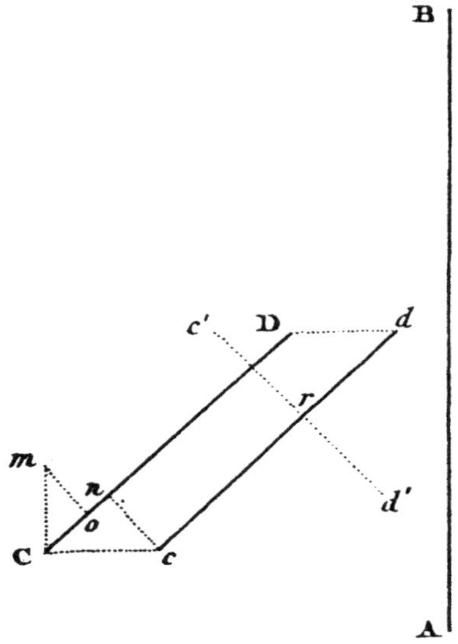

Fig. IV

XXI. Soit *AB* la quille d'un vaisseau, mu avec une vîtesse *C* de *A* vers *B* dans des eaux calmes; qu'on suppose en suite une pale verticalement plongée dans les eaux, mais obliquement à la quille, cette pale est représentée par la ligne *CD*; soit enfin cette pale poussée avec une vîtesse *v*, de *CD* en *cd*, sous une direction *Cc*, ou *Dd*, perpendiculaire au plan vertical de la quille, de maniere que la pale conserve constamment le parallelisme, il est clair que de cette façon la pale aura un double mouvement, l'un parallele à *AB* avec une vîtesse *C*, qui resulte du mouvement du vaisseau, & l'autre perpendiculaire au premier avec une vîtesse *v*; & il faut, avant toutes choses, examiner quels efforts la pale souffrira par la résistance des eaux en considérant ce double mouvement. De ces efforts connus, il faudra tirer l'effort utilement employé, pour entretenir le vaisseau dans son mouvement, & enfin le travail requis pour cette nouvelle maniere de voguer. Dans l'application de cette nouvelle théorie il faudra des deux côtés du navire un nombre égal de ces pales obliques, & je ferai voir que toutes ces pales obliques pourront agir sans aucune interruption; je réprésenterai la somme des pales tout d'un coup par la surface dont *CD* fait une section horizontale, & j'appellerai cette surface *S*, pendant que le plan qui mesure la résistance du vaisseau est exprimé par *n*.

XXII. Soit à présent la vîtesse v, avec laquelle on pousse la pale oblique perpendiculairement vers le plan vertical de la quille exprimée par Cc, & la vîtesse c, avec laquelle tout le sistême avance dans la direction parallele à AB, exprimée par Cm, soit le sinus total $= r$, le sinus de l'angle $nCc = s$, son cosinus $= \sqrt{rr - ss} = \sigma$, là-dessus il faut résoudre le mouvement Cc en Cn, qui est inutile, & $nc = \frac{s}{r}v$, de même le mouvement Cm doit être résolu en Co, qui est inutile, & $om = \frac{\sigma}{r}c$, par là on voit que la pale est poussée perpendiculairement contre les eaux en deux sens directement opposés, sçavoir de n vers c, & de o vers m; ici il faut voir lequel des deux mouvemens perpendiculaires au plan de la pale prenant sur l'autre, si nc est plus grand que om, la configuration de la surface de la pale du côté du point m seroit tout à fait indifférente, quand même cette surface ne seroit pas plane, & si le mouvement nc étoit plus petit que le mouvement om, alors la surface qui regarde le point c seroit indifférente; pour notre maniere de naviguer le mouvement nc doit toujours être plus grand que son mouvement contraire, & enfin $nc - om$ exprimera le mouvement perpendiculaire absolu de la pale contre les eaux; ce mouvement perpendiculaire absolu est donc $= \frac{s}{r}v - \frac{\sigma}{r}c$: ensuite le quarré de cette vîtesse, multiplié par la surface S, marquera l'effort perpendiculaire des eaux contre la pale, lequel par conséquent sera $= \left(\frac{s}{r}v - \frac{\sigma}{r}c\right)^2 S$. Enfin cet effort perpendiculaire doit pour notre dessein être encore résolu en deux pressions latérales, l'une perpendiculaire au plan vertical de la quille, qui sera $= \frac{s}{r} \times \left(\frac{s}{r}v - \frac{\sigma}{r}c\right)^2 S$, & l'autre parallele à la quille qui sera $= \frac{\sigma}{r} \times \left(\frac{s}{r}v - \frac{\sigma}{r}c\right)^2 S$. La premiere marque la force avec laquelle il faut pousser la pale de CD en cd, & si nous la nommons π, nous aurons $\pi = \frac{s}{r} \times \left(\frac{s}{r}v - \frac{\sigma}{r}c\right)^2 S$. La seconde marque la force avec laquelle tout le sistême est poussé de A vers B, & celle-ci doit être égale à la résistance des eaux contre la proue du vaisseau; si donc cette résistance est nommée P, on aura $P = \frac{\sigma}{r} \times \left(\frac{s}{r}v - \frac{\sigma}{r}c\right)^2 S$. Comme la résistance du vaisseau P, est égale au quarré de la vîtesse c multipliée par la surface n, nous aurons $ncc = P = \frac{\sigma}{r} \times \left(\frac{s}{r}v - \frac{\sigma}{r}c\right)^2 S$. De cette équation nous pourrons tirer la relation entre v & c puisqu'elle nous donne $v = \left[\frac{\sigma}{s} + \frac{r}{s}\sqrt{\left(\frac{n}{S} \times \frac{r}{\sigma}\right)}\right]c$.

Enfin remarquons aussi que $\pi : P = s : \sigma$, ou $\pi = \frac{s}{\sigma} P$, on pourra employer chacune de ces équations suivant les différentes questions qu'on se formera.

XXIII. Voici à présent les résultats, qui nous sont nécessaires.

I. Le travail essentiellement requis doit être exprimé par $P \times c$, puisqu'il s'agit d'enlever une résistance P avec la vîtesse c.

II. Le travail actuel par la même raison est $= \pi \times v$, puisqu'on exerce continuellement une pression π contre la pale, & que la pale est poussée avec la vîtesse v.

III. Ainsi le travail actuel est au travail essentiellement requis comme $\pi \times v$ est à Pc, ou comme $\frac{s}{\sigma} Pv$ à Pc, ou comme sv à σc, ou enfin, (si on substitue pour v sa valeur déterminée au précédent §) comme[30] $1 + \frac{r}{\sigma}\sqrt{\frac{n}{S} \times \frac{r}{\sigma}}$ est à 1.

C'est cette derniere analogie, qui faisoit l'objet principal de ces recherches.

XXIV. J'admire ici la conservation perpetuelle des travaux & de leurs effets, puisque le travail actuel devient parfaitement égal au travail essentiellement requis aussitôt qu'il n'y a point d'effet inutile. L'effet inutile est ici la force vive qui résulte du mouvement qu'on donne aux eaux par le coups de pale; mais si l'on suppose la surface S infinie, on rétranche entiérement cet effet inutile parce qu'alors on ne fait que presser la pale contre les eaux, sans donner à ces eaux aucun mouvement, & la vîtesse v devient simplement $= \frac{\sigma}{s} \times c$. Ceci est confirmé par notre derniere analogie, puisque faisant $\frac{n}{S} = 0$, elle nous donne le travail actuel au travail essentiellement requis comme 1 à 1; ce qui marque une parfaite égalité entre ces deux travaux. Cette égalité entre le travail actuel, & le travail essentiellement requis subsistera en ce cas toujours, quelque valeur qu'on donne à $\frac{r}{\sigma}$, c'est-à-dire, quelqu'inclinaison qu'on donne aux pales avec la quille du vaisseau. Mais aussitôt qu'on donne à la surface S une grandeur déterminée par celle de n, le travail actuel sera composé de deux parties, l'une utile & égale au travail essentiellement requis, l'autre inutile, & qui est en pure perte. Ces deux parties seront en raison de 1 à $\frac{r}{\sigma}\sqrt{\frac{n}{S} \times \frac{r}{\sigma}}$; & comme tout travail a son effet, avec cette simple différence, qu'il y a des effets inutiles par rapport à celui qu'on se propose, on n'a qu'à

30 The 1 was misprinted as $\int$ in the original.

diminuer autant qu'il est possible le travail inutile pour obtenir le plus grand effet possible.

Cette considération nous conduit à cette regle pour la surface S non infinie, qu'il faut rendre le cosinus de l'angle nCc, presqu'égal au sinus total r, c'est-à-dire, que la pale CD doit être dirigée presque perpendiculairement à la quille & n'être inclinée que très-peu; alors le travail utile sera au travail inutile presque comme 1 à $\sqrt{\frac{n}{S}}$. Ce qui fait entiérement le même rapport que celui que nous avons trouvé au §7. Part. prem. pour l'usage ordinaire des rames.

XXV. Nous voyons donc qu'en se servant de cette espece de rames, le travail inutile est par rapport au travail total sensiblement le même que dans l'usage des rames ordinaires, si on fait la surface S de part & d'autre la même, pourvu que le plan de la pale soit presque perpendiculaire à la quille du vaisseau; mais d'un autre côté nous verrons de plus en plus, que les rames obliques ont un avantage infini sur les rames ordinaires & directes, lorsque les rames doivent être appliquées à un grand vaisseau. Cependant comme il arrive souvent que les préceptes de la théorie, pour se mettre dans le plus grand avantage, sont sujets dans l'exécution à de certains inconveniens accidentels, j'examinerai les différentes circonstances que la diversité de l'angle formé par le plan de la pale avec la quille entraîne, je commencerai cet examen en supposant d'abord la surface S infinie.

(*a*) Plus cet angle est grand, plus il faudra donner de vîtesse à la pale, puisque $v = \frac{\sigma}{s}c$; cette vîtesse devroit être infinie si ledit angle étoit droit; pour un angle de 80^{d}, on auroit $v = 5.67c$; pour un angle de 70^{d}, on aura $v = 2.74c$, ou [pour un angle de 60^{d} on aura] $= 1.73c$. Si on suppose la vîtesse du vaisseau par exemple de quatre pieds par seconde, on aura pour les angles de 80, 70 & 60 degrés les vîtesses[31] v de 22.68; de 10.99; & 6.92 pieds par seconde.[32]

(*b*) Si on destine une seconde de tems à chaque excursion, les pales doivent parcourir un espace de 22.68 de 10.96 ou de 6.92 pieds, suivant que l'angle mCo est de 80, de 70, ou de 60 degrés; & si au lieu d'une seconde, on en prenoit une & demie, lesdites espaces deviendroient encore plus grandes en raison de 2 à 3.

31 The number 10.99 was printed as 10.96 in the original.
32 Mark this is only true if $S \gg n$.

(c) En supposant la vîtesse du vaisseau du premier rang de quatre pieds par seconde, on obtient $P = 2800$ livres, & comme (§22.) $\pi = \frac{S}{\sigma}P$; nous aurons pour les angles de 80, 70 & 60 degrés, pour π les valeurs de 494, de[33] 1019, & de 1618 livres. Ces poids marquent pour chacun des différents cas la pression totale, ou la somme de toutes les pressions que l'équipage doit exercer contre les pales; ainsi pour avoir la pression contre chaque pale, il faut diviser cesdits poids par le nombre des pales, en supposant que celles-ci agissent sans interruption, ce que nous verrons pouvoir se faire.

(d) Si la surface S n'est pas assez grande pour pouvoir être censée infinie, la vîtesse v, devient plus grande que $\frac{\sigma}{s}c$, le surcroit de vîtesse sera exprimé par $\frac{r}{s}\sqrt{\frac{n}{S} \times \frac{r}{\sigma}} \times c$. C'est ce surcroit de vîtesse, qui n'augmente aucunement la vîtesse du vaisseau, tant que les hommes agissent avec la même force sur les pales, & qui rend une partie du travail inutile; & comme les travaux, quand les pressions sont égales, sont en raison des vîtesses, il s'ensuit que le travail utile est au travail inutile comme $-\frac{\sigma}{s}c$ est à $\frac{r}{s}\sqrt{\frac{n}{S} \times \frac{r}{\sigma}} \times c$, c'est-à-dire, en raison composée réciproque de la raison sesquipliquée des cosinus de l'angle nCc, & de la raison sousdoublée de la surface S, comme nous avons déja dit dans le précédent article: supposons, par exemple, $S = 600$ pieds quarrés, ou $\frac{n}{S} = \frac{1}{4}$, en retenant $c = 4$ pieds par seconde, nous aurons en faisant successivement l'angle[34] mCn de 80, 70 & 60 degrés, la vîtesse nécessaire v égale à $22.68 + 11.60$ ou[35] $10.99 + 6.03$ ou $6.92 + 4.29$ ou bien[36] $v = 34.28$ ou 17.02 ou 11.21 pieds par seconde; si on employe une seconde de tems pour chaque saccade, chaque excursion des pales aura le même nombre de pieds; & comme il ne convient pas de faire ces excursions excessivement grandes, il faudra encore se relâcher un peu sur l'avantage qu'on obtient de la grandeur de l'angle de l'inclination des pales.

(e) Nous verrons, dans l'application de cette théorie, qu'il n'est pas difficile de donner à S la valeur de 600 pieds quarrés, & si mon dessein étoit de ne faire voir que le beau côté de ma nouvelle maniere de naviger, j'aurois pu la supposer plus grande sans tomber dans des inconveniens considérables,

33 Misprinted as 1022 in the original.
34 This is the angle β between the board and the line of the keel in Fig. 2, p. 60 h. v.
35 Misprinted as $10.96 + 6.01$ in the original.
36 17.02 was misprinted as 16.97 in the original.

mais j'aime mieux prendre les choses au pis; la même maxime m'engage à ne donner à l'angle *mCn*, que la valeur de 60^{d}, quoique j'eusse peut-être pu le supposer de 10^{d}, car l'étendue des excursions n'est pas un inconvenient fort essentiel, d'autant moins qu'on peut y employer aussi peu de tems qu'on le trouve à propos: le travail que je proposerai sera fort semblable à celui de scier du bois, dans lequel on fait facilement deux mouvemens dans une seconde de tems, & chaque mouvement de deux pieds. Or en supposant $S = 600$ pieds quarrés, & l'angle *mCn* de 60^{d}, nous avons vu dans la note (*d*), que le travail utile est au travail inutile comme[37] 10.96 est à 6.01 & par conséquent le travail utile au travail total comme 10.96 à 16.97. Donc dans ces suppositions le travail utile fait $\frac{1096}{1697}$, ou 0.645 du travail total[38], & il y a à remarquer, que c'est-là dans notre maniere de voguer presque l'unique perte qu'on fait, comme je le ferai voir dans la suite.

Après avoir bien pesé toutes les circonstances, je suis bien persuadé que le travail utile fera au moins 0.550 du travail entier, pendant que nous n'avons trouvé dans notre premiere Partie §25 que 0.446 pour les galeres voguant à toutes rames.

Il suit donc de là que notre nouvelle navigation est plus profitable, quoique appliquée aux grands vaisseaux, que n'est l'usage ordinaire des rames sur les galeres, & on a certainement tout lieu d'être content d'un tel résultat. Encore est il certain que le travail utile sur les galeres exprimé par 0.446, est pris tout au plus haut, car il est fondé sur la grandeur du plan de résistance de 16 pieds, pendant que M. Bouguer ne donne à ce plan que 10 pieds, & si nous avions suivi l'estime de M. Bouguer, nous n'aurions trouvé pour le travail utile sur les galeres, que 0.279, ce qui ne feroit qu'environ la moitié du travail utile qui provient de l'emploi de nos rames obliques.

(*f*) Si le travail utile est 0.550 du travail entier, le vaisseau prendra par le travail ordinaire de 500 hommes la vîtesse $5.55 \times \sqrt[3]{0.550}$ ou 4.54 & par un travail forcé 5.63. Ces vîtesses sont sans doute considérables puisqu'elles font plus, que les $\frac{4}{5}$ des vîtesses possibles entiéres; & comment pourroit-on ne pas s'en contenter, quand on fait réflexion que 500 hommes sur notre grand vaisseau ne font pas plus, que 53 hommes sur les galeres, qui tout au plus souffrent pour des vîtesses égales $\frac{16}{150}$ de la résistance du grand vaisseau, ce seroit donc n'être pas content, que 53 hommes donnassent par un travail

37 In fact, this refers to 70^{d}, cf. note 31.
38 This is the number for 70 degrees and not the one for 60 degrees.

ordinaire aux galeres une vîtesse à faire plus de quatre pieds & demi par seconde.[39]

XXVI. J'ai prevenu dans la note (*e*), du précédent article, que j'ai imaginé une maniere de rendre l'action des pales contre les eaux continuelles, cela demande une explication préalable. Nous avons vu, que pendant que la pale *CD* est poussée en *cd*, le vaisseau sera poussé suivant la direction *AB*; pour obtenir le même effet en repoussant la pale, & en l'éloignant des bords du vaisseau, il n'y a qu'à imaginer un mecanisme, qui fasse tout d'un coup un peu tourner la pale, & lui faire prendre la position $c'd'$, à la place de sa position précédente *cd*; je proposerai un moyen assez facile pour obtenir ce renversement précisement aux momens qu'on commence à pousser ou à repousser la pale. Dans ce mecanisme l'angle crc' doit être double de l'angle *nCc*, & par conséquent de 60^{d}, pour l'hypothèse de la note (*e*).

XXVII. Les rames, qui soutiennent les pales seront suspendues verticalement, la pale enfoncée dans l'eau, & le reste dans l'air, chaque rame aura au bout deux tourillons qui entreront dans des viroles, ces tourillons serviront à soutenir la rame, ils seront parfaitement horizonteaux, & paralleles à la quille, & ne permettront à la rame d'autre mouvement que celui de balancer dans un plan vertical perpendiculaire à la quille; les viroles destinées à recevoir les tourillons seront menagées dans des solives ou pièces de bois, qui sortent hors du vaisseau assez pour permettre aux rames leur jeu; le maniement consistera à approcher & éloigner alternativement les rames des bords du vaisseau moyennant une longue manivelle. Les hommes destinés à ce travail pourront être également placés sur le tillac, ou dans les entreponts, il y en aura plusieurs appliqués à la même manivelle, qu'ils tiendront horizontalement, & avec laquelle ils tireront & pousseront alternativement la rame; la distance depuis les tourillons jusqu'à la manivelle ne sera que le tiers, ou le quart de toute la longueur de la rame; de cette façon chaque allée & venue sera d'environ trois pieds, qui se feront dans une seconde de tems avec un effort d'environ vingt livres pour chaque homme, ce qui fait, à mon avis, le travail le plus naturel des hommes.

39 This is obtained assuming a 100% efficiency; if the customary efficiency is assumed one would obtain 146 men (see the Introduction p. 63 h. v.).

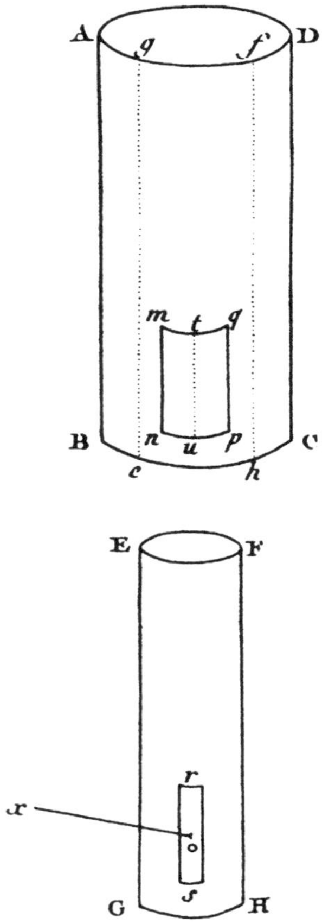

Fig. V

XXVIII. Il ne nous reste plus qu'à donner quelqu'idée de la façon de renverser à propos les pales, de façon que la surface qui pousse les eaux soit toujours tournée du côté de la pouppe. On pourra pour cet effet composer la rame de deux pieces, une partie de la piece inférieure entrera dans la supérieure, de maniere que celle d'enbas tourne librement sur son axe. Soit *ABCD* le bout de la piece supérieure creusée pour recevoir le bout de la piece inférieure *EFGH*, qu'on fasse dans la piece supérieure une rainure *mnpq*, & que l'amplitude de l'arc *mq* ou *np* soit d'un peu plus de 60^{d}, qu'on ajoute à la piece inférieure *EFGH* une espece d'arrêt formé en bosse représenté par *rs*, dont la hauteur *rs*, soit tant soit peu plus petite, que la hauteur de la rainure. Cet arrêt est destiné à toucher & heurter alternativement contre les bords *mn* & *pq* de la rainure, & il avancera un peu hors de la rainure, la manivelle qui doit servir à pousser & repousser la rame pourra être appliquée en *o* au millieu de l'arrêt *rs*, de maniere que la direction de la manivelle ne passe pas par l'axe de la rame, mais qu'elle la rase; soit la manivelle *ox*, & que les hommes la tirent vers eux, alors l'arrête *rs* viendra aussitôt à heurter le bord *mn*, & *y* restera comme collée pendant tout le tems qu'on tire la rame; mais aussitôt qu'on commence à la repousser, l'arrêt *rs* se mettra contre l'autre bord *qp*, & *y* restera pendant tout le tems qu'on employe à repousser la rame, & ce mouvement de l'arrêt *rs*, servira à renverser alternativement la pale. On voit que ce mecanisme demande que la verticale *tu*, qui partage également l'amplitude de *m*, ou *np* prolongée reponde précisement à l'axe des tourillons, par lesquels la rame est suspendue. Je n'ay exposé ce mecanisme que pour faire voir que je ne suppose rien, qui ne soit facile dans l'exécution; les artistes sçauront peut-être substituer à cet artifice un autre beaucoup plus convenable & je demande à ceux-ci de l'indulgence sur la maniere de m'être énoncé sur cet article. J'ai ajouté la planche seconde[40] pour expliquer mieux mes pensées.

XXIX. J'ai dit préalablement dans la note (*f*) du §25, que 500 hommes pourront donner au vaisseau par cette manœuvre une vîtesse à faire $4\frac{1}{2}$ pieds[41] par seconde; c'est-là sans doute un résultat, dont ont a tout lieu d'être content, mais que nous n'avons trouvé que par la méthode analytique; procédons à présent synthétiquement; quand notre grand vaisseau fait $4\frac{1}{2}$ pieds par seconde, il souffre une résistance continuelle de 3500 livres[42] s'il y a 50 rames, 25 de chaque côté, chaque pale doit surmonter une résistance de 70 livres; si la pale est inclinée[43] de 60^{d}, il faut pour avoir la force avec laquelle on attire ou repousse la pale, multiplier 70 livres par le rapport du sinus de 30^{d} à son cosinus, c'est-à-dire par $\frac{50\,000}{86\,602}$, ce qui fait à-peu-près 40.4 livres[44], si on fait $\dfrac{n}{S} = \frac{1}{4}$, c'est-à-dire si on fait la somme de toutes les

40 In his letter of June 8th, 1771 to Johann-Albrecht Euler, Bernoulli complains:

> *«...j'en ai trouvé toute l'impression extremement remplie de fautes; les formules irreconnaissables, les nombres alterés; la seconde table y manque; Mr de Lalande me marque qu'elle a été detournée &c. et tout cela après un delay de 18 ans.»*

There is therefore no second sheet of figures and the *Figs. 6, 7, 8* are non-existent.

41 Misprinted as $4\frac{1}{4}$ in the original.

42 By using formula $P = nc^2\frac{7}{6}$ and rounding off 3544.

43 This refers truly to 60^{d}.

44 Misprinted as 44 livres in the original.

pales = 600 pieds quarrés ou chaque pale de 12 pieds quarrés, l'équation $v = \left(\frac{\sigma}{s} + \frac{r}{s}\sqrt{\frac{n}{S} \times \frac{r}{\sigma}}\right)c$ du §22, fait $v = 12.33$; ainsi en employant une seconde pour chaque agitation, la pale fera chaque fois une excursion de 12.33 pieds. Si on fait la distance depuis l'axe des tourillons jusqu'à l'endroit, où la manivelle est appliquée à la rame égale[45] à $\frac{100}{411}$ parties de la distance du même axe des tourillons jusqu'au milieu des pales, la manivelle parcourra à chaque seconde un espace de trois pieds, & l'effort requis sera $= \frac{411}{100} \times 44$ livres = 180 livres; & comme il y aura dix hommes à chaque manivelle, le travail de chaque homme consistera à élever à chaque seconde un poids de 18 livres à la hauteur de trois pieds, & un tel travail ne fait encore que les $\frac{9}{10}$ du travail ordinaire d'un homme, j'ai reservé cette dixieme partie pour reparer quelques pertes de très-peu d'importance qu'on pourroit encore faire de ses forces, car pour les pertes essentielles, nous y avons déja eu égard en donnant à chaque pale des excursions de 12.33 pieds, pendant qu'elles seraient plus petites, si on donnoit plus d'étendue aux pales, ce qui diminueroit le travail requis.

XXX. Finissons cette matiere par un petit parallele entre la maniere ordinaire d'employer les rames sur les galeres, & ma nouvelle maniere de voguer, & par quelques autres réflexions. Le principal avantage de nos rames, & de la façon de les mettre en œuvre est sans contredit celui de pouvoir être appliquées fort commodement à toutes sortes de navires, depuis les plus petits jusqu'au plus grands. Il me semble que c'est-là tout ce que l'Académie demandoit; le second avantage est de pouvoir facilement employer à cette manœuvre autant d'hommes qu'on voudra, sans qu'ils s'embarrassent le moins du monde, puisqu'on pourra multiplier le nombre des manivelles tant qu'on voudra. Le troisieme avantage est que chaque homme peut-être mis dans une sorte de travail qui lui est le plus naturel, & qui demande que les hommes fassent des mouvemens d'environ trois pieds à chaque seconde; le vogue-avant sur les galeres fait un mouvement beaucoup trop grand, & le rameur le plus proche de l'apostis fait un mouvement beaucoup trop petit. Le quatrieme avantage est, qu'il n'y a ici aucune interruption dans le travail, pendant que sur une galere de trois mouvemens il n'y a qu'un seul d'utile; les hommes menagent donc l'inertie de leur propres corps. Le cinquieme avantage est que l'inertie des rames ne cause ici aucune perte ou très-peu, car les balancemens sont d'une nature à se conserver d'eux-mêmes.

45 Misprinted as $\frac{400}{411}$ in the original.

Cet article demanderoit cependant une grande discussion, si on vouloit le traiter avec autant d'exactitude qu'il peut l'être. Le sixieme avantage, qui est encore très grand, est qu'on peut-ici remédier beaucoup mieux que par l'usage des rames ordinaires à cet inconvient, si essentiel & si opiniâtre, qui consiste dans l'insuffisante étendue des pales; par la seule continuité de l'action, on triple tout d'un coup la surface des pales, & rien n'empêche, dans notre nouvelle navigation, d'aggrandir considérablement chaque pale. Tout ces avantages manifestes m'encouragent à proposer avec beaucoup de confiance l'usage des nouvelles rames dont je viens de donner la description, & ne me laissent absolument point douter de leur succès. Il ne sera pas difficile d'en faire des expériences préalables sur des bateaux, & pour peu que celles-ci soient favorables, le succès sera certain sur les grands vaisseaux. Mais tant d'avantages ne peuvent guere s'obtenir sans tomber en même tems dans quelques petits désavantages, que je suis bien éloigné de vouloir pallier. J'avoue donc qu'on fait une petite perte du travail par l'obliquité des pales qu'on est obligé de leur donner pour éviter les trop grandes vîtesses avec lesquelles il faudroit remuer les pales en les faisant trop perpendiculaires à la quille. J'avoue encore qu'on employe inutilement le petit travail qu'il faudra employer pour renverser les pales, en les faisant incliner tantôt d'un côté, tantôt de l'autre. On pourroit remédier à ce petit défaut, en donnant assez de mouvement aux pales pour sortir des eaux, & pour pouvoir les renverser dans l'air, mais une telle manœuvre rendroit déja le travail un peu interrompu, & feroit en même tems, que les pales ne seroient pas toujours entiérement plongées dans les eaux, ce qui diminueroit la vraie surface S. On se contentera de la précaution d'augmenter la surface des pales plutôt en hauteur qu'en largeur. Voilà tout ce que j'ai pu trouver de désavantage, & je suis bien persuadé, que ces deux petites pertes ne valent pas à beaucoup près celles qu'on fait par l'usage ordinaire des rames, lors-même qu'on est à portée de les employer avec toute la commodité qu'on peut désirer. Je vois même qu'on peut sans grand inconvenient rendre les pales plus perpendiculaires à la quille que nous ne les avons supposées dans notre évaluation, si on ne sçauroit donner assez de mouvement aux pales en les tirant vers les bords, rien n'empêche de les en éloigner tant qu'on voudra; j'aurois pu donc supposer les pales inclinées de 70^{d}, au lieu[46] de 60^{d}, & par-là j'aurois diminué d'environ la moitié, la petite perte qu'on fait à cet égard; mais j'ai mieux aimé m'en tenir au plus sûr que de pousser trop loin les hypothèses du plus avantageux. Je dirai encore deux mots sur la nature de la manœuvre; le jeu de nos rames n'est pas si arbitraire que celui des rames ordinaires, mais

46 See note 48 of the Introduction, p. 63 h. v.

on n'en pourra pas moins gouverner le vaisseau selon que les circonstances l'exigent par les différentes combinaisons que les rameurs peuvent faire de plusieurs façons, les uns en arrêtant les rames, les autres en tenant les pales hors de l'eau, & d'autres encore en redoublant de force. Au reste on voit bien que lorsqu'il s'agit simplement de faire du chemin, on doit être fort attentif, que les rames ne soient pas un seul moment en repos, tant que les pales sont sous l'eau; le vaisseau en seroit aussitôt rétardé; c'est à quoi il faut sur tout faire attention dans ces instans, ou il s'agit de faire rebrousser les rames, ce sont des momens, ou les hommes appliqués à la même manivelle doivent agir avec un grand concert. Je dirai à cet égard au hazard d'être traité de vetilleur, qu'il ne sera peut-être pas inutile de menager des obstacles des deux côtés, qui non seulement arrêtent tout d'un coup les rames, mais qui les rejettent même vers l'endroit d'où elles étoient venues; par un tel ou semblable artifice, on obtient en même tems un autre avantage, c'est celui de ne rien perdre absolument du travail qu'on fait par l'inertie des rames, pourvu que lesdits obstacles soient propres à faire rebondir les rames, c'est-là une réflexion qu'on ne doit pas négliger.

XXXI. Voilà sans doute la meilleure maniere de suppléer à l'action du vent sur les grands vaisseaux par le travail des hommes, & je me flatte que tous ceux qui seront entrés dans nos principes en conviendront avec moi. Nous avons indiqué une maniere d'obtenir les $\frac{4}{5}$ parties de toute la vîtesse possible, & cela d'une maniere qui n'a aucun inconvenient par rapport aux autres avantages qu'on doit tâcher de conserver. Les cinquante nouvelles rames avec tout leur appareil n'occuperont, pas plus d'espace que les rames sur une simple galere, elles seront ôtées & remises avec beaucoup de facilité & de promptitude.

L'application & la manœuvre de ces nouvelles rames ne demandent pas le moindre changement à la construction ordinaire des grands vaisseaux, pendant que la commodité des rameurs fait un des principaux points de vue pour la construction des galeres; si quelqu'un s'étoit attendu a une plus grande vîtesse, il auroit eu certainement grand tort, & il quittera bientôt ce préjugé en considérant seulement que le grand vaisseau que nous avons pris pour exemple souffre une résistance presque dix fois plus grande, que n'en a une galere, & même quinze fois plus grande, suivant l'estime de M. Bouguer: si l'on vouloit donc prétendre, qu'un équipage de 500 hommes donnât à notre grand vaisseau une plus grande vîtesse que celle que nous avons démontrée lui pouvoir être donnée, ce seroit prétendre que cinquante-trois rameurs, suivant notre estime, ou trente-trois, suivant l'estime de M. Bouguer, fissent parcourrir à une grande galere plus de $4\frac{1}{2}$ pieds par seconde, en n'employant qu'un travail ordinaire & soutenable pendant huit heures par jour, ou plus de $5\frac{1}{2}$ pieds avec un travail extrêmement forcé, car cette vîtesse peut aussi être imprimée au grand vaisseau avec un tel travail.

On ne peut donc plus s'empêcher de réconnoître notre nouvelle navigation pour la meilleure qu'on puisse raisonnablement esperer, à moins qu'on ne trouve à rédire à ce quinzieme de vîtesse qu'on y perdra encore sur la vîtesse totale possible; mais voici une réflexion, qui nous fera bientôt perdre toute espérance d'obtenir jamais un plus grand succès de quelque maniere qu'on s'y prenne; c'est qu'on aura toujours indispensablement besoin de quelque chose, qui tienne lieu de pale, puisqu'on n'a pas hors du vaisseau un point fixe qui puisse servir d'appui, & on ne pourra jamais donner une étendue infinie à ce qui est substitué aux pales; d'où il résultera constamment une perte de forces inévitable par sa nature. Outre cette perte on en fera nécessairement quelques autres, il y aura toujours quelque frottement, & quelques mouvemens inutiles, qu'il faille continuellement réproduire dans la matiere, qui sera substituée aux rames, sans parler des pertes qu'on fait lorsqu'on applique les hommes à un genre de travail peu conforme à leur constitution, par lequel ils sont souvent bien éloignés de produire tout cet effet, que nous leur avons supposé, & duquel ils sont très-capables, s'ils sont duement employés. Je ne crains donc plus de proposer notre derniere maniere de naviger par le travail des hommes, comme fort approchante de la meilleure qu'il soit possible d'imaginer, ce que j'ajouterai pourra faire voir avec combien d'attention j'ai examiné tout ce qui appartient à notre question, & peut-être en même tems utile pour plusieurs circonstances particulieres dans lesquelles on peut se trouver.

XXXII. Imaginons-nous d'abord un grand plan vertical flottant dans les eaux à une certaine distance depuis la proue perpendiculaire à la quille, & qu'on tire ce plan moyennant une corde d'une longueur suffisante vers la proue du vaisseau, il arrivera que le vaisseau, & ce plan s'approcheront l'un de l'autre; & faisant abstraction de l'inertie du vaisseau, la vîtesse de celui-ci sera à la vîtesse du plan, comme la racine du plan supposé est à la racine du plan, qui mesure la résistance du vaisseau. Ainsi la vîtesse du plan supposé sera encore, $= c\sqrt{\frac{n}{S}}$, si nous entendons, par c la vîtesse du vaisseau par S le plan supposé, & par n le plan qui mesure la résistance du vaisseau. Si on considéroit le plan S comme infini, sa vîtesse seroit comme o, & les hommes ne perdroient absolument rien de leur travail, que j'exprimerai par l'unité, mais si le plan S n'est pas considére comme infini, on aura pour le travail utile $\frac{c}{c+v}$, & pour le travail inutile $\frac{v}{c+v}$, ou bien $\frac{\sqrt{S}}{\sqrt{S}+\sqrt{n}}$ pour le premier, & $\frac{\sqrt{n}}{\sqrt{S}+\sqrt{n}}$ pour le second. Comme rien n'empêche ici de donner une très-grande étendue au plan S, & même de multiplier le nombre de ces plans à son plaisir, il sera facile d'anéantir presque entiérement la qua[nt]ité $\frac{\sqrt{n}}{\sqrt{S}+\sqrt{n}}$, qui marque la perte du travail, ce n'est pas-là la plus grande difficulté; mais ce travail presque entiérement utile ne durera que jusqu'au moment que le vaisseau aura atteint ce plan contre lequel on l'aura tiré. Il faudra donc de nouveau éloigner le plan du vaisseau; cela pourroit se faire moyennant une chaloupe, & il sera très-facile de faire que pendant ce transport le plan présente son tranchant aux eaux de maniere que la chaloupe puisse entraîner le plan presque avec autant de vîtesse que les rameurs peuvent donner à la chaloupe, en la supposant entiérement libre. On pourroit aussi composer ces plans de doubles battans, qui se replient l'un contre l'autre par la force des eaux pendant tout le tems qu'on employe à ce transport; dans cet intervalle il faudroit lâcher la corde qui va du vaisseau au plan jusqu'à ce que celui-ci soit suffisamment éloigné du vaisseau; après quoi on lâcheroit la corde qui va de la chaloupe au même plan, l'équipage du vaisseau tireroit l'autre corde, les deux battans se rouvriroient pour ne faire qu'un seul, & même plan, & le vaisseau s'avanceroit vers le plan. Par une telle ou autre semblable manœuvre, on feroit assez peu de perte du travail; il y auroit pourtant toujours celle qui résulte du défaut de la surface du plan, perte qu'on ne peut jamais éviter entiérement; outre cette perte on feroit encore celle du travail des hommes qui rameroient sur la chaloupe, comme ne faisant qu'un travail accessoire. Quant à la premiere perte, on a ici au moins autant de commodité pour la diminuer que par toute autre méthode, & la seconde perte peut être facilement negligée, à cause du peu de proportion qu'il y a entre les rameurs sur la chaloupe, & les hommes qui travaillent sur le vaisseau; mais le plus grand inconvenient consisteroit dans la perte du tems que chaque allée de la chaloupe demanderoit; cependant comme cet intervalle seroit en même tems un repos parfait pour l'équipage du vaisseau, cet équipage pourra redoubler ses efforts par un travail forcé pendant tout le tems de leur travail actuel; cette alternative de repos, & de travail forcé remettroit l'équipage dans le cas d'un travail ordinaire, & soutenable pendant plusieurs heures de suite, & l'effet d'un tel travail alternativement interrompu seroit le même aussi à-peu près que celui d'un travail ordinaire non interrompu, mais on pourra aussi prévenir cette interruption en employant deux chaloupes; dont chacune fasse alternativement avec l'autre la fonction que nous avons exposée.

XXXIII. Je me contenterai d'avoir indiqué cette nouvelle maniere de naviguer; je serois trop long, si je voulois examiner en détail tout ce qu'une telle idée demande pour l'exécution, & pour sa perfection, & après tout je ne pourrois dire que des choses qu'il sera très-facile de voir, surtout à ceux qui auroit bien voulu lire avec quelque attention tout ce que nous avons déja dit dans ce Mémoire; pour moi, après avoir pesé toutes les circonstances, je suis entiérement persuadé, qu'on pourra de cette maniere donner au vaisseau au moins les $\frac{3}{4}$ de la vîtesse possible totale, & on pourroit même esperer un plus grande succès, si au lieu des plans que nous avons supposés, on vouloit employer des corps creux conoïdiques de la moindre résistance, & fort longs dont la base seroit tournée du côté du vaisseau, & la pointe vers la chaloupe, on pourroit de cette maniere faire les intervalles du travail fort petits, les rameurs sur la chaloupe ne donneroient que quatre ou cinq coups des rames pour avancer ces corps, & puis l'équipage les retireroit avec beaucoup de vigueur vers le vaisseau; il seroit même facile de faire ensorte que ces corps coniques creux se fermassent toutes les fois qu'ils seroient tirés par la chaloupe, & s'ouvrissent lorsqu'ils seroient tirés par le vaisseau; rien n'empêcheroit d'attacher par une même corde une file de ces sortes de corps, lesquels offriroient une résistance extrêmement grande quand ils seroient tirés vers le vaisseau, & cette résistance seroit fort augmentée par l'inertie des eaux renfermées dans le creux de ces corps coniques pendant que la résistance seroit extrêmement petite durant le retour de ces corps. Enfin une longue expérience ne manqueroit pas de mettre cette maniere de naviguer dans une grande perfection, quand on auroit une fois commencé à s'en servir; j'y vois quelques inconveniens, mais je n'en vois point d'essentiels qu'on ne puisse espérer d'éviter après quelques essais qu'on auroit faits. Il est sûr du moins que dans les occasions, où l'on ne prétend que de simplement remorquer un vaisseau, on le fera avec beaucoup de facilité, & avec un succès incomparablement plus grand que par la maniere ordinaire de le faire. Je vois aussi qu'une semblable manœuvre fournira à l'equipage un moyen d'exercer une plus grande force quand il s'agiroit, par exemple, de remettre un navire à flot qui seroit échoué sur un banc de sable; il y a des cas où l'on ne se propose pas de donner au vaisseau une grande vîtesse, mais simplement de le remettre en mouvement, & dans ces cas on doit s'appliquer à pouvoir exercer une grande pression; quand on a des points fixes, on peut obtenir une pression aussi grande qu'on se propose avec une pression aussi petite qu'on veuille employer, mais en manquant de points fixes, comme sur les vaisseaux, c'est une question toute différente de notre sujet principal, quelle est la maniere de pousser le plus vigoureusement qu'il est possible un navire; l'action de pousser un corps en repos n'est qu'une force morte; mais sur un vaisseau on ne sauroit venir à bout d'exercer une telle action qu'avec une force vive, ou avec un certain travail qui demande une pression, & une vîtesse à la fois; parce qu'à la place de points fixes, on n'a que des points, qui cedent avec de certaines résistances: je trouve, par exemple, pour la galere, dont M. Chazelles à fait l'espérience, que la chiourme, par un travail ordinaire, ne peut enlever au de-là d'un poids de 1800, ou 2000 livres, c'est-à-dire, que si la galere étoit attachée par une corde, qui passât sur une poulie, & qu'à l'autre bout il y eût un poids de 1800 à 2000 livres suspendu, un tel poids arrêteroit la galere, & feroit équilibre avec les forces de la chiourme; l'effet de ce travail ne consisteroit donc qu'à tenir suspendu un poids de 1800 à 2000 livres, qui seroit produit par 260 hommes, de maniere que chaque homme n'exerceroit qu'une pression continuelle d'environ sept livres contre le vaisseau quoiqu'en repos, pendant que le même homme pourroit faire une pression continuelle infinie, si la pale & la longueur du manche pouvoient être censées infinies.

XXXIV. S'il est sûr que nous avons indiqué toutes les mesures qu'on peut prendre pour tirer le plus grand succès qui soit possible du travail des hommes destinés à suppléer à l'action du vent sur les grands vaisseaux, & s'il est vrai que le plus grand succès possible ne surpasse pas sensiblement le succès réel qu'on obtient par les manieres de voguer les grands vaisseaux, que nous avons exposées, j'ose me flatter d'avoir satisfait entiérement à la question proposée par l'Académie. On ne doit pas être moins persuadé dans la mécanique du plus grand avantage possible, que nous avons démontré, qu'on l'est dans la géometrie de renfermer le plus grand espace possible dans une péripherie circulaire; mais la généralité de nos principes est telle que nos réflexions sont encore vraies, quand on se serviroit, sur les vaisseaux de tout autre principe de mouvement; car de même que les hommes sont une source de forces vives, toutes les choses dont on pourroit se servir pour suppléer à l'action du vent renferment sous des apparences différentes une certaine quantité de forces vives; telle est l'action du feu, d'un air condensé, d'un air échauffé, celle des vapeurs, de la poudre à canon &c. Le vent lui-même est compris dans cette classe; pour tirer le parti le plus avantageux de tous ces principes, il n'y a qu'à éviter tous les effets inutiles autant qu'il est possible; alors toute la quantité de forces vives renfermées dans les choses naturelles dont on veut se servir pour suppléer à l'action du vent sera transmise dans les eaux poussées, & déplacées par la proue du vaisseau, ce qui fait seul tout l'effet utile, comme étant inséparable d'avec la navigation. Je finirai ce Mémoire par quelques remarques sur ce sujet.

XXXV. Supposons qu'on ait sur le vaisseau en sa puissance une certaine quantité de forces vives, qu'on veuille employer à faire aller un vaisseau sur mer, & examinons, quel sera le plus grand effet qu'on en pourra tirer. Soit cette quantité de forces vives exprimée par MA, c'est-à-dire, telle qu'une masse M acquiert en tombant librement de la hauteur verticale A: soit encore le plan d'une résistance directe égale à celle de la proue $= n$, la vîtesse du vaisseau $= c$, la hauteur verticale génératrice de cette vîtesse sera[47] $= \frac{cc}{60}$; là-dessus on aura la résistance du vaisseau égale au poids d'un prisme d'eau, dont la base est n, & la hauteur $\frac{cc}{60}$, & le poids contiendra autant de livres, qu'il y a d'unités en $\frac{70ncc}{60}$; je le fais[48] donc $= \frac{7}{6}ncc$; supposons enfin que la force vive MA se consume uniformement dans un tems t exprimé en secondes, cela fera l'espace parcouru uniformement par le vaisseau pendant ce tems ou $s = ct$, & la force vive, que le vaisseau aura communiquée aux eaux $= \frac{7}{6}ncc \times s$, ou $= \frac{7}{6}nc^3t$. Cette formule exprime l'effet que j'appelle utile, qui résulte de la somme des forces motrices contenues dans la force vive MA, & si toute cette force vive est utilement employée, je dis qu'on aura $\frac{7}{6}nc^3t = MA$; ce sera là le cas du plus grand avantage possible, & on ne pourra jamais obtenir un plus grand effet.

Voici à présent quelques proprietés que cette équation nous enseigne.

α Les vîtesses c sont en raison réciproque des racines cubiques des tems, pendant lesquels la force vive MA se consume: on obtiendra, par exemple, une vîtesse double en consumant ladite force vive dans la huitieme partie du tems.

β Puisque $s = ct$, on aura aussi $\frac{7}{6}nccs = MA$; ce qui marque que les vîtesses du vaisseau c sont en raison réciproque des racines des espaces que le vaisseau parcourt pendant que la force vive se consume. On pourra donc, par exemple, moyennant une certaine quantité de forces vives, conserver au vaisseau une vîtesse double, si on ne veut lui faire parcourrir que le $\frac{1}{4}$ de l'espace.

47 See Introduction p. 50 h. v., formula (7).
48 See Introduction p. 50 h. v., formula (9).

γ Comme on peut aussi faire $\frac{7}{6}n \times \frac{s^3}{tt} = MA$, on voit que les espaces parcourus par le vaisseau, pendant que la force vive se consume, sont comme les racines cubiques des quarrés des tems. Il est donc possible de faire parcourir à un vaisseau d'une grandeur quelconque un aussi grand espace qu'on voudra, moyennant une force vive aussi petite qu'on voudra employer; mais ce sera en ne consumant la force vive que dans un tems suffisant pour cela: cela suppose que les eaux n'ayent absolument aucune autre résistance que celle qui est proportionelle aux quarrés des vîtesses, & on remarquera aussi qu'il ne s'agit pas ici de donner au vaisseau un certain mouvement, mais de l'entretenir dans un mouvement déja acquis.

δ Enfin, comme la force motrice du vaisseau que je nomme P, est égale à la résistance exprimée par $\frac{7}{6}ncc$, on aura encore $Ps = MA$, & par conséquent P réciproquement proportionel à s; on pourra donc, moyennant une même force vive, obtenir une force motrice aussi grande qu'on voudra; mais le vaisseau ne parcourra qu'un petit espace pendant tout le tems que cette force motrice dure.

XXXVI. Pour voir la vérité de tout ce que je viens de dire, on n'a qu'à s'imaginer que le force vive MA consiste dans un long ressort bandé dont un bout soit appuyé contre la pouppe du vaisseau, pendant que l'autre bout s'appuye contre un poteau, ou un autre point fixe. Il est évident que par une telle maniere d'employer la force vive renfermée dans le ressort bandé, on en tire tout l'usage possible. Eclaircissons les regles par un exemple: supposons un pied cubique rempli de poudre à canon: il y a plusieurs expériences qui prouvent que la force vive renfermée dans une telle quantité de poudre est pour le moins = 200 000 000 livres élevées à la hauteur d'un pied[49]. Cet exemple fait donc $MA = 200\,000\,000$. Si on pouvoit employer cette force vive potentielle toute utilement pour faire aller un vaisseau du premier rang qui donne $n = 150$, qu'on voulut employer une telle quantité de poudre à chaque heure ou à chaque 3600 secondes, il faudroit substituer dans notre équation $\frac{7}{6}nc^3t = MA$, pour t, 3600 secondes pour n, 150 pieds quarrés, & pour MA, le nombre 200 000 000, après quoi on trouve-

49 This was discussed in the *Hydrodynamica* (*Werke* 5, Section X, Annex after §46, pp. 350–360) where experiments to measure the energy yield of a given quantity of powder were described. These in turn were copied from D. Bernoulli St. 14a, *Dissertatio de actione fluidorum* (CP II, 1727 (1729), pp. 304–342 – *Werke* 4) and had their origin in experiments conducted by general Günther in St. Petersburg in 1727 (see L. Euler, *Briefwechsel II* (mit Johann I Bernoulli ...), Basel: Birkhäuser, 1998, pp. 86–87, notes 3 and 4).

roit[50] $c = \sqrt[3]{\dfrac{6 \times 200\,000\,000}{7 \times 150 \times 3600}} = \sqrt[3]{317} = 6.8$, c'est-à-dire, qu'on pourroit, par une telle dépense en poudre à canon, entretenir un vaisseau du premier rang dans un mouvement à faire 6.8 pieds dans une seconde, & cet effet repond dans la Table fondamentale à celui de 900 hommes supposés pareillement d'employer utilement toutes leurs forces avec un travail naturel. Voici un autre exemple qui servira à montrer ce que l'on pourroit attendre des vapeurs & du feu[51]. J'ai lu dans une description de la grande machine hydraulique de Londres, qu'elle élevoit à chaque minute 80 pieds cubes d'eau à la hauteur de 124 pieds, comme c'est ici mesure d'Angleterre, je donnerai 58 livres à un pied cube d'eau, & je reduirai la hauteur de 124 pieds à celle de 116 pieds de Paris, ainsi nous aurons $MA = 80 \times 58 \times 116 = 538\,240$, $t = 60$, & par conséquent $c^3 = 51.261$, ou $c = 3.71$; c'est une vîtesse à faire environ 3 pieds $8\frac{1}{2}$ pouces par seconde. Je crois à la vérité qu'il seroit facile d'imaginer une machine pour obtenir réellement cet effet, car je trouve que dans la machine de Londres, on perd encore une très-grande partie de l'effet qu'on pourroit, tirer de l'action des vapeurs, mais je n'en suis pas moins persuadé que ces deux exemples calculés sur des principes infaillibles suffiront pour nous faire perdre toute espérance de pouvoir substituer sur les grands vaisseaux avec quelque succès considérable les forces motrices renfermées dans les choses naturelles aux travaux des hommes, à moins que ce ne soit dans des cas particuliers, qui ne demandent de semblables forces que pour peu de tems. Il y a après cela une autre réflexion à faire, c'est qu'il est difficile de faire un tel usage de ces forces motrices, qu'une très grande partie de leurs effets ne se perde inutilement. Le principe de l'effet utile consistera toujours dans la réaction d'une matiere poussée, & jettée en direction contraire à celle du vaisseau; cette matiere ne sçauroit être que de l'eau, & la force vive qu'on imprimera à ces eaux sera toujours un effet perdu; cette

50 Misprinted as $c = \sqrt{\dfrac{6\,,\,200\,000\,000}{7\,,\,150\,,\,3600}} = \sqrt{317}$ in the original.

51 The engine Bernoulli has read about is a Newcomen type engine which operated at the York Buildings (a site now covered by Charing Cross station) from 1726 to 1731; it drove a pump to raise the water of the Thames to a reservoir from whence it was distributed to the town. It was extensively described a.o. by the German author Weidler (*Io. Friderici Weidleri Tractatus de Machinis Hydraulicis toto terrarum orbe maximis Marlyensi et Londinensi...* (2nd edition), Wittemberg 1733) who is mentioned by Bernoulli in the *Hydrodynamica*; Weidler cites original sources (in english, with translation in latin) and displays detailed engravings of the machinery. Contemporary sources claimed that it lifted 3 tuns (i.e. 630 imp. gal.) of water a minute over 102 feet, according to H.W. Dickinson (in: *Watersupply of Greater London*, Leamington Spa and London: Newcomen Society at the Courier Press, 1954). The numbers adopted by Bernoulli are somewhat different, but the power is in both cases about 9000 pound feet/sec (French units).

force vive perdue sera cependant d'autant plus petite, qu'on poussera, ou qu'on jettera une plus grande masse d'eau en tems égaux, & si cette masse étoit infinie, on ne perdroit rien à cet égard. Le recul d'un canon toujours chargé de la même quantité de poudre deviendroit deux fois plus grand, si on le chargeoit de quatre boulets au lieu d'un, & si on pouvoit faire un boulet d'un poids infini, toute la force de la poudre seroit *à cet égard*, employée utilement pour faire reculer le canon; je dis *à cet égard*, parce qu'on ne laisseroit pas de faire encore une autre perte très-considérable, c'est qu'aussitôt que le boulet seroit sorti du canon, il n'y auroit plus de réaction, & le reste de la force de la poudre enflammée se perdroit encore inutilement, & ce reste seroit toujours très-grand.

XXXVII. Les réflexions que je viens de faire suffisent pour voir en quoi consistent les pertes qu'on fait des forces motrices en question, elles sont toutes comprises (car je ne parlerai pas de quelques pertes entiérement accidentelles, telles que celles qui proviennent du frottement) dans le mouvement qu'on donne aux corps mobiles, qui servent en quelque façon d'appui, & dans la grande quantité des forces motrices qu'on laisse inutilement échapper, & cette derniere perte fait le plus grand défaut de toutes les machines à feu qu'on a encore imaginées. La nature de ces deux pertes nous enseigne d'abord tout ce qu'on peut faire pour nous approcher du plus grande effet qu'on peut tirer de semblables forces motrices; mais je vois assez que cet effet ne sauroit jamais être assez grand pour mériter beaucoup d'attention. Je conclus donc que si nous avons donné la meilleure maniere d'employer le travail des hommes pour suppléer à l'action du vent, nous aurons en même tems montré de toutes les manieres praticables, qui soient possibles, celle dont on tirera le plus de profit.

PRINCIPES HYDROSTATIQUES

ET MÉCHANIQUES,

Sur la queſtion propoſée pour la ſeconde fois par l'Académie Royale des Sciences :

Quelle eſt la meilleure maniere de diminuer le Roulis & le Tangage d'un Navire, ſans qu'il perde ſenſiblement par cette diminution aucune des bonnes qualités que ſa conſtruction doit lui donner.

Piece qui a remporté le prix de l'Académie Royale des Sciences en 1757.

Par M. DANIEL BERNOULLI, Aſſocié Etranger de l'Académie Royale des Sciences, Membre des Académies de Berlin, de Petersbourg, de Bologne, &c.

Qui dubiis auſus committere flatibus alnum
Quas natura negat, præbuit arte vias.

St. 48

Principes Hydrostatiques et Méchaniques, sur la question proposée pour la seconde fois par l'Académie Royale des Sciences:
Quelle est la meilleure maniere de diminuer le Roulis et le Tangage d'un Navire, sans qu'il perde sensiblement par cette diminution aucune des bonnes qualités que sa construction doit lui donner

Prix Paris VIII (pièce IV) 1757 (1771), pp. 2–96

Qui dubiis ausus committere flatibus alnum
Quas natura negat, praebuit arte vias[1]

§. I. Quoique l'Académie Royale des Sciences ait adjugé le prix de l'année 1755 à M. Chauchot[2], très-savant & très-habile Constructeur de Vaisseaux, elle n'a pas laissé de proposer une seconde fois la même question pour le Prix de 1757. Ce sujet étoit, *La maniere de diminuer le plus qu'il est possible le Roulis & le Tangage d'un navire, sans qu'il perde sensiblement, par cette diminution, aucune des bonnes qualités que sa construction doit lui donner*.

L'Académie donne en même tems à connoître la raison qui l'a engagée à ce nouveau procédé: elle a voulu couronner l'excellence & la beauté de la piece de M. Chauchot; mais en même tems, elle a eu égard à l'importance d'un sujet qu'elle juge *susceptible de recherches plus profondes*. Ces recherches plus profondes ne peuvent consister sans doute que dans un examen exact & précis de tout ce qu'une bonne Méchanique & Hydrostatique peuvent nous dicter sur ce sujet.

§. II. De toutes les recherches qu'on peut faire sur notre question, la plus essentielle est celle qui concerne la stabilité des navires. C'est aussi ce qui a engagé M. Bouguer, cet illustre Géometre, & l'homme du monde le plus entendu dans tout ce qui regarde la navigation & la construction des vaisseaux, à traiter cette question avec une scrupuleuse attention & toute la sa-

1 Bernoulli took the Latin phrase from the book by Bouguer who used it as *motto* at the head of his first chapter. It means: «Who dared commit a ship to the uncertain winds, he supplied by Art the ways Nature denies.»

2 See the Introduction and note 3, p. 73 h. v.

gacité possible, dans son *Traité du Navire, de sa construction & de ses mouvemens*[3]. M. Chauchot, à son exemple, n'a pas manqué de citer là-dessus quelques formules, & d'y ajoûter quelques réflexions, mais sans cette précision & ce détail que l'importance de la question exige. Je ne saurois donc me dispenser de reprendre tout de nouveau cette matiere, mais toujours dans la vue d'appliquer mes recherches à notre sujet principal; & comme depuis très longtems j'ai tâché de me la rendre familiere, j'ose me flatter de n'en rien dire dans ce Mémoire de superflu ou d'étranger à la question de l'Académie.

CHAPITRE PREMIER.

Sur la stabilité des corps flottans.

§. III. Tout corps flottant dans des eaux calmes, affectera une certaine position déterminée; il n'y a que les corps sphériques, dont le centre de gravité se trouve placé au centre même des corps, qui soient absolument indifférens à toute position[4]. La moindre force peut faire quitter à ces corps leur premiere position, & ils ne la reprendront point, quoique la force cesse d'agir. Dans les corps cylindriques ou sphéroïdiques, qui auroient leur centre de gravité placé dans l'axe, il n'y auroit que l'axe même qui prendroit toujours une certaine position déterminée; mais rien ne détermineroit leur position autour de leur axe. Tous les autres corps flottans prennent une certaine position déterminée en tous sens[5]. Il est vrai que la moindre force leur fait encore quitter cette position; mais la force cessant d'agir, les corps reprennent toujours leur position d'équilibre. C'est cet effort de se rapprocher de sa position d'équilibre qui forme la stabilité des corps flottans, comparable en tout à celle d'un corps arrondi posé sur un plan horizontal[6], qui auroit le centre

3 See §1.3 Visibility and impact of the memoir (p. 74 h. v.) and note 12, p. 75 h. v.

4 A floating homogeneous sphere can indeed take any orientation and still be in equilibrium. Bernoulli states the inverse proposition, viz. that a floating body in equilibrium in any position must be a sphere; this is unproven to this day, as appears from S. M. Ulam, *A collection of mathematical problems*, New York 1960 and from E. N. Gilbert, *How things float*: Am. Math. Monthly 98, (Nr 3) p. 201, 1991.

5 Taken literally this is not true: a cube e.g. has several orientations around which it shows stable equilibrium. The point of the remark is of course that any such position is an isolated point or line which has its own basin of attraction.

6 Appell describes a method (due to Dupin, P. V. Acad. Sci. Paris of August 30th, 1814) to map a problem of floating equilibrium into a problem of a rounded object resting on a horizontal plane which works out this idea of Bernoulli: P. Appell, *Traité de mécanique rationelle III*, Paris 1903, Ch. XXXI, p. 215.

de gravité au-dessous du centre de la courbure pour le point d'attouchement & pour le plan vertical, dans lequel on rouleroit le corps. Ordinairement la force réquise pour éloigner les corps plongés, de leur position naturelle, est d'autant plus grande que les corps s'en éloignent davantage; cependant cette augmentation de force ne va que jusqu'à un certain degré. Cette force dépend encore de plusieurs autres circonstances, ce qui fait que les corps plongés ont plus ou moins de stabilité, quoique rapportée au même plan & à une inclinaison égale. Je me propose donc, avant toutes choses, d'examiner les circonstances qui déterminent la stabilité des corps plongés; & comme les navires sont fort sujets à s'éloigner extrêmement de leur position droite, je pousserai à cet égard mes recherches plus loin qu'on ne fait ordinairement; je considererai des inclinaisons finies quelconques. Il faut considérer d'abord deux points dans les corps plongés; l'un est le centre de gravité du corps, & l'autre, le centre de gravité de sa partie submergée, considérée comme homogene[7]; on regardera ensuite la pesanteur de tout le corps comme concentrée dans le premier de ces deux points; de-là il résulte une force égale au poids du corps appliquée à son centre de gravité, agissant verticalement de haut en bas. Dans l'autre point on supposera une force égale qui agit verticalement de bas en haut; cette seconde force provient de la poussée de l'eau qui soutient le corps. Lorsque le corps plongé se trouve dans sa position naturelle, les deux dits centres de gravité se trouveront nécessairement dans une même ligne verticale; mais supposons le corps, par des forces purement horizontales, arrêté dans une autre position, il faudra alors, par les deux dits centres de gravité, tirer des verticales, & la distance entre ces verticales marquera le levier, sur l'extrêmité duquel le poids du corps agit & fait effort pour remettre le corps dans sa position naturelle. Je passe légerement sur ce principe comme clair par lui-même. Si on nomme donc P le poids du corps & r la distance entre les deux dites verticales, le produit Pr marquera le *momentum* de la force, qui tend à rapprocher le corps plongé de sa position naturelle, & qui fera son effet aussitôt que les forces horizontales étrangères cesseront d'agir. Ainsi notre question se trouve réduite à celle de déterminer la quantité r pour chaque position du corps, dont le volume submergé est supposé constamment égal. Je remarquerai cependant ici, que cette supposition ne trouve plus lieu aussitôt que le corps est abandonné à lui-même

7 We use, in the notes and comments, the term «centre of buoyancy» for this point. For *moment of the force* we shall substitute freely the term «torque» in the comments for the sake of clarity, although it is anachronistic. Naval architects call it the righting moment.

& qu'il commence à balancer, & que ce corps tirera alternativement plus & moins d'eau[8].

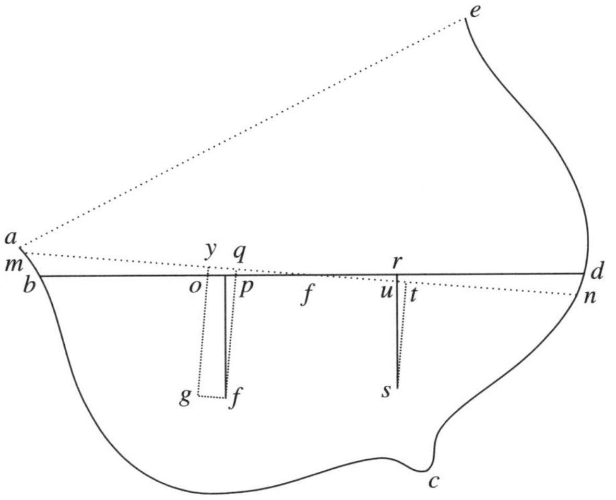

Fig. 1 (redrawn after the original)

§. IV. Commençons par une simple tranche uniformement épaisse, plongée verticalement dans l'eau, & perpendiculaire à l'axe de rotation[9]. Supposons la tranche d'une figure[10] quelconque, telle que *ace*, qui peut représenter la plus grande section verticale d'un navire perpendiculaire à sa longueur. Supposons aussi que dans la position naturelle de la tranche la ligne *ae* soit horizontale, & que de cette position naturelle la tranche soit parvenue dans la position oblique *ace*: soit alors *bd* la ligne d'eau, & *bcd* la partie submergée. De cette maniere l'angle entre les prolongées *ea* & *db* sera l'angle de rotation, que j'appellerai σ, & que je mesurerai par un arc de cercle qui a

8 The combined movements (rocking and bobbing) of a floating body were studied in D. Bernoulli St. 32, *De motibus oscillatoriis corporum humido insidentium*: CP XI, 1739 (1750), pp. 100–115. Those are «normal modes» and hence, in the limit of small deviations from equilibrium, they can be treated as independent.

9 The problem of a thin slab, constrained to stay vertically e.g. by two enclosing vertical walls, is especially simple. Such a slab can only rotate and bob up and down (sideways movements can be excluded by a choice of initial conditions as no horizontal force is present). At equilibrium the vertical position is fixed by Archimedes' law and the only remaining movement is an angular one. In his first publication on the subject St. 30, *De statu aequilibrii corporum humido insidentium*: CP X, 1738 (1747), pp. 147–163, D. Bernoulli started also by considering a slab.

10 In *Fig. 1* the intersection of *mn* and *bd* should be labeled x, to agree with the text.

l'unité pour rayon. Il est question de trouver une équation entre la différentielle $d\sigma$ & l'incrément de force correspondant. Soit donc pour la position infiniment proche de la précédente la ligne d'eau mn, & comme la partie submergée est supposée constamment égale, il faut que le triangle élémentaire[11] bxm soit égal au triangle opposé dxn, ce qui fait $bx = dx$; le petit angle bxm ou dxn sera exprimé par $d\sigma$. Soit le centre de gravité de la tranche ace au point s, les centres de gravité de la partie submergée homogene pour les lignes d'eau bd & mn aux points f & g. Si après cela on tire des points s & f, les lignes sr & fo perpendiculaires à la ligne d'eau bd, l'interceptée ro sera, en vertu du §. III, la longueur du levier sur lequel le poids de la tranche fait effort pour rapprocher cette tranche de sa position naturelle relativement à la ligne d'eau bd; & si pareillement on tire des points s & g, les lignes st & gy perpendiculaires à la ligne d'eau mn, l'interceptée ty sera dans le même sens le levier relativement à la ligne d'eau mn; ainsi la différence entre ty & ro, marquera l'incrément du levier pendant que le poids appliqué à ces leviers reste constamment le même. Il s'agit donc de déterminer la différentielle $ty - ro$: or tirant encore la ligne fo perpendiculaire à bd, on aura $ty - ro = yq - op + ut$; & comme la petite ligne gf est nécessairement parallèle au niveau d'eau mn, on aura $yq = gf$, & par conséquent $ty - ro = gf - op + ut$: nous allons maintenant déterminer analytiquement ces trois élémens. Soit $bd = q$, la surface $bcd = M$, on trouvera par les premiers principes de la Statique $gf = \frac{q^3 d\sigma}{12M}$. Considérant ensuite l'égalité des angles bxm, ofp & ust, & dénotant la verticale fo par y, la verticale sr par z, on aura $op = y\,d\sigma$ & $ut = z\,d\sigma$, & enfin la quantité cherchée

$$gf - op + ut = \Big(\frac{q^3}{12M} - y + z\Big)d\sigma\ .$$

Si on indique par s la hauteur verticale du centre de gravité de la tranche par-dessus le centre de gravité de la partie submergée homogene, ladite quantité devient $\Big(\frac{q^3}{12M} - s\Big)d\sigma$. Voilà pour chaque rotation élémentaire $d\sigma$ l'accroissement du levier; & si on multiplie cette quantité par le poids de la tranche, on aura l'accroissement du *momentum* de la force. Soit donc l'épaisseur de la tranche $= \alpha$ & la quantité $M\alpha$ exprimera la quantité d'eau

11 In the arguments presented in §4 Bernoulli neglects, without saying so, all terms of order $d\sigma^2$; his statement $bx = dx$ e.g. is only strictly valid if the ship would have straight parallel sides; gf is not parallel to the level line mn but to the median of the triangle mxb.

déplacée, au poids de laquelle le poids de la tranche est égal. Ainsi nous exprimerons ce poids par $M\alpha$, & de cette maniere nous aurons ledit accroissement du *momentum* de la force

$$= \left(\frac{q^3}{12} - Ms\right)\alpha \, d\sigma \ ,$$

& le *momentum* total sera exprimé par l'intégrale de ladite quantité ou par $\int \left(\frac{1}{12}q^3 - Ms\right)\alpha \, d\sigma$.

§. V. La solution que nous venons de donner de notre problème, s'étend à tous les corps prismatiques, l'épaisseur de la tranche indiquée par α pouvant être quelconque; mais s'il étoit question d'autres corps, il faudroit prendre des abscisses x dans une ligne perpendiculaire au plan *ace*, ensuite substituer dx pour α, & faire une seconde intégration, puisqu'il est clair que le *momentum* total sera exprimé par

$$\int dx \int \left(\frac{1}{12}q^3 - Ms\right) d\sigma \ .$$

Cette derniere expression est à la vérité générale & comprend absolument tout ce qu'on peut desirer sur cette matiere; mais comme il n'y a aucune loi de continuité, qui puisse exprimer la valeur de s pour chaque tranche d'un navire, verticale & parallele à la tranche principale, la formule devient tout-à-fait inutile pour cet usage, & il faudra avoir recours à d'autres expédiens; mais avant que d'entrer dans ces discussions, nous examinerons d'abord les corps prismatiques, & il est indifférent que toutes les tranches de ces corps soient uniformément chargées de matiere ou non, pourvu que ces tranches demeurent dans des plans verticaux.

§. VI. Pour trouver la valeur de

$$\int \left(\frac{1}{12}q^3 - Ms\right)\alpha \, d\sigma \ ,$$

il faut connoître pour chaque angle σ les quantités q & s, & il seroit extrêmement difficile de determiner une relation entre ces quantités, si on ne vouloit demeurer dans des suppositions bien simples touchant les configurations des sections *ace*.

Je supposerai premierement la figure *ace* former un grand segment de cercle: c'est la supposition la plus simple, & en même tems celle qui approche le plus de la section principale d'un navire. Dans ce cas la quantité q devient constante; mais pour trouver la quantité s, considérons la tranche dans sa position naturelle, & supposons alors la hauteur du centre de la figure par dessus le centre de gravité $= f$, & puis la hauteur de ce dernier centre par dessus le centre de gravité la partie submergée $= g$: si après cela la tranche a pris une autre situation quelconque, on voit facilement qu'on aura toujours la hauteur verticale du centre par-dessus le centre de gravité $= f \cos\sigma$, pendant que la hauteur du centre par-dessus le centre de gravité de la partie submergée est constamment $f + g$; d'où l'on tire

$$s = f + g - f \cos\sigma \ ;$$

ainsi nous aurons

$$\int \Big(\frac{1}{12}q^3 - Ms\Big)\alpha\, d\sigma =$$
$$\int \Big(\frac{1}{12}q^3 - Mf - Mg + Mf \cos\sigma\Big)\alpha\, d\sigma =$$
$$\Big(\frac{1}{12}q^3 - Mf - Mg\Big)\alpha\sigma + Mf\alpha \sin\sigma \ ;$$

mais la nature du cercle est telle, que la quantité

$$\frac{1}{12}q^3 - Mf - Mg$$

est toujours $= o$; donc toute la quantité qui exprime le *momentum* de la force devient simplement $= Mf\alpha \sin\sigma$. Cette formule nous montre que ladite force est proportionnelle au sinus de l'angle d'inclinaison. Elle augmente donc continuellement jusqu'à une inclinaison de 90 degrés; mais les augmentations deviennent continuellement plus petites. Cette remarque mérite quelque attention, parce que les roulis d'un navire ne sauroient manquer d'être à-peu-près pareils à ceux qu'auroient les corps que nous considérons ici. Or les vaisseaux sont souvent fort inclinés par un vent de côté permanent, & alors c'est cette position, quoi qu'oblique, qui doit être regardée comme naturelle; en ce sens on peut dire que la stabilité devient d'autant plus petite, que la position naturelle est plus oblique. Il me semble cependant qu'il faudroit tâcher de donner d'autant plus de stabilité aux navires, qu'ils sont plus inclinés par quelque cause uniforme, parce qu'alors un nouveau degré d'inclinaison peut être dangereux. Il suit aussi de-là que les balancemens d'un navire considérablement incliné dans sa position moyenne, seront plus lents que ceux d'un navire dont la position moyenne est droite: ces derniers balancemens differeront encore entre eux en ce qu'ils seront d'autant plus lents qu'ils seront plus grands. Notre expression marque aussi que la stabilité de ces corps est proportionnelle à la quantité f, c'est-à dire, à la plus grande hauteur du centre par dessus le centre de gravité, sans que la hauteur du centre de gravité de la partie submergée y entre pour rien pour ces corps figurés en segmens de cylindres. Enfin, la stabilité est encore proportionnelle à la quantité d'eau déplacée par le corps: le corps aura donc d'autant plus de stabilité, qu'il tirera plus d'eau, pourvu que son centre de gravité ne change pas. En considérant donc la figure *ace* comme un grand segment de cercle creux, s'il est question de quelle maniere il faut le charger de matiere pour lui donner la plus grande stabilité, je répons qu'il faut tâcher de faire la quantité Mf la plus grande qu'il est possible. Si donc par exemple la hauteur du centre par dessus le centre de gravité étoit de cinq piés, toute la charge qu'on ajoûtera au-dessus de ces cinq piés diminuera la stabilité, & l'augmentera si c'est au-dessous. De-là on peut conclurre qu'en pompant les eaux d'un navire, on diminuera le plus souvent sa stabilité. Mais d'un autre côté le mouvement des eaux qui coulent du même côté que le navire s'incline, diminue pareillement la stabilité.

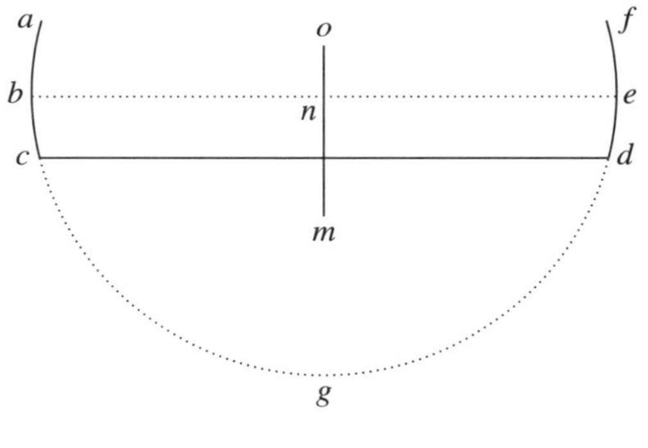

Fig. 2 (redrawn after the original)

§. VII. La matiere qui nous occupe est si essentielle à notre sujet principal, qu'elle mérite toute notre attention; examinons donc encore l'effet de quelques autres figures. Voici une observation fort importante: soit[12] *abcdef*

12 The point o is meant to be the centre of the circle through *acgdf*. The point n marks only the water level on the vertical axis.

(Fig. 2) la section du corps prismatique plongée dans l'eau jusqu'en *be*; supposons que les bords *ac* & *fd* soient des arcs d'un même cercle, & que jamais les inclinaisons ne soient assez grandes pour que l'une des extrémités *c* ou *d* sorte hors de l'eau: je dis que l'on pourra facilement déterminer la stabilité, quelle que soit la figure de la base *cd*, en joignant les deux bords circulaires par l'arc intermédiaire *cgd*, de maniere que tout l'arc *acgdf* soit un arc circulaire. Comme l'eau n'exerce aucun effort dans l'eau, il est clair que pourvu que le plan *cgdc* soit d'eau, le plan proposé *acdf* aura la même stabilité qu'auroit tout le plan *acgdf*, si la partie supérieure *acdf* restant chargée comme elle l'est, la partie inférieure *cgd* étoit toute d'eau. Soit donc le centre commun des arcs *abc*, *cgd* & *daf* au point *o*, le centre de gravité de tout le plan *acgdf*, chargé comme nous l'avons dit, au point *m*; ces deux points *o* & *m* sont dans une même ligne verticale pour la position naturelle; qu'on suppose $om = \varphi$ & l'espace $bcgde = \mu$; on aura, en vertu du §. VI, la stabilité[13] pour la figure plongée *acdf* égale à la quantité $\mu\varphi\alpha \sin\sigma$.

Cette formule nous fait voir que pour une même quantité submergée, la stabilité est ici beaucoup plus grande qu'elle n'est dans les hypotheses du précédent article, le rapport des deux stabilités étant celui de $\mu\varphi$ à Mf; or puisque M n'est qu'une partie de μ, de même que f de φ, la quantité Mf doit être beaucoup plus petite que la quantité $\mu\varphi$, lorsque les arcs *bc* ou *cd* sont petits, & par conséquent la largeur *bc* assez grande pour que l'espace *bcde* soit $= M$. La quantité $\mu\varphi$ croit en raison des cubes de *be*, ou en raison réciproque des cubes de *bc*, ou du tirant d'eau. Ainsi de deux navires qui ont la solidité de la carene égale, celui qui aura le moindre creux aura la plus grande stabilité. Je ne m'arrêterai pas à tirer de notre formule tous les corollaires qu'elle offre; remarquons seulement que la force réquise pour incliner ces corps prismatiques, dont la section est représentée par *acdf*, en supposant *cd* droite ou courbe: que dis-je, cette force est encore proportionnelle au sinus d'inclinaison ou à $\sin\sigma$.

De tout ce que nous venons de dire, il s'en suit que le creux d'un navire ne sauroit augmenter ou diminuer fort sensiblement sa stabilité. Il me paroît important d'examiner l'effet que le plus ou moins de creux doit faire dans un navire; voici cet effet.

Supposons la base *cd* s'éloigner de la ligne d'eau *be*, ce qui augmente le creux: pour savoir si cette augmentation du creux augmente ou diminue la stabilité, il n'y a qu'à examiner si le surcroît du creux survenu est rempli d'une matiere plus ou moins pesante que l'eau: dans le premier cas, la stabilité est augmentée; & dans l'autre, diminuée. Mais un autre effet plus

13 $T = \mu\varphi\alpha \sin\sigma$ is obviously the torque and not the stability.

sensible consistera dans la retardation ou l'accélération des balancemens: plus le tirant d'eau est grand, plus les balancemens sont retardés; & dans les balancemens, il ne faut pas seulement avoir égard aux excursions, on doit encore considérer la durée de chaque balancement. Il faut également tâcher de diminuer & de retarder les balancemens des navires lorsque la chose est possible.

§. VIII. La maniere singuliere dont je me suis servi dans le précédent article pour déterminer la stabilité dans le cas que les deux bords *ac* & *fd* sont des arcs d'un même cercle, quelle que soit au reste la figure du fonds *cd*, & l'expression $\mu\varphi\alpha \sin\sigma$ m'engagent à faire ici encore une remarque qui marquera en même tems la justesse de nos méthodes & répandra quelque nouveau jour sur nos formules. Voici cette remarque.

Soit la hauteur *bc* ou *ed* de la partie submergée *bcdeb* très-petite[14]; alors le centre de gravité de tout le systême que nous avons supposé au point *m*, sera le même que le centre de gravité du segment homogene *bcgde*; en prenant donc le point *m* en ce sens, la Geométrie & la Méchanique nous apprennent que la quantité $\mu\varphi$ (c'est-à-dire l'espace *bcgde* multiplié par la distance du point *m* au centre *o*) est toujours égale[15] à $\frac{1}{12}\overline{be}^3$. Cette propriété convient admirablement bien avec la formule générale marquée à la fin du §. IV

$$\int\left(\frac{1}{12}q^3 - Ms\right)\alpha\, d\sigma\ ,$$

puisque dans ce présent cas q est constant & $s = o$, & que cette intégrale devient $= \frac{1}{12}q^3\alpha\sigma$; mais à cause de la petitesse de *bc*, les inclinaisons doivent toujours être très-petites; on peut donc censer $\sigma = \sin\sigma$. Ainsi $\frac{1}{12}q^3\alpha\sigma = \mu\varphi\alpha \sin\sigma$.

§. IX. Après avoir déterminé généralement la stabilité, pourvu que les bords *ac* & *fd* soient circulaires, je mettrai encore ici le calcul en supposant les bords *ac* & *fd* droits & verticaux. La différence des résultats ne sauroit manquer de répandre quelques nouvelles lumieres sur cette matiere; mais quoique nous ayons donné à la fin du §. IV la formule générale pour déterminer la stabilité, il faut pourtant avouer que chaque nouvelle figure de-

14 See the Introduction, p. 88 h. v.

15 Obviously $q = \overline{be}$; the formula was printed as $\frac{1}{12}$ *cub. be*.

mande un nouveau calcul; la méthode cependant est toujours la même; je ne ferai que mettre ici le résultat pour le nouvel exemple[16].

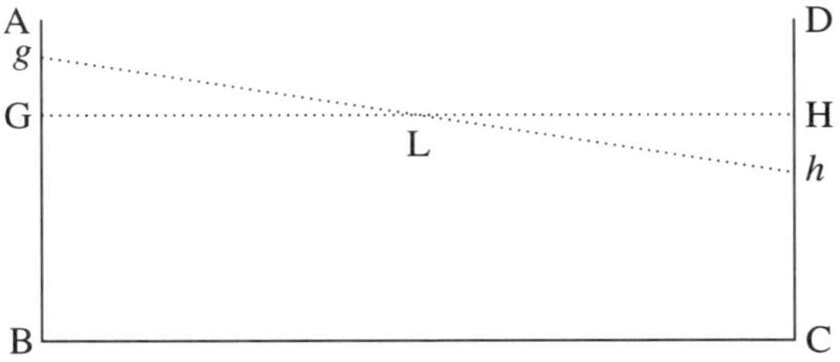

Fig. 3 (redrawn after the original)

Soit donc *ABCD* (Fig. 3) la section verticale du prisme flottant; que dans la position naturelle du parallelepipede, *AB* & *DC* soient verticales, *BC* horizontales & *GH* la ligne d'eau; soit ensuite $GH = 2a$, GB ou $HC = c$, la hauteur du centre de gravité du navire par dessus le centre de gravité de la partie submergée $= d$. Qu'on suppose à présent la section *ABCD* inclinée de maniere que *gh* soit la ligne d'eau; l'angle *GLg* sera l'angle d'inclinaison σ; mais dans ce cas il vaut mieux de déterminer l'obliquité de la position par *Gg*, & nous supposerons $Gg = x$. Sur ces dénominations, on trouvera[17] le *momentum*[18] =

$$\left(\frac{aa}{3c} - d + \frac{xx}{2c}\right) \times \frac{x}{\sqrt{aa+xx}} \times M\alpha \ .$$

Afin d'approcher la formule de celles des §§. VI & VII, nous remarquerons que $\frac{x}{\sqrt{aa+xx}}$ est $= \sin\sigma$. En substituant cette valeur, on pourra donner à la force de stabilité, cette expression

$$M \times \left(\frac{aa}{3c} - d + \frac{xx}{2c}\right)\alpha \sin\sigma \ .$$

§. X. Comme cette formule n'a plus rien qui puisse demander quelque explication, je ne m'arrêterai pas à marquer ce qu'il convient de faire pour rendre

16 See the Introduction, p. 85 h. v.
17 See p. 86 for the original formula (12).
18 Error, the third term $\frac{xx}{2c}$ should be, here and in the sequel, $\frac{xx}{6c}$.

ces parallelepipedes flottans plus stables. Je ne ne m'attacherai qu'à une seule circonstance, de laquelle j'ai fait mention au milieu du §. VI; c'est qu'ici la stabilité n'est plus proportionnelle au sinus de l'inclinaison, comme lorsque les bords sont des arcs circulaires d'un même cercle; elle augmente en plus grande raison que ledit sinus. Plus le corps est incliné ici, plus il résiste pour être incliné davantage; les augmentations de force pour produire toujours un nouveau degré d'inclinaison, croissent ici en beaucoup plus grande raison que dans la précédente hypothese. On obtiendra facilement ici que le corps ne s'incline jamais au-delà d'un certain degré, bien loin de courrir risque de renverser; & cette seule qualité rend l'hypothese du parallelepipede infiniment recommandable. Si on faisoit $d = \frac{aa}{3c}$, on auroit le point que M. Bouguer, mon illustre guide, appelle *métacentre*; si $a = 25$ & $c = 10$, la hauteur du métacentre devient $= 20\frac{5}{6}$ comme M. Bouguer le détermine pour l'arche de Noë. Mais qu'arriveroit-il alors? L'arche renverseroit-elle? Voilà ce que l'on pourroit croire, quand on ne calcule la stabilité que pour des inclinaisons infiniment petites; mais nous voyons tout le contraire par notre méthode. La stabilité seroit exprimée par[19] $M\frac{xx}{2c}\alpha\sin\sigma$. Elle seroit nulle en faisant $x = o$ & $\sigma = o$; mais elle commence à devenir réelle avec la moindre inclinaison.

Je mettrai ici pour les deux hypotheses, sçavoir, pour celle des bords circulaires & des bords droits & verticaux, les stabilités pour des inclinaisons égales, en supposant de part & d'autre la même quantité M, comme aussi la même stabilité initielle; c'est-à-dire, en faisant $\mu = M$ & $\varphi = \frac{aa}{3c} - d$. De cette maniere, les stabilités pour les deux hypotheses seront en raison[20] de

$$\left(\frac{aa}{3c} - d\right)\sin\sigma\ ,$$

&

$$\left(\frac{aa}{3c} - d + \frac{xx}{2c}\right)\sin\sigma\ .$$

Pour ne pas nous éloigner de la nature, nous ferons $c = a$, & $d = \frac{1}{4}a$; & comme nous voulons prendre a pour le sinus total, nous ferons $a = 1000$ & $d = 250$; un tel exemple donnera les stabilités ou les forces ré-

19 Error, should be $\frac{Mxx}{6c}\alpha\sin\sigma$; cf. note 18.

20 See the Introduction p. 87 h. v.

quises pour produire de certaines inclinaisons en raison de $\frac{250}{3}\sin\sigma$ à $\left(\frac{250}{3}+\frac{xx}{2000}\right)\sin\sigma$, en entendant par x la tangente de l'inclinaison pour le sinus total 1000. Voici à présent une petite Table de 5 en 5 degrés jusqu'à l'inclinaison de 45 degrés. La premiere colonne marque les inclinaisons, la seconde les forces pour les bords circulaires, & la troisieme pour les bords droits & verticaux[21].

Inclinaisons.	Forces pour les bords circulaires.	Forces pour les bords droits & verticaux.
0 degrés.	0.00	0.00
1	1.45	1.45
2	2.90	2.92
3	4.35	4.42
4	5.81	5.98
5	7.26	7.59
10	14.47	17.17
15	21.57	30.86
20	28.50	51.15
25	35.22	81.17
30	41.33	124.66
35	47.80	188.40
40	53.56	274.70 †
45	58.93	412.48

Nous voyons par cette Table combien il y a à gagner pour la stabilité en approchant la coupe principale verticale & perpendiculaire à la longueur, de la figure rectangulaire; mais surtout près de la surface d'eau: car les parties qui restent toujours submergées, ne sauroient changer la stabilité qu'en tant qu'elles changent la hauteur du centre de gravité, par-dessus le centre de gravité de la partie submergée, ce qui le plus souvent n'est pas de grande importance.

21 The number marked † is misprinted and should be 279.86. This is the original Table where the third column is erroneous; the correct Table is in the Introduction p. 87 h. v.

§. XI. Venons à présent aux corps plongés, & ne nous arrêtons plus à une seule section de ces corps. Lorsque toutes les sections paralleles de ces corps sont les mêmes & égales en tous sens, la solution pour ces corps prismatiques, ou cylindriques, ne demande rien de nouveau; il n'y a qu'à entendre par α la longueur de ces corps, & nos formules serviront encore pour ces corps; il n'est pas même nécessaire que le centre de gravité de chaque coupe soit pareillement placé, ni que le poids soit égal pour toutes les tranches également épaisses; car le poids entier du corps pourra toujours être considéré comme concentré dans une seule & même coupe, & le centre de gravité du corps comme placé dans la même coupe à une pareille hauteur. Il est vrai que les navires different considérablement de ces corps; cependant on voit déja que la principale coupe en largeur d'un navire reglera presque entierement la stabilité d'un bord à l'autre. Cette stabilité pourra faire comme les deux tiers ou les trois quarts de la stabilité qu'auroit le corps cylindrique de la même longueur: la section principale en longueur du navire reglera pareillement sa stabilité dans le sens de la proue à la pouppe. Je ne laisserai pas d'examiner cette matiere avec plus de précision dans le dessein de chercher toutes les circonstances qui peuvent contribuer à la stabilité d'un navire, tant à l'égard des roulis que des tangages. Mais pour ne nous point engager dans des difficultés inutiles, je ne considererai que des corps semblables en quelque façon aux navires; & dans ces corps je ne ferai attention qu'aux inclinaisons qui se sont faites, soit en largeur, soit en longueur, sans entrer dans les recherches sur les inclinaisons intermédiaires, qui ne font qu'un mouvement composé des deux autres mouvemens.

§. XII. Ceux qui auront lu avec quelque attention nos remarques sur cette matiere, verront d'abord que la stabilité de tous les corps flottans dépend uniquement de deux choses: la premiere est la section horizontale du corps faite au niveau des eaux, & la seconde est la hauteur du centre de gravité du corps par-dessus le centre de gravité de la partie submergée, considérée comme homogene, laquelle hauteur doit être multipliée par tout le poids du corps plongé, ou par tout le poids de l'eau déplacée. Examinons chacun de ces deux points à part.

Nous avons vu, au §. V, que la stabilité des corps est

$$= \int dx \int \left(\frac{1}{12}q^3 - Ms\right) d\sigma = \int dx \int \frac{1}{12}q^3 d\sigma - \int dx \int Ms\, d\sigma \,.$$

Quant au premier membre, on remarquera qu'il peut arriver que les mêmes coupes pourront tirer dans les différentes inclinaisons tantôt plus, tantôt moins d'eau, & en ce cas la quantité

$$\int dx \int \frac{1}{12} q^3 d\sigma$$

ne seroit plus exactement vraie, puisqu'elle est fondée sur ce que chaque tranche déplace continuellement la même quantité d'eau. Il peut cependant arriver qu'un navire différemment incliné tire plus ou moins d'eau par le milieu, & qu'il arrive le contraire vers les extrêmités. En ce cas, il faudroit une autre solution de notre problême; il faudroit considérer la figure de la section du corps au niveau de l'eau. Dans cette figure on tireroit une ligne parallele à l'axe de la rotation élémentaire. Cette ligne doit partager ladite section de maniere que les deux coins solides formés par la rotation soient égaux. Voici, après cela, comme il faudroit procéder. Soit la surface plane du coin d'un coté $= S$, & de l'autre $= S'$, la distance du centre de gravité de cette surface jusqu'à la ligne qui partage la section du corps faite au niveau de l'eau, d'un côté $= q$, & de l'autre $= q'$; l'angle élémentaire de rotation $= d\sigma$. On sait que le solide[22] du coin sera d'un côté $= Sq\,d\sigma$ & de l'autre $= S'q'\,d\sigma$, & ces deux quantités doivent être égales. Soit après cela la distance du centre de gravité du coin à ladite ligne qui partage la section entiere $= p$, d'un côté, & de l'autre $= p'$; je dis qu'on aura la stabilité du corps, en tant qu'elle est formée par la section du corps avec le niveau de l'eau

$$= \int Spq\,d\sigma + \int S'p'q'\,d\sigma = \int (p + p')Sq\,d\sigma\ .$$

Dans cette expression, on suppose toutes[23] les quantités p, p', S & q déterminées par l'inclinaison σ, & elle a cet avantage qu'elle ne demande qu'une seule intégration. On voit assez que la quantité $\int Spq\,d\sigma$ marque le *momentum* du coin d'eau élémentaire qui survient d'un côté, & $\int S'p'q'\,d\sigma$ celui du coin opposé & retranché du contrepoids. Quant à l'autre membre de la formule, qui exprime la stabilité d'un corps plongé quelconque, on peut, au lieu de considérer le double centre de gravité de chaque coupe, examiner

22 *Solide* means volume. The argument relies on the theorem of Guldin giving the volume of a solid obtained by rotation of a plane figure.

23 This argument, with the wedges dipping in or out, was first developed in St. 30, *De statu aequilibrii corporum humido insidentium*, for infinitesimal angles. Those same expressions are also found in P. Bouguer, *Traité du navire, de sa construction et de ses mouvements*, Livre II, Paris: Jombert, 1746, Section II, Ch. III, art. III. Here Bernoulli expands the argument and makes it valid for finite angles. In the subsequent §13 he reverts to infinitesimal angles and retrieves his old formula.

tout d'un coup la place du centre de gravité de tout le corps, comme aussi celle du centre de gravité de la partie submergée homogene. Si on nomme ensuite encore s la hauteur verticale du premier par-dessus le second, & P tout le poids du corps plongé, ou de l'eau déplacée, on aura $\int Ps\, d\sigma$ pour l'expression de l'autre membre qu'il faut retrancher du premier membre, ou l'y ajoûter, suivant que le centre de gravité du corps est plus haut ou plus bas que le centre de gravité de la partie submergée.

La relation entre les deux membres de la formule qui exprime la stabilité des corps est telle, que dans les corps qui tirent peu d'eau & qui forment une grande section au niveau de l'eau, le premier membre l'emporte ordinairement beaucoup sur l'autre; & que c'est tout le contraire lorsque cette section est très-petite.

§. XIII. Toutes les réflexions que je viens de faire seroient trop abstraites, si elles n'étoient éclaircies par quelques exemples applicables à notre sujet: il convient donc de descendre à quelques exemples. Celui qui répand le plus de jour sur cette matiere, est de supposer l'inclinaison assez petite pour que les quantités finies n'en reçoivent aucun changement considérable.

Dans cette restriction nous pourrons nous servir de la formule du §. V,

$$\int dx \int \left(\frac{1}{12}q^3 - Ms\right) d\sigma ,$$

qui devient

$$= \int \frac{1}{12}q^3\sigma\, dx - \int Ms\sigma\, dx .$$

Ici l'élément dx marque l'épaisseur de chaque tranche, q la largeur variable, M la surface plane de chaque tranche variable, & s la hauteur verticale du centre de gravité de la tranche par dessus le centre de gravité de sa partie submergée, & enfin σ la petite inclinaison commune à toutes les tranches. Quant au dernier membre, on sait par la théorie sur le centre de gravité, qu'il est égal à $P\, d\sigma$, en entendant par P, le poids de tout le corps, que nous mesurons par la quantité d'eau déplacée, & par d la hauteur du centre de gravité de tout le corps par-dessus le centre de gravité de la partie submergée; mais le premier membre $\int \frac{1}{12}q^3\sigma\, dx$ ou bien $\frac{\sigma}{12}\int q^3 dx$ dépend uniquement de la figure de la section horizontale du corps au niveau de l'eau. Si donc cette section étoit un parallelogramme rectangle, dont la longueur fût égale à b, on auroit simplement $\int \frac{1}{12}q^3\sigma\, dx = \frac{1}{12}q^3 b\sigma$. Mais comme dans les vaisseaux ladite section approche beaucoup d'une ellipse, il est bon de faire le calcul de cette intégrale, en supposant la section une ellipse dont le petit axe soit $= a$, & le grand axe $= b$. Si on prend les abscisses x depuis le centre de l'ellipse sur le grand axe, on aura $q = \frac{2a}{b}\sqrt{(\frac{1}{4}bb - xx)}$, & $q^3 = \frac{8a^3}{b^3}(\frac{1}{4}bb - xx)^{\frac{3}{2}}$, & par conséquent

$$\frac{\sigma}{12}\int q^3 dx = \frac{\sigma}{12} \times \frac{8a^3}{b^3} \times \int \left(\frac{1}{4}bb - xx\right)^{\frac{3}{2}} dx\ .$$

Les regles connues du calcul intégral donnent cette quantité[24]:

$$= \frac{a^3\sigma}{6b^3}\left[4x\left(\frac{1}{4}bb - xx\right)^{\frac{3}{2}} + \frac{3}{4}bb\int dx\left(\frac{1}{4}bb - xx\right)^{\frac{1}{2}}\right].$$

En faisant $x = \frac{1}{2}b$, ladite quantité devient

$$= \frac{a^3\sigma}{8b}\int dx\sqrt{\left(\frac{1}{4}bb - xx\right)}\ ;$$

dont le facteur sommatoire exprime le quart de cercle qui a la ligne $\frac{1}{2}b$ pour rayon. Nous substituerons donc à ce facteur[25] $\frac{11}{14} \cdot \frac{1}{4}bb$ ou bien $\frac{11}{56}bb$, & par-là nous aurons $\frac{11}{448}a^3 b\sigma$, qui fait la quantité cherchée pour la demi-section elliptique; prenant donc le double, nous aurons la quantité entiere $= \frac{11}{224}a^3 b\sigma$.

Comparons à présent cette quantité avec celle qui convient à la figure rectangulaire, savoir avec $\frac{1}{12}q^3 b\sigma$, ou (en donnant la même largeur aux deux figures) avec $\frac{1}{12}a^3 b\sigma$; le rapport sera comme $\frac{11}{224}$ à $\frac{1}{12}$ ou comme 132 à 224, ou environ[26] comme 3 à 5. Nous voyons par cet exemple, que la simple figure de la section du vaisseau faite par la surface de l'eau, peut contribuer considérablement à la stabilité, sans qu'on ait changé la longueur, la largeur ni la profondeur, ni même la solidité, ou le volume de la partie submergée.

Pour avoir une idée plus exacte de la stabilité entiere d'un navire, dont la section à fleur d'eau est supposée elliptique, nous évaluerons aussi le second membre $P\,d\sigma$; P marque la solidité de la carene, elle est à-peu-près

24 The factor $4x$ is misprinted in the original as x.

25 The integral is equal to $\frac{1}{4}\pi(\frac{b}{2})^2$ and Bernoulli substitutes $\pi = \frac{22}{7}$, as he usually does in this memoir.

26 The ratio is $\pi\frac{3}{16} = 0.5890\ldots$

égale à la moitié (ou plus exactement à $\frac{11}{21}$, lorsque la carene est formée par la demi-révolution d'une ellipse sur son axe) du parallelepipede circonscrit; si nous nommons donc c la profondeur de la carene, nous aurons sa solidité[27] P à-peu-près $= \frac{1}{2}abc$. Quant à la hauteur du centre de gravité du vaisseau par-dessus celui de la carene, que nous avons nommée d, il est difficile de la déterminer avec beaucoup de précision; le centre[28] de gravité de la carene sera environ $\frac{1}{3}c$ au-dessous de la surface d'eau, ou un peu davantage; le centre de gravité de tout le navire sera peut-être de $\frac{1}{5}$ ou de $\frac{1}{6}$ de la profondeur c au-dessous de l'eau, suivant le lest, la charge, leur pesanteur spécifique, & leur arrangement; enfin je supposerai $d = \frac{1}{6}c$. Dans cette supposition nous aurons $P\,d\sigma = \frac{1}{12}abcc\sigma$, & la stabilité entiere du navire, en supposant sa section à fleur d'eau elliptique, sera $= \left(\frac{11}{224}a^3b - \frac{1}{12}abcc\right)\sigma$.

Si la plus grande longueur de la carene est supposée de 140 piés, sa plus grande largeur à fleur d'eau de 40 piés, & la profondeur de 18 piés, la stabilité du vaisseau pour les petites inclinaisons d'un bord à l'autre sera environ $= (440\,000 - 151\,200)\sigma$. Ainsi le premier membre est presque trois fois plus grand que le second, & de-là on voit que la stabilité dépend beaucoup plus de la section du vaisseau à fleur d'eau que de tout le reste. La stabilité entiere sera donc en ce cas $= 288\,800\sigma$. Dans les grandes inclinaisons moyennes les navires s'inclinent de côté d'environ 15 degrés; ce qui donne σ environ $= \frac{1}{4}$. Si on considere cette inclinaison assez petite pour induire peu d'erreur dans nos formules, la stabilité sera = 72 200; c'est-à-dire que le *momentum* de la force pour tenir le vaisseau incliné de 15 degrés, sera égal au poids de 72 200 piés cubes d'eau, agissant sur un levier d'un pié de longueur. Mais si nous supposons un levier égal à la demi-largeur du vaisseau, le poids agissant sur ce levier, sera celui de 3610 piés cubes d'eau, ou de 129 tonneaux, en prenant 28 piés cubes d'eau pour le poids d'un tonneau; ce qui fait environ la quatorzieme partie de tout le poids du vaisseau. Si la section à fleur d'eau étoit supposée faire un rectangle, la stabilité absolue en deviendroit double; car nous avons vu que le premier membre de la formule, qui exprime la stabilité, en seroit augmenté en raison de 3 à 5, pendant que le second membre n'en est presque point changé; & ce second membre, faisant environ le tiers du premier, les stabilités pour la section elliptique & pour la section rectangulaire, seroient comme 3 – 1 à 5 – 1, ou comme 2 à 4.

27 The exact value is $P = \frac{\pi}{6}abc$.

28 If the hull, up to the waterline, were indeed a half ellipsoid the centre of buoyancy would be $\frac{3}{8}c$ under the water level.

§. XIV. Nous venons de donner, dans le précédent article, des mesures absolues pour les stabilités relatives aux roulis; considérons en deux mots celles qui regardent les tangages. Il n'y a, pour cet effet, qu'à convertir les quantités a & b pour le premier membre des formules; car le second membre restera le même pour le roulis & pour le tangage. Cette seule considération nous marque que le premier membre augmente en raison quarrée de la largeur du vaisseau à sa longueur, qui est ordinairement comme 2 à 7; de sorte que le premier membre augmente en raison de 4 à 49: & comme le second membre, pour la section du vaisseau à fleur d'eau elliptique, n'est qu'environ le tiers du premier membre à l'égard du roulis, il n'en sera plus qu'environ la trente-septieme partie à l'égard des tangages. De-là nous voyons qu'on peut, pour les tangages, négliger absolument le second membre; & que la stabilité du vaisseau pour les inclinaisons de la proue à la pouppe, dépend presque uniquement de la section du vaisseau à fleur d'eau. Si donc cette section est une ellipse, la stabilité de la proue à la pouppe sera simplement exprimée, pour les petites inclinaisons[29], par $\frac{11}{224}ab^3\sigma$; & si la section étoit un rectangle, elle seroit exprimée par $\frac{55}{672}ab^3\sigma$. La formule pour la section elliptique donne pour l'exemple du vaisseau rapporté dans le précédent article, en faisant $P = \frac{1}{2}abc$, la stabilité telle que s'il y avoit attaché à la proue ou à la pouppe un poids qui fût égal[30] à $\frac{55}{36}\sigma P$. Il faudroit donc appliquer à la proue un poids qui fût environ la huitieme partie du poids de tout le vaisseau, pour ne l'incliner que de 5 degrés. Je trouve que le vent le plus impétueux ne pourra jamais faire la dixieme partie de cet effet, ni par conséquent incliner le vaisseau de la pouppe à la proue au-delà d'un $\frac{1}{2}$ degré.

§. XV. J'ajoûterai encore quelques théorêmes sur les corps plongés qui seroient formés par la rotation d'une figure quelconque autour d'un axe horizontal. Toutes les sections perpendiculaires à l'axe de rotation, sont alors des arcs de cercle; & si l'inclinaison d'un tel corps plongé se fait autour d'un axe parallele audit axe de rotation, chaque arc de cercle demeurera constamment le même, quelle que soit l'inclinaison du corps. Voici maintenant une propriété remarquable sur notre sujet.

Si on multiplie un segment de cercle quelconque par la distance du centre de gravité du segment au centre du cercle, le produit sera toujours égal à la douzieme partie du cube de la corde du segment. Si nous nommons donc q

29 The exact value is $\frac{1}{12}ab^3c$; see Introduction p. 90 h. v.

30 See the Introduction p. 90 h. v., formula (14). In the original this was misprinted as $\frac{556}{36}P$.

la corde du segment submergé, nous voyons que l'expression $\int \frac{1}{12}q^3dx$ (*voyez* §. V), n'est autre chose pour ces corps que le volume constant de la partie submergée de tout le corps multiplié par la distance du centre de gravité du même volume à l'axe de rotation de la figure génératrice. Soit donc le poids de tout le corps, ou celui du volume d'eau déplacée $= P$, la distance du centre de gravité de la partie submergée homogene à l'axe de rotation de la figure[31] génératrice $= f$, nous aurons

$$\int \frac{1}{12}q^3dx = Pf \ .$$

Ces facteurs P & f demeurent constamment les mêmes pour toutes les inclinaisons du corps plongé; & le premier membre de la formule générale pour la stabilité des corps plongés, deviendra ici $Pf\sigma$, quelle que soit la figure génératrice.

Examinons maintenant le second membre de cette formule. Nous avons vu que ce second membre est toujours égal à $\int Ps\,d\sigma$, en entendant par s la hauteur du centre de gravité de tout le corps par-dessus le centre de gravité de la partie submergée homogene; & il est clair que dans les corps formés par la rotation d'une figure autour d'un axe horizontal, la hauteur verticale s est $= f - g\cos\sigma$, en entendant par g la distance du centre de gravité du corps à l'axe de rotation de la figure génératrice. Nous avons donc

$$\int Ps\,d\sigma = \int Pf\,d\sigma - \int Pg\,d\sigma\cos\sigma = Pf\sigma - Pg\sin\sigma \ ;$$

& cette derniere expression fait le second membre de la formule générale sur la stabilité des corps plongés, qu'il faut retrancher du premier membre, qui est pour les corps en question toujours $= Pf\sigma$. Ainsi la formule générale pour la stabilité des corps en question devient simplement $Pg\sin\sigma$.

§. XVI. Voilà une propriété bien remarquable sur la stabilité des corps formés par la rotation d'une figure quelconque autour d'un axe horizontal. Cette stabilité est la même que si le corps étoit suspendu en l'air par ledit axe horizontal, & qu'il fût incliné d'un angle σ; car le poids P seroit le même; son action seroit concentrée au centre de gravité, & il agiroit sur un levier $g\sin\sigma$; & enfin son *momentum* seroit aussi $Pg\sin\sigma$. Ici le centre

31 The same proposition was already stated in D. Bernoulli St. 32, *De motibus oscillatoriis corporum humido insidentium*: CP XI, 1739 (1750), § 12, ex. III. Having decided not to cite himself, Bernoulli demonstrates it again.

de gravité de la partie submergée homogene n'entre plus en considération; toutes les inclinaisons, grandes ou petites, font les *momenta* des forces réquises, pour tenir les corps inclinés, proportionnels aux sinus des inclinaisons. Si la partie submergée fait la moitié du corps formé par la révolution entiere de la figure génératrice, l'axe de rotation sera à fleur d'eau. Ce cas ne s'éloigne pas beaucoup de l'exemple des navires; & comme le métacentre ne peut être que fort proche de la surface d'eau, étant pour ces corps dans l'axe de rotation, le centre de gravité du vaisseau tout chargé doit être placé plus bas que la surface d'eau, pour que le navire obtienne une stabilité tant soit peu considérable. Mais comme la hauteur des mâts, des voiles, l'accastillage, & tout ce qui est au-dessus de l'eau, ne permet pas de donner une profondeur considérable audit centre de gravité, on pourra augmenter la stabilité du navire en haussant l'axe de rotation; c'est-à-dire, en faisant ensorte que la carene fasse une plus petite portion que la moitié du solide formé par la révolution entiere. La stabilité dans le sens de la longueur du navire n'est si grande, que parce que la carene ne forme que comme une très-petite portion du solide entier, en considérant la carene comme formée par la révolution d'un demi-cercle autour d'un axe fort éloigné horizontal & perpendiculaire à la quille. Une carene ainsi formée, auroit l'axe de rotation élevé par-dessus la surface de l'eau de 127 piés, en donnant 18 piés de profondeur à la carene, & 140 piés de longueur.

§. XVII. Voilà ce que j'avois de plus essentiel à dire sur la stabilité des corps plongés. La matiere est très-riche, & j'ai supprimé bien des choses pour n'être pas trop long; je n'ai fait que choisir ce que j'ai cru avoir le plus de rapport à notre sujet principal.

Pour mettre à profit la théorie que je viens d'établir, je vais examiner la relation qu'il y a entre les deux especes de stabilités, & les roulis & les tangages; après quoi j'exposerai tout ce qu'on peut faire pour augmenter l'une & l'autre stabilité.

CHAPITRE II.

Examen de ce que la stabilité des vaisseaux peut contribuer pour diminuer leur roulis & leur tangage.

§. XVIII. Une force quelconque, qui agit sur un vaisseau, si elle est entierement uniforme & permanente, ne sauroit faire d'autre effet, que celui de plier le vaisseau jusqu'à un certain degré, & ce degré dépend uniquement de la stabilité du vaisseau: plus la stabilité est grande, moins le vaisseau sera incliné par la force qui agit sur lui. Pour la construction ordinaire des vaisseaux, les sinus d'inclinaison seront à-peu-près en raison réciproque[32] des forces qui les produisent. A cet égard la stabilité renforcée est déja d'une grande utilité; car il est bon de conserver la position droite du vaisseau, tant qu'il est possible. Un vaisseau qui, dans les routes obliques, plie sous ses voiles, pourra, par la force du vent, être quelquefois incliné jusqu'à quinze, & même, si on ne diminuoit alors l'étendue des voiles, jusqu'à 20 degrés. Un vaisseau, ainsi incliné, sera encore sujet aux mêmes balancemens, que dans sa position droite; & il est évident que la mâture sera beaucoup plus fatiguée, par ces balancemens, lorsque la position moyenne est fort oblique, qu'elle n'est lorsque cette position moyenne est droite. Car si ces balancemens étoient par exemple de vingt degrés de chaque côté, & si l'inclinaison moyenne étoit pareillement de vingt degrés, le vaisseau dans l'une des extrêmités seroit incliné jusqu'à quarante degrés, & la mâture souffiroit beaucoup par le poids des mâts & des voiles. D'ailleurs, le vaisseau seroit déja dans un péril d'être renversé; le moindre accident pourroit être funeste, d'autant plus que les forces qui tendent à relever le vaisseau, ne croissent à-peu-près qu'en raison des sinus des inclinaisons; de sorte qu'un nouveau degré de force qui fait balancer le vaisseau, fait d'autant plus d'effet, que le vaisseau est déja incliné. Telle force qui feroit incliner le navire depuis sa position droite jusqu'au vingtieme degré, pourra le faire incliner depuis vingt degrés jusqu'à quarante-trois degrés dix minutes; & une petite force, qui dans le premier cas le feroit incliner d'un degré, ajoûtera à un vaisseau déja incliné de quarante degrés, une inclinaison d'un degré dix-neuf minutes.

Ces réflexions prouvent déja qu'il ne sauroit y avoir que de l'avantage à augmenter la stabilité des vaisseaux, tant qu'il est possible de le faire sans nuire à ses autres qualités réquises, puisque l'inclinaison moyenne, d'autant plus nuisible qu'elle est plus grande, en sera sûrement diminuée.

32 *Raison réciproque* is to be understood here as *raison directe*.

§. XIX. Examinons à présent les balancemens par eux-mêmes. Ils ne sauroient être produits que par des causes variables; ces causes, quelles qu'elles soient, sont nécessairement périodiques; elles augmenteront & diminueront alternativement. Un seul accès de ces forces motrices ne sauroit ébranler considérablement la masse énorme d'un vaisseau, pour le faire rouler; ce n'est qu'après un certain nombre de nouveaux accès répliqués coup sur coup, que le vaisseau roulera avec force. Je trouve que les roulis dépendent en partie de la force de chaque accès, & en partie de l'intervalle de tems entre deux accès consécutifs. Quant à la force, chacun voit, que plus elle est grande, plus elle doit faire d'effet; mais il n'est pas également facile à déterminer comment les intervalles d'un accès à l'autre concourent pour former les roulis. C'est ici une question qu'on n'a pas encore traitée, & que je vais tâcher de réduire aux loix de la méchanique.

§. XX. Les accès de force motrice sont, ou tout-à-fait irréguliers, ou uniformément périodiques. Cette derniere espece peut encore être sous-divisée par les durées de chaque accès: cette durée est ou plus petite, ou égale, ou plus grande que la durée d'un balancement naturel du vaisseau; car tout vaisseau incliné, & puis entierement abandonné à lui-même, fait des allées & venues à-peu-près isochrones, que je nomme *balancemens naturels*.

Lorsque les accès de force motrice sont irréguliers, ils agiteront le navire: mais ces agitations seront assez petites; car ces forces accélereront tantôt un balancement déja imprimé au vaisseau, & tantôt le retarderont, en agissant dans un sens contraire à son balancement; & quand même le hazard feroit que trois, quatre, ou cinq coups de suite conspirassent à augmenter les balancemens du vaisseau, ceux-ci ne pourront pas encore devenir excessifs, à moins que la force de chaque accès ne soit excessivement grande. Dans tous les cas, la stabilité du vaisseau sera un vrai remede qu'on peut opposer à ces efforts: plus la stabilité sera grande, plus les balancemens provenans de ces efforts seront petits. Cette réflexion nous conduit donc encore pour ces cas à la conclusion d'augmenter la stabilité des navires, tant que les autres circonstances le permettent.

§. XXI. Considérons à présent les accès de force motrice, lorsqu'ils sont réguliers & uniformément périodiques; je ne doute pas qu'ils ne soient à-peu-près tels assez souvent; car lorsque le vent n'est pas variable, qu'il est parvenu à agiter les eaux de la mer avec toute la force qu'il peut leur imprimer, & que la mer est libre & bien profonde jusqu'à une grande étendue, il est certain que les agitations des eaux ne sauroient manquer d'être régulieres. Alors ces agitations consisteront à faire des allées & venues conformes aux balancemens d'un pendule simple, & il est question d'examiner l'effet que ces agitations des eaux, devenues régulieres, pourront faire sur le navire.

Je prens ici d'abord pour principe, que dans tout systême de corps qui agissent les uns sur les autres, si ces corps font des mouvemens réciproques, réguliers & permanens, ils feront toujours des balancemens harmonieux & synchrones, qui commenceront & finiront tous dans les mêmes instans. C'est-là une vérité que j'ai reconnue avec une évidence entiere, toutes les fois qu'il m'est arrivé d'examiner de pareils mouvemens réciproques composés; & cela m'est arrivé très-souvent. Le systême que nous avons ici devant nous, sont le navire & les eaux qui l'environnent, car je suppose un vent fait & uniforme à tous égards. Or le navire, quelques balancemens qu'il fasse, ne sauroit changer les mouvemens réciproques des eaux agitées. Ceux ci maîtriseront donc entierement le navire, qui par conséquent fera dans ces cas ses balancemens harmonieusement avec ceux des eaux agitées; de sorte que la durée de chaque balancement soit égale à celle de chaque accès moteur.

§. XXII. La propriété que nous venons d'indiquer sur la durée de chaque balancement nous fournit une maniere de déterminer en quelque façon l'étendue de chaque balancement, & d'indiquer les circonstances dont cette étendue dépend, & dont elle dépendroit, quand même un navire balanceroit sans trouver aucune résistance. Nous comparerons les balancemens d'un navire avec ceux d'un pendule simple, & nous les supposerons isochrones les grands avec les petits: nous supposerons aussi que les accès de force motrice résultante de l'action des eaux, agissent sur le vaisseau indépendamment de ses inclinaisons; ce qui sera d'autant plus vrai, que les balancemens sont petits.

Pour expliquer à présent mes idées sur cette matiere avec plus de clarté, je considererai un pendule simple, long d'un peu plus de douze piés, & qui batte à chaque coup deux secondes[33]. Supposons une force qui oblige ce pendule à battre à chaque seconde[34], il faudra que la force accélératrice entiere devienne dans chaque point quatre fois plus grande; & comme la force de la pesanteur qui agit sur le corps du pendule reste la même, il faut que la force étrangere qui est survenue, & qui oblige le pendule de doubler ses battemens, soit trois fois plus grande que celle qui anime le pendule par sa pesanteur: il n'y a donc plus qu'à considérer l'intensité de ladite force étrangere. Supposons-la, par exemple, telle qu'elle soit, égale à la force de la pesanteur qui anime le pendule, lorsque celui-ci se trouve à douze degrés

33 See the Introduction p. 91 h. v.
34 See the Introduction p. 92 h. v.

de distance avec la verticale qui passe par le point de suspension. Je dis que dans ce cas le pendule fera des excursions de quatre degrés de chaque côté, puisqu'alors la force étrangere est trois fois plus grande que la force de la pesanteur qui anime le pendule, & qu'elle demeure telle durant tout le balancement; car nous supposons la force étrangere être de la nature de celles qui produisent des mouvemens isochrones, & agir également sur le pendule incliné de 12^{d} & de 4^{d}. Si le pendule faisoit toute autre excursion, ses balancemens ne pourroient jamais se faire harmoniquement avec les accès de la force qui l'agite. Dans ce cas les accès de la force qui agite le pendule, conspirent avec la force de la pesanteur qui anime le pendule; mais lorsque lesdits accès se suivent avec moins de rapidité que ne feroient les balancemens naturels du pendule, il arrive nécessairement le contraire, c'est-à-dire que la force qui agite le pendule, doit toujours agir d'une maniere opposée à la pesanteur du corps du pendule; c'est ce que je vais expliquer par un autre exemple.

Supposons que la force étrangere oblige notre pendule à faire chaque vibration en quatre secondes de tems. En ce cas, la force entiere qui anime le pendule, ne doit plus faire que le quart de la simple force de la pesanteur. Il faut donc que la force étrangere ôte les trois quarts à la force naturelle de la pesanteur du corps. De là nous connoissons que le pendule doit faire des excursions de 16^{d} de chaque côté; car puisqu'à 12^{d} d'inclinaison la force de la pesanteur est supposée égale à la force qui l'agite, il s'ensuit qu'à 16^{d} d'inclinaison, la force de la pesanteur sera à la force étrangere, comme 4 à 3; & que par conséquent celle-ci fasse les trois quarts de la premiere. Ce n'est qu'avec ces excursions de 16^{d} de chaque côté, que les balancemens du pendule peuvent devenir réguliers & prendre un état de permanence. Mais ici la force étrangere seroit toujours opposée à la force de la pesanteur; c'est-à-dire, que le corps du pendule sera secondé par la force étrangere pendant tout le tems qu'il emploie à monter; & qu'au contraire, cette force étrangere s'opposera au mouvement du corps pendant tout le tems qu'il emploie à descendre.

§. XXIII. Par les deux exemples que j'ai examinés dans le précédent article, on comprend aisément la solution générale sur la question de l'excursion du pendule animé en même tems par la pesanteur & par une force étrangere qui soit de la nature des forces qui produisent des balancemens isochrones. Soit donc à présent l'intensité de la force étrangere telle, que dans le moment qu'elle est la plus grande, elle soit égale à la force de la pesanteur du corps du pendule sous un angle s. Soit le tems d'un balancement naturel du pendule t, & que ce pendule soit obligé par la force étrangere à faire chaque vibration dans un tems $\frac{1}{n}t$, je dis que le pendule fera de chaque côté des excursions, dont l'angle avec la verticale[35] sera $= \frac{s}{nn-1}$.

Il suit de cette formule, que plus les accès de la force étrangere se succedent rapidement, plus les excursions du pendule seront petites; mais lorsque les intervalles entre deux accès sont à-peu-près égaux à la durée d'un balancement naturel du pendule, ce pendule fera des excursions excessives. C'est ici une propriété a laquelle je souhaite qu'on fasse attention. Enfin, lorsque lesdits intervalles surpassent la durée d'un balancement naturel du pendule, il arrive que la force étrangere & la force qui résulte de la pesanteur des corps agissent d'un sens contraire, parce que l'angle de l'excursion réelle devient négatif; & alors l'excursion elle-même devient d'autant plus petite, que les accès de la force étrangere se succedent plus lentement.

Tout cela sera éclairci par l'expérience physique qui suit. Qu'on prenne par exemple un pendule simple à secondes, & qu'on le tienne suspendu entre deux doigts; qu'on fasse avec la main des mouvemens réciproques horizontaux, en imitant le plus qu'il est possible ceux d'un pendule d'une longueur quelconque; on verra que suivant la rapidité de la main, qu'on pourra supposer faire toujours les mêmes excursions, le pendule s'accommodera aux mêmes durées de vibrations de la main, & que les excursions du pendule pourront ou se faire du même côté que celles de la main, ou du côté opposé; mais on remarquera surtout, que lorsque chaque allée & venue de la main se fait précisément dans une seconde de tems, alors les plus petits mouvemens de la main feront faire au pendule des excursions excessives, qui même ne sont limitées que par la résistance de l'air & par l'imparfaite flexibilité du fil.

§. XXIV. Les propriétés remarquables que nous venons d'indiquer sur l'étendue des balancemens d'un pendule, peuvent toutes être appliquées aux roulis d'un navire, puisqu'il est clair qu'on peut substituer au corps du pendule tout autre corps soutenu d'une façon à pouvoir faire des balancemens. Les hypotheses que nous avons faites ne répondent pas à la nature de la chose avec une précision entiere; ce que je ne dissimulerai pas. Les roulis *naturels* d'un vaisseau ne sauroient être parfaitement isochrones entre eux: les accès de la force qui agite le navire auront aussi rarement des intervalles de tems parfaitement égaux; cette force n'agira pas précisément avec cette loi que demande l'isochronisme, & elle n'agira pas tout-à-fait également sur le navire plus ou moins incliné dans les momens où ces inclinai-

35 See the Introduction p. 92 h. v. (formula (15)).

sons sont grandes; mais tout cela ne doit pas nous empêcher d'avoir de l'indulgence pour nos principes, puisqu'il est certain que les roulis auront toujours quelque penchant marqué pour cet état que notre théorie exige, pendant tout le tems que les agitations des eaux sont fort régulieres & permanentes; ce qui fait le cas dont nous parlons. J'appliquerai donc les balancemens du pendule, que j'ai expliqués & que j'ai pris pour exemple uniquement pour être intelligible, aux balancemens d'un navire.

§. XXV. Nous avons vu au §. XXIII, que l'étendue des balancemens de chaque côté est exprimée par $\frac{s}{nn-1}$. Voyons d'abord quel sens il faut attacher à cette formule lorsqu'il est question des roulis d'un navire. Pour cet effet, il faut premierement considérer la force qui agite le vaisseau dans le moment qu'elle est la plus grande; j'appellerai cette force φ. Ensuite il faut examiner quelle inclinaison le navire prendroit si ladite force φ demeuroit constamment la même, sans augmenter ni diminuer. L'angle de cette inclinaison est exprimé dans notre formule par s. On voit bien que cet angle s sera d'autant plus petit que la stabilité du navire est plus grande; & que, pour les petits angles, l'un est réciproquement proportionnel à l'autre. Nous avons vu que dans les petites inclinaisons la formule qui exprime la stabilité des corps plongés, peut toujours être réduite à $\pi\sigma$; ainsi la stabilité pour l'angle s sera πs, & de-là nous aurons $\varphi = \pi s$ ou $s = \frac{\varphi}{\pi}$. Substituons cette valeur dans la formule $\frac{s}{nn-1}$, & nous aurons $\frac{\varphi}{(nn-1)\pi}$. Soit après cela, comme au §. XXIII, le tems d'un *balancement naturel* du vaisseau $= t$, & le tems entre deux accès de force qui agitent le vaisseau $= \theta$, nous aurons $\theta = \frac{1}{n} t$, & par conséquent $n = \frac{t}{\theta}$, ou $nn - 1 = \frac{tt - \theta\theta}{\theta\theta}$; & si nous substituons encore cette valeur dans notre formule $\frac{\varphi}{(nn-1)\pi}$, nous aurons enfin

$$\frac{\theta\theta\varphi}{(tt-\theta\theta)\pi}.$$

§. XXVI. Voilà la formule qui exprimera toujours à-peu-près l'étendue des balancemens du navire permanens & causés par des efforts réguliers. Elle nous apprend donc que la quantité des roulis dépend de la force qui agite le navire, de la stabilité du navire, & du rapport qu'il y a entre les tems t & θ, c'est-à-dire, entre la durée d'un *balancement naturel* du navire, & la durée de chaque accès de la force qui agite le navire. Nous voyons aussi que les cas les plus à craindre sont ceux où les balancemens des eaux qui causent les roulis sont à-peu-près de la même durée avec les *balancemens naturels* du navire. Je n'examinerai pas encore toutes les conséquences qu'on peut tirer de cette formule. On voit généralement que plus la quantité π, qui est proportionnelle à l'intensité de la stabilité, est grande, plus l'étendue des roulis sera petite. Il est vrai que le tems t dépend le plus souvent de la stabilité, puisque si la stabilité est plus grande, les balancemens naturels du navire en doivent être accélérés. On peut dire que tout le reste étant égal, le tems t est à-peu-près réciproquement proportionnel à la racine quarrée de la stabilité; mais on remarquera qu'on peut changer le tems t, sans rien changer du côté de la stabilité, en éloignant ou en approchant la matiere, qui fait toute la charge du vaisseau, de l'axe de rotation. La relation qu'il y a entre les quantités π, t & θ, pourroit faire, dans de certains cas, qu'en augmentant la stabilité, on rendît les agitations du navire plus grandes. C'est donc une question bien importante, s'il convient d'augmenter la stabilité du navire. Là-dessus on peut faire les réflexions suivantes. Comme le tems θ est incertain, puisqu'il dépend de la nature des agitations des eaux de la mer, & par conséquent des mers même où l'on se trouve, & de plusieurs circonstances accidentelles; il arrivera toujours, quel que soit le tems t, ou que θ soit plus grand que t, ou qu'il soit plus petit, ou qu'il lui soit égal. Si θ est plus grand que t, il est clair que la quantité $\frac{\theta\theta\varphi}{(tt - \theta\theta)\pi}$, devient d'autant plus petite, que l'intensité de la stabilité π est plus grande, puisqu'on peut supposer la quantité $tt\pi$ à-peu-près constante, & on remarquera qu'il est indifférent que la valeur du dénominateur soit affirmative ou négative pour la quantité de l'excursion, & que cette différence marque simplement si les forces φ & π agissent du même côté ou du côté opposé; mais si θ est plus petit que t, on augmentera au contraire les roulis en augmentant la stabilité, parce que le dénominateur devient plus petit, en supposant toujours la quantité $tt\pi$ constante. On sera peut-être surpris d'entendre qu'il peut y avoir des cas où une augmentation de stabilité cause une augmentation de roulis. Je réponds à cela, que la stabilité diminue à la vérité toujours l'inclinaison causée par une force uniforme & permanente; mais que l'excursion des roulis causée par des accès de force, ne se regle pas toujours sur ladite inclinaison. Si θ étoit infiniment plus grand que t, l'étendue des roulis seroit exprimée par $\frac{\varphi}{\pi}$, & on diminueroit les roulis à mesure qu'on augmenteroit la stabilité. Si au contraire θ étoit infiniment plus petit que t, la même étendue de roulis seroit exprimée par $\frac{\theta\theta\varphi}{tt\pi}$, & la stabilité ne changeroit presque pas l'étendue des roulis; d'ailleurs les balancemens seroient en ce cas très-petits par eux-mêmes, à cause de la petitesse supposée de $\frac{\theta\theta}{tt}$. Le premier cas mérite donc plus d'attention que le second; outre cela, il semble que le premier cas sera beaucoup plus fréquent que l'autre; car lorsque les accès, qui font rouler le vaisseau, sont réguliers, je ne doute pas que ces accès ne soient isochrones avec les mouvemens réciproques des lames, & ceux-ci sont le plus souvent beaucoup plus tardifs que ne sont les balancemens naturels d'un navire. Après tout, il n'y a que le seul cas à craindre où les tems t & θ sont à-peu-près égaux; & comme on ne sait pas lequel prévaudra sur l'autre, c'est sans doute prendre le parti le plus sûr que d'augmenter la stabilité tant qu'il est possible, puisque pour la même quantité $tt - \theta\theta$, les roulis en sont certainement diminués. Je ferai voir ci-dessous de quelle façon particuliere on pourra se précautionner contre ce cas, qui est le plus fâcheux. Au reste, on voit assez que nos principes sont toujours les mêmes, quelle que soit la source des forces qui agitent le navire. Après avoir examiné tous les cas qui peuvent arriver, & après avoir fait voir, que le meilleur parti & le plus sûr pour diminuer les balancemens d'un navire, est toujours d'augmenter sa stabilité, il est tems de considérer ce dernier article avec tout le détail réquis.

CHAPITRE III.

Exposition des moyens qu'on peut employer pour diminuer le roulis & le tangage des navires fondés sur leur stabilité.

§. XXVII. Il s'agit, dans ce Chapitre, d'examiner quels sont les moyens convenables d'augmenter la stabilité du navire[36], & de diminuer par-là ses balancemens. Je ne parle pas de cette augmentation de stabilité qu'on obtiendroit en augmentant les dimensions principales du navire. Tout le monde sait qu'un grand vaisseau a plus de stabilité qu'un petit. Nos formules montrent à cet égard que la stabilité croit en raison biquarrée des dimensions homologues pour les vaisseaux semblables, suivant toutes les circonstances. Mais l'inertie de toute la matiere rélative aux balancemens du vaisseau, croit comme la cinquieme puissance des dimensions homologues. D'où il suit que la longueur du pendule simple isochrone avec les *balancemens naturels* du

36 See the Introduction p. 95 h. v.

navire, suit la raison des dimensions homologues; & la durée de chaque balancement, la raison sous-doublée des dimensions homologues. Ainsi un grand vaisseau pliera moins sous les voiles, dans les routes obliques, qu'un petit, en raison réciproque des dimensions homologues; parce que le *momentum* de l'effort du vent pour faire plier le vaisseau, en supposant la mâture & les voiles garder la même proportion en tout sens, ne suit que la raison de la troisieme puissance des dimensions homologues, pendant que la stabilité croit en raison de la quatrieme puissance. On pourroit donc employer plus de voiles à proportion sur un grand vaisseau que sur un petit. Mais aussi, d'un autre côté, le centre de gravité du navire est à proportion plus haut dans les grands navires que dans les petits, tant à cause de la charge & de l'artillerie, que parce que la structure n'est pas entierement semblable; & cette circonstance diminue la stabilité des grands vaisseaux, surtout celle qui est rélative aux roulis, car la stabilité dans le sens de la longueur ne sera pas sensiblement diminuée par cette circonstance. Quant à la proportion des forces, qui agissent sur des navires de différente grandeur pour les faire balancer, & dont nous avons supposé, au §. XXV, l'intensité $= \varphi$, on voit, pour peu qu'on y fasse attention, que c'est un problème absolument indéterminé que d'indiquer cette proportion. Ces forces pourront consister quelquefois dans l'action d'une petite masse d'eau, qui exercera tout son pouvoir tant sur le petit que sur le grand navire, de maniere que ces forces seroient égales; d'autres fois ce seront de grandes masses d'eau, qui agissent sur toute la carene, ou sur une grande partie de la carene; outre cela la direction moyenne des efforts passera tantôt plus loin, tantôt plus près de l'axe de rotation du navire, &c. C'est donc là une question sur laquelle on ne peut rien affirmer de positif.

Mais voyons aussi quelle sera la proportion entre les balancemens des deux navires, lorsqu'ils se trouvent en même tems, dans les mêmes circonstances proches l'un de l'autre. Car quoique la théorie sur l'étendue des balancemens devenus tout-à-fait réguliers, que j'ai exposé depuis le §. XXI jusqu'à la fin du chapitre, ne sauroit jamais convenir aux balancemens des navires avec une précision entiere, parce qu'il y aura toujours un reste d'irrégularité dans les circonstances, il n'est pourtant pas douteux que ces balancemens ne suivent plus ou moins les loix de cette théorie, suivant le plus ou le moins de régularite dans leurs causes.

Supposons donc deux navires de grandeur différente, mais entierement semblables en toutes choses. Soit la longueur de l'un à la longueur de l'autre, comme 1 à n, & que l'étendue des balancemens soit exprimée, pour le petit navire, par $\frac{\theta\theta\varphi}{(tt - \theta\theta)\pi}$, qui est la formule que nous avons trouvée à la fin du §. XXV. Il s'agit d'examiner comment cette formule sera changée pour le grand vaisseau. Pour cet effet, on remarquera que le tems θ est le même pour l'un & l'autre navire; que le tems t doit être, en vertu de ce que nous avons marqué au commencement de ce §, changé en $t\sqrt{n}$, & π en $n^4\pi$; & enfin qu'on ne peut rien affirmer de positif sur le changement de la force φ. J'appellerai donc φ' cette force pour le grand vaisseau, quelle qu'elle puisse être. De cette façon, l'étendue des balancemens du grand navire sera exprimée par $\frac{\theta\theta\varphi'}{(ntt-\theta\theta)n^4\pi}$. Ainsi les balancemens pour le petit & pour le grand navire seront comme $\frac{\theta\theta\varphi}{(tt-\theta\theta)\pi}$ à $\frac{\theta\theta\varphi'}{(ntt-\theta\theta)n^4\pi}$, ou comme $\frac{\varphi}{tt-\theta\theta}$ à $\frac{\theta\theta\varphi'}{n^5tt-n^4\theta\theta}$. On voit assez, par ce rapport, que les balancemens du grand vaisseau doivent naturellement être plus petits que ceux du petit navire, parce que le plus souvent le dénominateur $tt-\theta\theta$ sera beaucoup plus petit que l'autre dénominateur $n^5tt-n^4\theta\theta$. Si on veut supposer t beaucoup plus petit que θ, & je crois qu'on se trouve souvent dans le cas à le pouvoir faire, ces balancemens seront en raison de $n^4\varphi$ à φ'. Je crois aussi que pour l'action des grandes lames régulieres on peut supposer à-peu-près $\varphi' = n^3\varphi$, parce que les forces agissent sur la surface de la carene proportionnelle à nn, & sur des leviers qui sont naturellement proportionnels à n. Dans cette seconde supposition, les balancemens seront en raison de n à 1, c'est-à-dire, en raison réciproque des dimensions homologues. Mais je vois aussi qu'on pourroit se trouver, par un concours de hazards, dans des circonstances que le grand navire fît de plus grands balancemens que l'autre; cela pourroit arriver lorsque la quantité ntt seroit à-peu-près égale à la quantité $\theta\theta$. C'est aux marins entendus & accoutumés aux observations, à juger d'un tel paradoxe; du moins suis-je bien assuré que souvent les grands navires feront d'aussi grands balancemens que d'autres navires considérablement plus petits, & que particulierement les grandes lames régulieres & uniformes doivent faire cet effet.

§. XXVIII. Voyons maintenant ce qu'on peut faire pour augmenter la stabilité du navire, & pour diminuer par-là ses balancemens, sans changer aucune de ses dimensions principales. Nous avons démontré dans le chapitre sur la stabilité des corps flottans, que cette stabilité dépend uniquement de deux choses; savoir, de la section horizontale du corps à fleur d'eau, & de la hauteur verticale du centre de gravité du corps par-dessus le centre de gravité de la partie homogene submergée. Considérons donc ces deux choses l'une après l'autre; car en les confondant, on obtient souvent des quantités qui se détruisent, & les deux sources de la stabilité ne sont plus reconnoissables par la formule qui exprime la stabilité absolue. La section horizontale du navire à fleur d'eau fait à-peu-près une ellipse. La stabilité qui en résulte dans le sens de la largeur du navire, est en raison composée du cube du petit axe de l'ellipse & du grand axe; & la stabilité dans le sens de la longueur, est au contraire en raison composée du cube du grand axe & du petit axe (§. XIII). De-là on voit combien il y auroit à gagner à augmenter la largeur du navire pour diminuer les roulis, & la longueur pour diminuer les tangages. Si cependant on veut conserver la plus grande largeur, de même que la plus grande longueur dans la section du navire à fleur d'eau, il y aura encore un autre moyen d'augmenter la stabilité; c'est de diminuer moins la largeur de ladite section vers la proue & vers la pouppe: plus la section approchera du rectangle circonscrit, plus la stabilité du navire augmentera, & cela non-seulement par rapport aux roulis, mais encore par rapport au tangage. J'ai démontré au milieu du dit §. XIII, qu'en changeant l'ellipse en son rectangle circonscrit, chacune des deux stabilités, entant qu'elles résultent de la coupe du vaisseau à fleur d'eau, est augmentée en raison de 3 à 5. Voilà tout d'un coup une grande augmentation de stabilité, mais qui devient encore plus grande du côté des roulis, en considérant la stabilité absolue, c'est-à-dire, en considérant les deux termes qui l'expriment ensemble, puisque j'ai fait voir, à la fin du §. XIII, que la stabilité entiere & absolue seroit doublée, par ce changement, pour les roulis, quoiqu'elle ne soit pas sensiblement augmentée au-delà de la proportion de 3 à 5 pour les tangages.

On devroit, au lieu de considérer la coupe horizontale du vaisseau à fleur d'eau comme une ellipse, la regarder plutôt comme composée de deux demi-ellipses sur un petit axe commun, & dont les demi-grands axes seroient comme 5 à 7, parce qu'on ne fait pas le fort du vaisseau précisément au milieu, mais environ à $\frac{5}{12}$ de toute la longueur, depuis la proue, & $\frac{7}{12}$ depuis la pouppe. J'avoue cependant que je ne vois pas ce qui a pu engager les constructeurs de vaisseau à suivre cette maxime, qui ne fait qu'augmenter la résistance des eaux contre la proue: c'est peut-être pour donner aux vaisseaux plus de soutien vers la proue que vers la pouppe, & pour occasionner par-là cette pente de la quille dans le sens de la proue vers la pouppe, que l'expérience a montré être utile, & que plusieurs raisons me font envisager comme telle. Mais on peut obtenir une telle pente par la simple disposition de la charge. M. Bouguer[37] fait voir pourtant que le navire en

37 P. Bouguer, *Traité du navire, de sa construction et de ses mouvements*, Livre II, Paris: Jombert, 1746.

gouverne mieux. Quoiqu'il en soit, & quelque parti qu'on prenne là-dessus, l'augmentation de la stabilité que nous conseillons, est la même pour les deux demi-ellipses & pour la simple ellipse entiere.

§. XXIX. Il me semble que dans la construction des vaisseaux, on ne dirige pas assez directement ses vues du côté de la figure que prendra la coupe du vaisseau à fleur d'eau, qui fait cependant un des points les plus essentiels, puisque la charge qu'on peut mettre sur le vaisseau, aussi-bien que la stabilité de celui-ci, dépendent presque uniquement de cette section. Par-là il arrive qu'une différence assez légere dans la conformation de la carene, jette une diversité assez considérable sur la figure de ladite coupe. Ayant examiné quelques exemples sur ces figures, j'ai remarqué qu'elles n'étoient assujetties à aucune loi fixe. J'ai vu seulement que la figure étoit toujours plus large vers la proue que vers la pouppe, quoique tantôt plus tantôt moins. Je ne saurois approuver cette maxime; car quand même l'expérience auroit entierement démontré qu'il faut rendre la proue plus grosse que la pouppe, & renfler la carene vers l'avant, rien n'empêche de se relâcher sur cette maxime, peut-être mal fondée, aux environs de la flottaison. Je ne veux pas qu'on retrécisse l'avant de la figure, mais qu'on en élargisse l'arriere. En un mot, il me semble que cette figure devroit approcher de la figure d'un rectangle arrondi par les quatre coins.

Si on considere la dite coupe comme composée de deux demi-ellipses, faites sur un petit axe commun, quelle que soit la proportion entre les deux demi grands axes, l'aire de la figure sera toujours égale à $\frac{11}{14}$ du produit de la longueur par la largeur[38]. Si donc la coupe a 120 piés de longueur sur 29 piés de largeur, (c'est un exemple que M. Bouguer donne dans son *Traité du Navire*, p. 242), l'aire de la coupe devroit être de 2734 piés quarrés. Cependant M. Bouguer, par une regle qu'il cite, ne la trouve que de 2655 piés quarrés. La section étoit donc encore plus petite que n'est l'aire des deux demi ellipses. Il est vrai que la regle alléguée par M. Bouguer[39], pêche toujours en défaut à cause du grand nombre de petits segmens curvilignes qu'on néglige; & je suis sûr que pour la construction ordinaire des vaisseaux, on trouvera l'aire de la section pour le moins avec une précision égale, en multipliant simplement la plus grande largeur par la plus grande longueur, & en prenant les $\frac{11}{14}$ du produit.

38 $\frac{11}{14}$ approximates the exact value $\frac{\pi}{4}$.
39 The reference to Bouguer is explained in the Introduction, pp. 94–95 h. v.

§.XXX. J'ai déja indiqué, que par le moyen que nous venons d'exposer, on augmentera également la stabilité dans le sens d'un bord à l'autre, & dans celui de la proue à la pouppe; & qu'ainsi on diminuera également le roulis & le tangage. Celui que je vais ajoûter est à-peu-pres de la même nature, quoiqu'il paroisse ne regarder que la diminution des roulis. Cet autre moyen consiste à faire les flancs du vaisseau, aux environs de la flottaison, plus droits & verticaux, qu'on n'a coutume de faire. Entrons, sur cet article, dans le détail qu'il mérite. Dans les routes obliques, le vaisseau incline considérablement, par la force du vent sur les voiles. Il faut alors considérer le vaisseau, ainsi incliné, comme dans sa position naturelle, que j'appellerai oblique. Le vaisseau ne sera pas moins sujet aux balancemens, dans cette position forcée, qu'il l'étoit dans sa position droite. Les balancemens fatigueront davantage la mâture, & seront plus à craindre par le péril de renverser. La stabilité est donc plus nécessaire pour les positions naturelles obliques que pour les droites. Mais lorsque la position naturelle est droite, si le vaisseau est balancé, il faut absolument que la force qui le repousse vers sa position naturelle, augmente à chaque nouvel incrément élémentaire d'inclinaison, sans quoi le navire seroit bientôt renversé. On conviendra sans doute, qu'il est bon de donner auxdites augmentations de force élémentaire le plus d'étendue qu'il est possible. Plus un vaisseau incliné trouve d'obstacle à s'incliner davantage, plus ses balancemens seront diminués. On peut entendre par le mot de *stabilité momentanée* le surcroit de la force réquise pour ajoûter au vaisseau déja incliné un nouveau petit degré d'inclinaison constant, que nous avons exprimé ci-dessus par $d\sigma$. Dans ce sens on peut dire, qu'il résulte de la construction ordinaire du navire, que sa *stabilité momentanée*, diminue à mesure que l'inclinaison du navire augmente; car, suivant cette construction, on se trouve assez dans le cas du §.XV, pour lequel le calcul nous a conduit au beau théorême marqué à la fin dudit §.XV, qui donne la stabilité (ce mot y est pris dans un autre sens, & il y exprime le *momentum* de la force qui repousse le corps vers sa position naturelle, lorsque l'angle de son inclinaison est $= \sigma$) égale à $Pg \sin\sigma$, en entendant par P le poids du corps, & par g la hauteur de l'axe de la figure génératrice par-dessus le centre de gravité. Comme donc les quantités P & g sont constantes, la différentielle de ladite quantité devient $= Pg\, d\sigma \cos\sigma$; & cette formule nous apprend que la *stabilité momentanée* du navire diminue à mesure que l'inclinaison augmente, puisqu'elle est proportionnelle au cosinus de l'inclinaison actuelle. Cette diminution de stabilité peut sans doute occasionner de grands roulis. Il faut donc faire tout son possible pour que cette nouvelle espece de stabilité devienne, au lieu de diminuer, d'autant plus grande que le navire sera plus incliné. Les roulis vont souvent jusqu'à 35 degrés; un leger accident pourroit alors achever de renverser le vaisseau, s'il n'est retenu par une grande stabilité; d'ailleurs, le navire ne fera ces roulis énormes & les inclinaisons moyennes excessives, qu'à cause même de ladite diminution de stabilité, qu'on ne sauroit prendre trop de soin de prévenir.

§. XXXI. Examinons la cause qui fait que la stabilité en question diminue; c'est que la section du navire à fleur d'eau demeure à-peu-près toujours la même, pendant que la hauteur verticale du centre de gravité du navire par-dessus le centre de gravité de la carene homogene augmente. Cette derniere circonstance, qu'on ne sauroit prévenir, à moins qu'on ne vienne à bout de placer le premier centre de gravité plus bas que le second, fait que la quantité à retrancher pour avoir la stabilité absolue, devient d'autant plus grande, que le vaisseau est incliné davantage, & par conséquent que la stabilité absolue devient plus petite. Le moyen pour prévenir cette diminution de stabilité, sera de faire ensorte que la section horizontale du vaisseau à fleur d'eau augmente à mesure que le vaisseau incline. C'est dans cette vue que je conseille d'élever verticalement les flancs du vaisseau, tant au-dessus qu'au-dessous de la flottaison, puisque de cette façon la coupe en question augmente comme les secantes des inclinaisons, & que la stabilité provenante de cette section augmente comme les cubes de ces secantes (§. IV). Si les constructeurs de vaisseau trouvent un tel changement excessif, ils pourront du moins diminuer la rentrée des vaisseaux, & conserver un peu au-dessous de la flottaison toute la largeur des coupes verticales perpendiculaires à la longueur. C'est pour faire voir le grand succès qu'on peut se promettre d'un tel changement, que j'ai donné, au §. X, la petite Table qui montre assez la grande différence entre les stabilités, lorsque les inclinaisons commencent à devenir sensibles. On voit, par cette Table, que la premiere espece de stabilité est pour une inclinaison de 20 degrés, augmentée par ce changement, en raison de 28.50 à 51.15; & que la seconde espece de stabilité, depuis 20^d jusqu'à 25^d, est augmentée en raison de 35.22 – 28.50 à 81.17 – 51.15 ou en raison[40] de 7.72 à 30.02. On remarquera, dans cette Table, que les différences des nombres de la seconde & de la troisieme colonne marquent les *stabilités momentanées*. Ces différences diminuent dans la seconde colonne, comme les cosinus des inclinaisons; mais elles augmentent considérablement, au lieu de diminuer, dans la troisieme colonne. Je ne doute donc nullement qu'un tel

40 The number should read 6.72 instead of 7.72. But, as pointed out in the Introduction, §3.3.2 «Rectangular versus round» p. 87 h. v., the comparison is based on a wrong formula. One should compare to 50.53 – 36.05 = 14.48 as can be read from the corrected table instead of 30.02; the ratio is in fact 0.464 instead of 0.224.

changement, assez leger, ne diminue très-considérablement le roulis. La stabilité latérale étant augmentée, la longitudinale le sera aussi, puisque celle-ci est proportionnelle à la largeur du vaisseau (§. IV). De-là il suit que plus les roulis sont grands, plus les tangages seront petits (car ces deux sortes de balancemens peuvent fort bien coéxister en même tems), parce que les roulis augmenteront la stabilité longitudinale.

§. XXXII. Tant qu'on considere à part les deux membres de la formule qui exprime la stabilité, on voit facilement que nous avons indiqué actuellement tout ce qu'il est possible de faire pour augmenter la stabilité, rélativement au premier membre; car toutes les parties du navire qui ne plongent jamais dans l'eau, de même que celles qui n'en sortent jamais, n'ont aucune relation avec la stabilité qui résulte de la valeur du premier membre. Ce n'est qu'en tant qu'elles font varier les deux centres de gravité, qu'elles pourront augmenter ou diminuer la stabilité, comme le second membre de la formule l'indique. La stabilité rélative à ce second membre, est exprimée par $\int Ps\, d\sigma$, en entendant par P le poids du navire, ou le volume de la carene, & par s la hauteur du centre de gravité du navire par-dessus le centre de gravité de la carene homogene; & la *stabilité momentanée* est simplement exprimée par Ps. Ce second membre doit toujours être retranché du premier, tant que le premier centre de gravité est plus haut que le second; & il faudroit trop s'écarter des principes usités, pour qu'il fût plus bas. Ainsi, pour rendre la stabilité absolue plus grande, il faut diminuer la quantité Ps. C'est ce précepte général qu'il ne faut jamais perdre de vue en construisant le vaisseau. Il faut donc baisser le centre de gravité du navire tout chargé qu'il est, & hausser le centre de gravité de la carene tant qu'il est possible. Le centre de gravité du navire est ordinairement autour d'un cinquieme du tirant d'eau plus bas que la surface d'eau. Sur ces préliminaires, voici les maximes qu'on pourra suivre.

(*a*) Il faut ménager également la hauteur & la matiere dans toutes les parties qui sont au-dessus de la surface d'eau, autant que les circonstances peuvent le permettre. J'entens ici les mâts, les vergues, les voiles, les cordages, les châteaux, &c. Il semble que la hauteur des mâts n'est dictée que par la hardiesse de l'homme, trop souvent funeste au genre humain.

(*b*) Le lest doit servir en quelque façon de contrepoids à toutes ces parties indiquées dans la remarque précédente: il faut le proportionner en partie à la hauteur des mâts. S'il y avoit donc quelque chose à gagner au sujet de la remarque (*a*), on gagneroit en même tems sur le lest requis; le navire enfonceroit moins; la résistance des eaux contre le sillage en deviendroit plus petite: on seroit dédommagé, en partie, de ce qu'on se relâcheroit sur la hauteur des mâts.

(*c*) En se mettant dans les circonstances qui demandent moins de lest, le poids P diminue, & la quantité Ps en devient d'autant plus petite, comme nous le souhaitions.

(*d*) Il faut prendre du bon lest, pour pouvoir placer plus bas son centre de gravité pour le même poids. Enfin on se conformera aux principes connus de la Statique, dans l'arrimage & dans l'arrangement de la charge.

§. XXXIII. Il ne nous reste qu'à examiner la figure la plus convenable de la carene, en supposant sa surface horizontale donnée, comme ayant déja été déterminée. Plus on l'élargira vers la flottaison, & plus on la rétrécira vers la quille, plus on élevera le centre de gravité de la carene homogene. Plus aussi le navire tirera d'eau sans changer le volume de la carene, plus on augmentera le *momentum* du lest. Voilà les deux considérations principales à faire sur la carene, ou du moins sur cette partie de la carene qui, pendant les plus grands roulis, reste constamment submergée, puisque cette partie ne peut avoir aucun rapport avec le premier membre de la formule qui exprime la stabilité. Il résulte de ces considérations, que depuis la flottaison on doit conserver aux navires toutes leurs largeurs jusqu'à une certaine profondeur. Cette premiere réflexion s'accorde heureusement avec le conseil que j'ai donné au commencement du §. XXX, de faire les flancs du vaisseau, tant au-dessous qu'au-dessus de la flottaison, droits & verticaux. A une certaine profondeur, on commencera à retrécir sensiblement les coupes verticales & perpendiculaires à la longueur du navire. Vers les deux tiers de la profondeur, on les retrécira brusquement. On pourra enfin donner à la courbure un point d'inflexion contraire, & faire diminuer très-peu jusqu'à la quille ces petites largeurs destinées à recevoir un bon lest. Je dirai ci-dessous un autre avantage essentiel qu'on donnera à la carene par une telle construction, les loix de la théorie ne laissant pas douter du bon succès d'une telle construction à l'égard de notre sujet; mais je ne prétens pas qu'on les suive aux dépens des Loix de l'Architecture Navale absolument constatées, me contentant d'indiquer tous les principes qui méritent l'attention du constructeur. Voici quelques réflexions générales sur la figure de la carene. Si on partage la carene toute chargée, en tranches horizontales, les tranches inférieures seront plus pesantes qu'un volume égal d'eau, & les supérieures seront plus légeres, parce qu'on doit arranger, suivant l'ordre de la pesanteur, tout ce qui doit être mis sur le vaisseau, en tant qu'on est libre sur ce point. Il y aura donc une tranche moyenne qui sera égale en poids au volume d'eau égal. Cependant il faut considérer le volume de la carene comme donné, parce qu'il n'est question que de la figure de la carene. Il me semble qu'il convient d'estimer ladite place de la tranche moyenne, de conserver aux vaisseaux presque toute leur largeur, jusqu'aux environs de ladite tranche; mais après cela on diminuera subitement les largeurs, pour ne plus employer le reste du volume de la carene qu'en profondeur. Plus on voudra accorder de profondeur à la carene, plus on pourra donner de stabilité au navire, tout le reste demeurant égal. C'est pourquoi il faut plus de lest aux navires qui sont plats de varangues, qu'aux autres, qui sont bien taillés, & dont les fonds sont fins. On ne doit applatir les navires par leur fond, que lorsqu'on est dans la nécessité de donner au navire peu de tirant d'eau. Je voudrois même, lorsqu'on se trouve dans cette nécessité, qu'on augmentât un peu au-delà des regles ordinaires la largeur du navire, pour pouvoir lui donner plus de façon, sans changer la solidité de la carene, d'autant plus que le navire en recevra plusieurs autres avantages. En un mot, on prendra pour maxime de jetter beaucoup de volume vers le haut, & beaucoup de poids vers le bas; mais comme on s'écarteroit infiniment des loix de l'Architecture Navale, si on ne vouloit suivre que ce seul principe pour déterminer la figure de la carene, & que cependant on a la liberté de l'observer jusqu'à un certain point, je me contenterai de l'avoir indiqué, en recommandant la profondeur de la carene, pour augmenter par-là l'effet du lest, comme aussi de ne commencer les rétrécissemens que vers la région destinée à contenir les choses pesantes. J'ai déja remarqué que le dernier conseil s'accorde heureusement avec ce que j'ai dit, de faire les flancs du vaisseau, tant au-dessous qu'au-dessus de la flottaison, droits & verticaux; & comme la coupe horizontale du navire à fleur d'eau doit aussi, en vertu du §. XXVIII, approcher de la figure rectangulaire, il en résulte que près de la flottaison le corps du navire doit former à-peu-près un parallelepipede convenablement arrondi par les coins. J'excepte cependant de cette regle la proue & la pouppe, sur lesquelles je me réserve quelques petites réflexions particulieres, que j'exposerai vers la fin de ce Mémoire.

CHAPITRE IV.

Remarques sur l'axe de rotation autour duquel se font les balancemens du navire, avec quelques autres moyens de diminuer ces balancemens.

§. XXXIV. Un navire incliné se rapproche de sa position droite par un mouvement de rotation, & nous devons tâcher de connoître l'axe de cette rotation, c'est-à-dire la ligne qui, durant les balancemens, n'a aucun mouvement horizontal. Les auteurs ne conviennent pas sur la position de cet axe, & il me semble que c'est pour n'avoir pas distingué les cas qui peuvent arriver. Quelques-uns le font passer par le centre de gravité du navire, & d'autres le placent autrement. Qu'on me permette de traiter en peu de mots cette question[41].

On peut considérer, dans cette question, toute la matiere du corps comme concentrée dans un seul & même plan vertical, que je suppose être le plan de rotation. Pour les roulis, on pourra choisir la premiere coupe verticale du vaisseau perpendiculaire à la longueur. Pour connoître sur quel point ce plan, venant à faire des balancemens, tournera; je dis qu'il faut distinguer les balancemens & leurs causes. Supposons d'abord ledit plan détourné de sa position droite & naturelle dans des eaux parfaitement calmes, & que tout d'un coup les causes qui ont détourné le plan viennent à cesser; en ce cas le plan fera des balancemens que je considere se faire avec une liberté entiere, & que j'appellerai par conséquent balancemens *libres*. Je dis que dans ces balancemens le centre de rotation sera nécessairement le centre de gravité, & voici comme je le prouve.

Durant les balancemens *libres*, il n'y a absolument que la force de la pesanteur qui agisse sur les parties. Une telle force agissant verticalement, ne sauroit produire aucun mouvement horizontal absolu; il faut donc que la quantité de mouvement horizontal de droite à gauche soit précisément égale à celle qui se fait de gauche à droite: il faut, en un mot, que le centre de gravité n'ait aucun mouvement horizontal; donc le centre de gravité sera le centre de rotation.

On pourroit objecter ici que quand même le centre de gravité du plan auroit un mouvement horizontal, il ne s'en suivroit pas qu'il y eût dans le sistême une quantité de mouvement horizontal absolu, parce que le centre de gravité commun au plan & à l'eau qui l'environne, ne laisse pas de demeurer entierement en repos par rapport au mouvement horizontal. Mais on voit assez qu'il ne s'agit pas ici dudit centre de gravité commun, puisque pour donner un mouvement horizontal à un corps flottant, il faudra toujours employer une force horizontale; & que rien ne peut occasionner ici une telle force, tant qu'on fait abstraction de la résistance des eaux.

Voici un autre principe, qui mene à la même conclusion. Le *momentum* de la force qui agit sur le plan durant ses balancemens, est le même, sur quel point que ce plan se tourne, puisque l'expression de ce *momentum* est indépendante du centre de rotation. Il est donc clair que le centre de rotation se placera de lui-même là où le plan aura la moindre inertie pour recevoir

41 See the Introduction pp. 96–100 h. v.

le même degré de vitesse angulaire (je considere la vitesse angulaire, parce que l'amplitude angulaire des balancemens doit être la même, quel que soit le centre de rotation); & il n'est pas difficile de démontrer que c'est le centre de gravité qui a cette propriété réquise. Tout corps qui doit être tourné sur un certain axe par une force donnée, agissant sur un levier donné, ou bien appliquée à une même distance, depuis l'axe de rotation, aura la moindre inertie angulaire, & recevra dans un tems donné la plus grande vitesse angulaire, lorsqu'on fait passer l'axe de rotation par le centre de gravité du corps. Je n'ajoûte pas la démonstration de cette proposition méchanique, parce que, quoiqu'assez facile, elle ne laisse pas de demander plusieurs éclaircissemens, qui pourroient nous écarter trop de notre sujet principal.

§. XXXV. Mais il n'arrivera pas toujours, ou pour mieux dire, il arrivera rarement que le vaisseau fasse des balancemens *libres*. Les eaux ne sont point calmes; elles sont agitées, & n'agissent pas simplement par le principe de la pesanteur sur le navire; elles pourront agir d'une infinité d'autres manieres. En ces cas, il faut considérer la direction moyenne de toutes les impressions de l'eau contre le navire. Si cette direction moyenne prolongée passe au-dessous du centre de gravité du navire, le centre de rotation momentanée sera au-dessus, & réciproquement. On peut appliquer ici toute la théorie de la percussion excentrique, exposée dans le IXe Tome des Mémoires de l'Académie de Petersbourg[42], servant pour l'année 1737, p. 189. Le théorême principal de cette théorie est, que si un corps, dont on ne considere que l'inertie, est animé par une puissance[43] dont la direction ne passe pas par le centre de gravité du corps, il faut tirer une ligne par le centre de gravité du corps perpendiculaire à la ligne qui marque la direction de la puissance; que par le point d'intersection de ces deux lignes, il faut tirer une ligne perpendiculaire au plan de rotation; considérer ensuite le corps comme suspendu par cette derniere ligne horizontale, comme par un axe; du corps ainsi suspendu, il faut prendre le centre d'oscillation; & c'est ce point autour duquel le corps tournera, ou plutôt autour de la ligne parallele à l'axe de suspension, & qui passe par ledit centre d'oscillation. Quelques auteurs ont ensuite nommé le centre sur lequel le corps animé par la puissance excentrique tourne, le centre de rotation spontanée[44].

42 The name of the author is not mentioned, but it is in fact an earlier publication of D. Bernoulli St. 27, *De variatione motuum a percussione excentrica*: CP IX, pp. 189–206, 1737 (1744) – *Werke* 3, Mechanik, p. 40 (Introduction) and p. 145 (text).

43 The word *puissance* is derived from the Latin *potentia* and means a force, in the present meaning of the term.

44 See the Introduction p. 102 h. v.

C'est donc ce point sur lequel le navire tournera, entant qu'il est agité par l'impulsion des eaux. Or on sait, par les théorêmes du grand Huguens, que la distance du centre de gravité à l'axe horizontal de suspension, & la distance du même centre au centre d'oscillation, sont réciproquement proportionnelles. Plus donc la direction de l'impulsion moyenne passe près du centre de gravité, plus l'axe de rotation *spontanée* sera éloigné du centre de gravité, & réciproquement; & si la premiere distance est nulle, l'autre sera infinie. La vérité de cette conclusion est bien palpable; car lorsque l'impulsion moyenne des eaux passe par le centre de gravité du navire, son effet ne peut être que celui d'emporter horizontalement le navire, sans lui faire faire aucune rotation; ce qui marque que la distance du centre de rotation est infinie. Lorsque la direction de l'impulsion moyenne est à une grande distance du centre de gravité, le centre de rotation sera très-près du centre de gravité; l'impulsion ne causera presque aucun mouvement progressif, & son effet sera tout employé à faire tourner le navire. Tel est l'effet d'un coup de vent sur les voiles, parce que l'impulsion moyenne est extrêmement éloignée du centre de gravité du navire; c'est pourquoi ces coups de vent sont fort dangereux; aussi ne manque-t-on pas de caler vitement les voiles quand on les appréhende. Dans l'instant que les impulsions contre les navires cessent, celui-ci ne tournera plus que sur son centre de gravité; mais cependant le mouvement horizontal, que le centre de gravité aura acquis, continuera jusqu'à ce que des impulsions contraires commencent à agir contre le navire (car ces impulsions ne sauroient manquer d'être réciproques); à arrêter son mouvement horizontal; & enfin à lui imprimer un mouvement contraire. On voit par-là que le navire, outre les balancemens de rotation, fera toujours des allées & venues horizontales. Ce sont sans doute ces mouvemens horizontaux & réciproques, qui, mêlés avec le sillage moyen, occasionnent ces élans qu'on remarque; les allées & venues horizontales seront d'autant plus grandes, que la direction de l'impulsion moyenne passera plus près du centre de gravité, & les balancemens de rotation en seront d'autant plus petits.

§. XXXVI. Il résulte de ces principes, qu'outre les balancemens *libres* qui se font autour du centre de gravité, il y a une autre espece de balancemens, que j'appellerai *forcés*, qui se font autour du centre de rotation *spontanée*. Ce dernier centre sera extrêmement variable, à cause de la variabilité des forces momentanées, qui causent les balancemens *forcés*.

Une remarque essentielle à faire sur les balancemens *forcés*, est qu'ils ne font à chaque instant qu'un mouvement composé d'un mouvement de rotation autour du centre de gravité, & d'un mouvement parallele. Soit *pd*

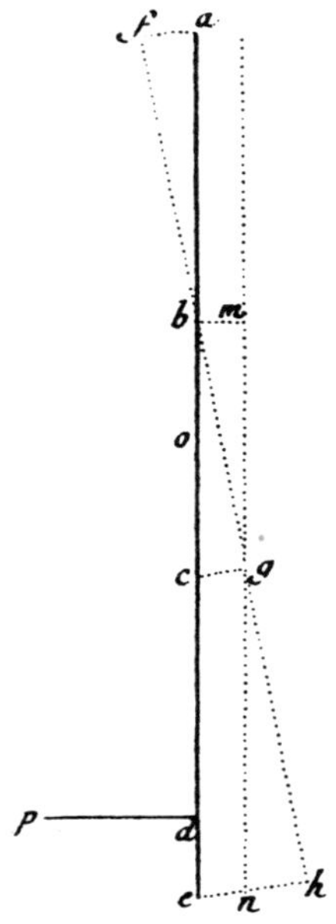

Fig. 4

(Fig. 4) la direction[45] de la force moyenne qui agit sur la carene; soit le centre de gravité du vaisseau en *c*, & que par ce point *c* on tire une ligne *ae* perpendiculaire à *pd*. Si on fait abstraction de l'action de la pesanteur, & si on considere simplement l'inertie de la matiere, la force *pd* fera prendre à la ligne *ae*, après un petit tems donné, la position *fh*; le point d'intersection *b* sera le centre de rotation spontanée. On voit d'abord que le mouvement, par lequel la ligne *ae* prend la situation *fh*, peut être résolu en deux mouvemens, par le premier desquels la ligne *ae* garde le parallelisme, & prend la situation *ln*, en jettant le centre de gravité de *c* en *g*; pendant que par le second, la ligne *ae* tourne autour du centre de gravité *g*, de maniere que *ln*, par cette rotation, prenne la situation *fh*. Ainsi *cg* mesurera la vitesse du mouvement parallele, & l'angle *fgl* mesurera la vitesse angulaire de la rotation. La proportion de ces vitesses sera la même pour la même direction *pd*; mais plus la force moyenne est grande ou petite, plus l'une & l'autre vitesse seront grandes ou petites. Pour déterminer le point *b*, je dis que ce point *b* est le centre d'oscillation, si le navire étoit suspendu par

45 In *Fig. 4* the letter *l* fails at the top of the vertical *nm*; the dotted lines *cg* and *en* should be horizontal.

le point d. Soit m une ligne constante, & $cd = x$; on aura, par la nature du centre d'oscillation, $bc = \frac{mm}{x}$. Soit aussi cg, qui exprime la vitesse du centre de gravité, $= c$; on voit que la même cg, rapportée au rayon bc, ou bien bm rapportée au rayon gm, exprime la vitesse angulaire. Mais comme la ligne bc se regle sur la ligne cd, que je regarde comme variable, il sera bon de rapporter les vitesses angulaires à un rayon constant a, par cette analogie, $bc : cg :: a : \frac{cg}{bc} \times a = \frac{ax}{mm}c$. Ainsi la vitesse du centre de gravité étant $= c$, la vitesse[46] angulaire ou rotatoire autour du centre de gravité, sera toujours exprimée[47] par $\frac{x}{mm}c$.

§. XXXVII. Nous voyons par-là que les balancemens du navire autour du centre de gravité, produits par les impressions des eaux agitées, sont proportionnels à la distance x, ou à la distance du centre de gravité depuis la direction de la force moyenne; & que si depuis le centre de gravité on prend une distance a, la vitesse de la rotation sera pour cette distance[48] $\frac{ax}{mm}c$. Ici il convient d'expliquer encore ce que c'est que la constante m que nous avons introduite; c'est la moyenne proportionnelle entre bc & cd. Soit donc oc égale à cette moyenne proportionnelle, ce point o aura pour tous les corps deux propriétés remarquables[49], la premiere est que si on suspend le corps par un axe perpendiculaire au plan de la figure passant par le point o (ou par un point quelconque pris dans la circonférence du cercle décrit autour du centre c par le rayon co), il fera ses balancemens brachystochrones, c'est-à-dire, de moindre durée que s'il étoit suspendu par tout axe parallele. La seconde propriété du point o est, qu'il est en même tems le centre des forces vives, lorsque le corps tourne autour d'un axe passant par le centre de gravité c; c'est-à-dire, que si toute la matiere étoit concentrée au point o, la force vive seroit la même que celle du corps. De cette derniere propriété on voit, que plus les extrêmités du corps, tournant autour du centre de gravité, sont chargées de matiere, plus le point o sera éloigné du point c, & plus par conséquent la ligne m devient grande. Tâchons à présent de mettre à profit la théorie que nous venons d'exposer.

46 m is the radius of gyration.
47 Misprinted as $\frac{ax}{mm}c$ in the original.
48 This is the linear speed, due to the angular velocity $\frac{x}{m^2}c$, for a point at a distance a from the spontaneous axis of rotation (b in Bernoulli's *Fig. 4*).
49 These two properties are not relevant for the problem.

§. XXXVIII. Nous avons dit que l'effet des impulsions réciproques des eaux contre la carene, est en partie de lui faire faire des allées & venues par un mouvement parallele, & en partie à faire tourner le navire autour du centre de gravité; le premier effet n'a absolument aucun inconvénient, & le second est le seul qu'on prétend de diminuer. Mais nous avons vu, dans le précédent article, que le mouvement de rotation résultant de l'impulsion des eaux, est exprimé par $\frac{ax}{mm}c$; pour faire que ce mouvement ait le moins de rapport à l'effet total, il faut que la quantité $\frac{x}{mm}$ soit la plus petite qu'il est possible; & de-là nous tirerons ces deux maximes, premierement d'augmenter la ligne m, & en second lieu de diminuer la ligne x, si l'on voit quelque moyen à cela. La premiere maxime demande d'éloigner la matiere du centre de gravité le plus qu'on peut; c'est ce que j'ai démontré dans le précédent article. Effectivement, si toute la matiere du navire étoit concentrée au centre de gravité, tout l'effet de l'impulsion tant soit peu excentrique consisteroit à faire tourner le navire sur son centre de gravité. En suivant cette maxime, on augmente en même tems l'inertie de la matiere; l'effet de chaque impulsion sera diminué, outre que les balancemens libres en seront rallentis. Mais y auroit-il aussi quelque moyen de diminuer la quantité x, c'est-à-dire, de faire passer la direction de l'impulsion moyenne près du centre de gravité du navire? On peut remarquer sur cela que si la carene étoit une portion de sphere dont le centre fût en même tems le centre de gravité du navire, non seulement la direction de l'impulsion moyenne passeroit par le centre de gravité, mais même la direction de chaque petite impulsion; parce que toute impulsion se fait perpendiculairement à la surface. Mais on voit aussi qu'un tel corps n'auroit plus aucune force de stabilité. Un autre principe seroit de faire ensorte que le *momentum* de toutes les impulsions dont les directions passeroient au dessus du centre de gravité, devint plus égal au *momentum* de toutes les impulsions dont les directions passeroient au-dessous. S'il y a quelque profit à espérer de ce principe, ce ne sera que sur un grand nombre d'observations & d'expériences faites, par des gens entendus, sur la nature des agitations de l'eau: la place du centre de gravité, la profondeur ou le creux de la carene, & sa figure, ne sont pas assez déterminés par les autres considérations, qu'on n'y puisse avoir égard au principe de rapprocher les deux *momentum* que je viens d'indiquer. Si les impulsions de l'eau sont tout-à-fait irrégulieres, elles ne pourront jamais causer de grands balancemens; & si elles sont régulieres & assujetties à certaines loix, il faudroit tâcher d'en connoître la nature, savoir à quelle profondeur elles ont le plus de force, & comment elles diminuent vers la quille & vers la surface de l'eau; comment elles peuvent différer entre elles d'un tems à l'autre & d'une mer à l'autre, suivant que les agitations des eaux sont plus ou moins fortes, & les lames plus ou moins grandes. Si on étoit un peu instruit sur ces points, & sur quelques autres d'une même nature, je ne doute pas qu'on n'en pût profiter, rélativement à notre derniere remarque. Je reprendrai cette matiere dans le chapitre suivant.

Un autre moyen pour diminuer les balancemens des navires, est celui de la résistance des eaux. M. Chauchot l'a exposé dans son Mémoire couronné. Ce nouveau principe me paroît fort convenable, à cause de la nature de la résistance des fluides, qui agit à fort peu près en raison quarrée des vitesses; elle ne sauroit donc mettre aucun obstacle au navire à se relever, après avoir achevé son excursion, parce qu'alors la vitesse est nulle; au contraire, lorsque le vaisseau est près de sa position naturelle, les eaux s'opposent avec le plus de force pour l'empêcher d'outrepasser cette position, & s'efforcent à l'y retenir. La résistance des eaux n'agit jamais qu'avec avantage, & il faut surtout en profiter à l'égard des roulis. Ceux-ci se font presque avec une liberté entiere, à cause de la figure arrondie de la carene & du peu d'éloignement qu'il y a depuis l'axe de la carene, considérée comme sphéroïdique, à l'axe de rotation qui passe par le centre de gravité du navire parallele à l'axe de la figure. Aussi voit-on que les roulis se font avec tant de liberté, qu'ils se continuent d'eux-mêmes pendant assez longtems, pendant que la seule résistance des eaux arrête tout d'un coup le tangage, qui ne sauroit se renouveller sans une nouvelle attaque.

§. XXXIX. Pour mettre ce moyen, fondé sur la résistance des eaux, à profit contre les roulis, il faut faire attention à la configuration de la carene & à la position de l'axe de rotation, qui doit passer par le centre de gravité du vaisseau; & alors on verra facilement sur quelle partie les eaux portent le plus de résistance. Je trouve en général, que c'est la hauteur de la quille, & les acculemens qu'on donne aux varangues, qui produisent la plus grande résistance des eaux dans les roulis[50]. La maitresse varangue n'a que très-peu d'acculement, mais ces acculemens augmentent à mesure que les varangues avancent vers la proue & vers la pouppe. Par cette construction, il provient comme une bande des façons, soutenue verticalement par la quille, dont les deux côtés sont comme deux plans verticaux. Cette bande des façons échancrée, choque directement les eaux pendant les roulis, conjointement avec les côtés verticaux de la quille.

50 *«En réalité c'est le frottement de l'eau sur la carène qui donne la plus grande partie de la résistance de l'eau.»* (A. L.)

On augmentera donc la résistance des eaux contre les roulis, en augmentant les acculemens, les hauteurs des façons, & la hauteur de la quille; par-là on diminuera en même tems la dérive dans les routes obliques. Le reste de la carene est trop arrondi, & l'axe de rotation trop près de l'axe de l'arrondissement, pour en attendre une grande résistance. En examinant la dite bande des façons, il paroît que les deux parties, depuis la maitresse varangue jusqu'aux deux extrêmités, sont assez inégales; celle de l'arriere étant plus longue & plus haute que celle d'avant, ne sauroit manquer de trouver plus de résistance. Il me semble que cette inégalité doit causer un mouvement de nutation horizontale dans les navires[51], lorsqu'ils roulent considérablement, & que cette nutation doit être nuisible au sillage. Je voudrois que les constructeurs examinassent scrupuleusement, si les raisons qu'ils alléguent pour donner une pente à la quille de la proue vers la pouppe, pour mettre la maitresse varangue plus près de la proue que de la pouppe, & pour donner plus d'acculement aux varangues de l'arriére qu'à celle de l'avant, sont bien réelles & suffisantes pour l'emporter sur les raisons contraires[52]. Si la quille étoit horizontale, & que les deux parties fussent égales & semblables, la bande des façons d'avant en seroit augmentée, les deux demi-bandes deviendroient égales, & on éviteroit ladite nutation. Je dirai encore, à l'occasion de cette résistance, que celle de l'air contre les voiles peut faire le même effet. Les Marins assurent que le navire roule plus avec un vent arriere, & qu'il tangue plus au plus près. C'est apparemment parce que l'air, dans le premier cas, s'oppose moins aux voiles pendant que le vaisseau roule, que pendant qu'il tangue; & que dans le second cas, c'est le contraire. Si cela est vrai, on peut encore se servir souvent des voiles pour modérer les balancemens du navire, surtout dans les calmes, pendant lesquels le navire roule quelquefois excessivement.

§. XL. Un remede bien sûr pour diminuer le roulis, est d'appliquer aux flancs du vaisseau deux ou trois bandes saillantes & horizontales un peu au-dessus de la flottaison. De telles bandes agiront par deux principes à la fois; car lorsque, pendant le roulis, elles atteignent les eaux, elles les frappent d'abord par leur saillant, & puis forment cette résistance dont je viens de parler; & aussitôt qu'elles sont submergées, il survient un nouvel accroissement de poussée d'eau de bas en-haut, qui n'agissant que d'un côté, emploie tout son effet à relever le navire. Je suis persuadé qu'avec une telle bande

51 In nautical terms: yaw.

52 *«La quille inclinée de la proue à la poupe présente l'inconvénient qu'en cas d'échouage du navire sur un banc de sable une grande partie du fond du navire sera en contact avec ce sable et il sera plus difficile de le déséchouer que dans le cas contraire.»* (A. L.)

de chaque côté, dont la saillie & la hauteur fût un peu considérable, on pourra parvenir à borner les roulis dans de certaines limites. Un navire, dans ses roulis excessifs, fait quelquefois jusqu'à 35^{d} d'excursion de chaque côté, & alors ces roulis sont certainement dangereux. Je crois qu'avec tous les moyens que j'ai exposés, on diminuera aisément jusqu'à 25^{d} les plus grands roulis, & ainsi la plus haute bande (si on en veut employer plus d'une) sera placée au-dessus de la flottaison, tout au plus de la hauteur de la cinquieme partie de la largeur, sans quoi elle pourroit devenir inutile à l'égard des roulis; c'est aussi jusqu'à cette hauteur, du moins, que je conseille de ne donner aucune rentrée aux coupes verticales du vaisseau faites perpendiculairement à la longueur.

§. XLI. Les bandes saillantes, dont je viens de parler, ne sauroient manquer d'être en même tems d'un grand secours contre le tangage, si on les prolonge jusqu'aux extrêmités des flancs, surtout si on vouloit augmenter leur largeur ou leur saillie à mesure qu'elles approcheroient de la proue & de la pouppe; ce qu'on obtiendrait si on faisoit leurs côtés opposés aux flancs entierement droits, puisque les côtés appliqués aux flancs auront toujours quelque concavité, malgré le conseil que j'ai donné ci-dessus d'approcher la section horizontale du vaisseau à fleur d'eau de la figure rectangulaire. Ces bandes ou ces aîles seroient naturellement hors de l'eau, & ne seroient par conséquent en aucune façon nuisibles aux autres qualités du navire, pendant que par leurs extrêmités élargies, elles s'opposeroient efficacement au tangage, aussitôt qu'elles atteindroient aux eaux, & cela par le double principe que j'ai exposé dans le précédent article. Je prie donc les Constructeurs de faire quelque attention à ce moyen, & de juger par eux-mêmes jusqu'à quel degré on peut l'embrasser, sans tomber dans des inconvéniens. C'est toujours avec cette réserve que je propose mes moyens.

§. XLII. Considérons enfin ces cas fâcheux dont j'ai parlé au §. XXIII & les suivans, jusqu'à la fin du chapitre, où j'ai fait voir que les roulis ne sauroient manquer d'être extraordinairement forts, lorsque le tems θ est à-peu-près égal au tems t; c'est-à-dire lorsque les *roulis forcés* sont à-peu-près de même durée avec les *roulis libres*. J'espere qu'on ne voudra pas traiter mes théorêmes sur cette matiere de spéculations purement hypotétiques, ou du moins mal appliqués à notre sujet. L'expérience physique, alléguée à la fin du §. XXIII, en prouve assez la réalité. D'ailleurs, personne n'ignore qu'un son fort met en vibrations très-sensibles une corde tendue à l'unisson, pendant que toute autre corde ne fait aucune vibration sensible. C'est ici absolument notre cas. Le son forme des ondulations de l'air, comparables aux ondes de la mer. Ces ondulations de l'air mettent la corde dans des vibrations, parce que les allées & venues de la corde sont isochrones avec les mêmes ondulations, & que la force motrice conspire toujours avec le mouvement de la corde. Les ondes de la mer feront le même effet sur un navire mis à l'isochronisme avec le mouvement réciproque des ondes, comme il l'est lorsque ses balancemens *libres* sont de même durée que les balancemens des eaux, ou bien que les balancemens *forcés* du navire. Aussi a-t-on remarqué souvent, qu'un navire rouloit extraordinairement lorsqu'on ne voyoit rien d'extraordinaire ni dans la force du vent, ni dans les agitations des eaux. N'auroit-on jamais remarqué que de deux vaisseaux inégaux l'un rouloit quelquefois plus extraordinairement que l'autre, quoiqu'ils fussent l'un & l'autre dans les mêmes circonstances? Je vois bien que ces roulis extraordinaires & excessifs doivent être rares, parce qu'il faut que non-seulement les balancemens des eaux soient synchrones avec les balancemens *libres* du navire, mais encore que les forces motrices soient pendant quelque tems uniformes & permanentes. Cependant tout cela peut arriver; & comme notre sujet demande de remédier aux roulis excessifs, nous devons examiner s'il n'y a absolument aucun moyen particulier pour cela.

§. XLIII. Remarquons d'abord que les moyens que nous avons déja donnés, ne manqueront pas de modérer aussi ces roulis extraordinaires; mais je suis persuadé que si on pouvoit mettre quelque achronisme entre les deux especes de balancemens, ou du moins troubler cette harmonie qu'il y a entre les accès des forces motrices & les balancemens du vaisseau, ceux-ci en seroient aussitôt considérablement diminués. Je crois qu'on y arrivera en remuant habilement & à propos le gouvernail, tantôt à droite, tantôt à gauche; c'est une manœuvre qu'on devroit étudier, si elle peut être utile; sur quoi je m'en rapporte aux experts. Je crois aussi qu'il sera toujours bon de connoître exactement pour chaque vaisseau la durée d'un balancement entierement libre; il en faudroit faire l'essai dans le port même. Deux bandes d'hommes opposées pourront, moyennant deux longues cordes attachées au mât, ébranler assez sensiblement le navire, par des efforts duement appliqués, pour distinguer les roulis; alors je voudrois qu'on en comptât le nombre pour trois ou quatre minutes de tems. Si on pouvoit parvenir à exciter de grands roulis, on pourroit alors reconnoître l'achronisme qu'il y auroit entre les grands roulis & les petits[53]. Ensuite de quoi on pourra toujours, dans l'occasion,

53 «*Pour provoquer le roulis de grande amplitude, il suffit de faire marcher cinq hommes sur le pont transversalement et changeant le sens de leur marche au rythme du roulis propre du navire.*» (A. L.)

comparer les balancemens actuels ou *forcés* avec les balancemens *libres*, en comptant les premiers pour une ou deux minutes de tems.

Si le navire est irrégulierement balotté, ces agitations, à mon avis, ne pourront pas être fort grandes, & elles ne sont pas dans le cas dont nous parlons; mais lorsque les roulis sont réguliers & uniformes en durée & en grandeur, on saura s'ils avancent ou retardent sur les roulis libres; & je présume que l'un & l'autre peut arriver. S'ils avancent, il faudra tâcher de retarder davantage les roulis *libres*; & s'ils retardent, il conviendra d'accélérer ces roulis: une très-petite accélération ou retardation des roulis libres, pourra faire ici quelque effet. On remarquera à ce sujet, qu'on retardera les roulis *libres*, tant en diminuant la stabilité, qu'en augmentant l'inertie du navire roulant. Quant à la stabilité, on ne pourra pas la faire changer autrement qu'en changeant la hauteur du centre de gravité par-dessus le centre de la carene homogene; si on augmente cette hauteur, on diminue la stabilité, & réciproquement. On peut hausser le centre de gravité par-dessus le centre de la carene, en plaçant plus haut tout ce qui est mobile, & qu'on pourra remuer commodément; & comme dans ces cas la grandeur des roulis n'est pas causée par aucun défaut de stabilité, un tel changement pourra se faire ici sans tomber dans de nouveaux inconvéniens. S'il s'agissoit d'accélérer les roulis libres, il faudroit faire tout le contraire.

Quant à l'inertie du navire roulant, on l'augmentera en éloignant tout ce qui est mobile de l'axe de rotation, qui passe par le centre de gravité; & comme l'inertie des parties augmente en raison quarrée de leurs distances à l'axe de rotation, un tel éloignement en fera d'autant plus d'effet. On hausse encore le centre de gravité en haussant les vergues & les voiles, aussi-bien qu'en mouillant ou *empesant* les voiles. La stabilité diminue & l'inertie augmente.

§. XLIV. Voyons enfin ce qui doit arriver, quand on augmente ou diminue la charge du vaisseau[54]. Soit la coupe horizontale du navire à fleur d'eau $= S$, & considérons-la comme demeurant la même, pendant qu'on augmente ou diminue la charge, conformément au conseil que j'ai donné de faire les flancs droits & verticaux. Soit la solidité de la carene $= aS$, & que dans cet état la distance du centre de gravité à la surface de l'eau soit $= b$ (je suppose ce point plus bas que la surface de l'eau), & la distance du centre de gravité de la carene homogene depuis la surface de l'eau $= c$. Qu'on augmente ensuite la charge jusqu'à rendre la solidité de la carene $= aS + \alpha S$. Pour

54 See the Introduction p. 104 h. v.

connoître l'effet que fera la nouvelle charge sur le navire, il faut savoir l'endroit où l'on placera cette nouvelle charge, ou plutôt son centre de gravité; supposons cet endroit plus bas que n'étoit auparavant le centre de gravité du navire, de la hauteur β. Sur cela les regles de la Statique donnent la hauteur de la surface d'eau par-dessus le centre de gravité, après l'augmentation de la charge, $= b + \alpha + \dfrac{\alpha}{a + \alpha}\beta$, & la hauteur de la surface de l'eau par-dessus le centre de gravité de la carene homogene, $= c + \alpha - \dfrac{c\alpha + \frac{1}{2}\alpha\alpha}{a + \alpha}$. Ainsi la hauteur du centre de gravité du navire par-dessus le centre de gravité de la carene, est avant l'augmentation de la charge $= c - b$, & après l'augmentation,

$$= c - b - \frac{c\alpha + \frac{1}{2}\alpha\alpha + \beta\alpha}{a + \alpha}\ .$$

Ainsi cette hauteur devient plus petite par l'augmentation de la charge tant que β est affirmatif. En considérant la quantité α comme fort petite, il faudroit que β fût $= -c$ pour que la distance entre les deux centres de gravité fût la même avant & après l'augmentation de la charge; c'est-à-dire, qu'en plaçant la nouvelle charge autant au dessus du centre de gravité du navire, que la surface de l'eau étoit élevée par-dessus le centre de gravité de la carene. On voit ici le grand effet du lest pour approcher le centre de gravité du navire du centre de gravité de la carene. Car supposons un navire avec toute sa charge & non lesté, & qu'on ait dans cet état $a = 10$ piés, $c = 6$ piés: qu'on leste ensuite le navire, & que la quantité du lest fasse la cinquieme partie du poids de tout le reste du navire chargé; on aura $\alpha = 2$ p.; & comme ce lest est placé au fond de cale, on pourra supposer $\beta = 14$ p. Ces suppositions, qui ne sont pas incongrues, font la quantité

$$\frac{c\alpha + \frac{1}{2}\alpha\alpha + \beta\alpha}{a + \alpha} = 3\tfrac{1}{2}\ \text{p.}$$

De sorte que le centre de gravité du navire s'approche de celui de la carene de trois piés & demi par le lest. Voilà le changement qui arrive à l'égard de la distance mutuelle des deux centres de gravité en question; mais comme le second membre de la formule qui exprime la stabilité est proportionnel à la solidité de la carene multipliée par la hauteur entre les deux dits centres, il faut, pour voir si la stabilité croît ou décroît par l'augmentation de la charge, multiplier $c - b$ par aS, & $c - b - \dfrac{c\alpha + \frac{1}{2}\alpha\alpha + \beta\alpha}{a + \alpha}$ par $aS + \alpha S$; le premier produit est $caS - baS$, & le second produit est

$$caS - baS - \frac{c\alpha + \frac{1}{2}\alpha\alpha + \beta\alpha}{a + \alpha} aS + c\alpha S - b\alpha S - \frac{c\alpha + \frac{1}{2}\alpha\alpha + \beta\alpha}{a + \alpha} \alpha S \ .$$

Le premier produit est plus grand que le second, de la quantité[55]

$$\frac{\frac{1}{2}a\alpha\alpha + a\beta\alpha + ba\alpha + b\alpha\alpha + \frac{1}{2}\alpha^3 + \beta\alpha\alpha}{a + \alpha} S \ .$$

Et comme ces produits doivent être retranchés de la valeur du premier membre de la formule, il s'en suit que l'augmentation de la charge augmente toujours la stabilité absolue, tant que β n'est pas négatif. Ainsi les eaux qui sont à fond de cale augmentent la stabilité; mais elles font plonger davantage le navire, & le sillage en est retardé. Si la quantité α est très petite, on peut censer le gain qu'on fait sur la stabilité, par l'augmentation de la charge, $= (b + \beta)\alpha S$. Lorsqu'on diminue la charge, il faut prendre negativement la valeur de α; & si l'endroit d'où l'on ôte la charge est plus haut que le centre de gravité, il faut aussi prendre négativement la valeur de β. Par cette double négation, on gagne considérablement sur la stabilité, en coupant les mâts, à cause de leurs hauteurs. Si on donne beaucoup de profondeur à la carene, si on emploie du bon lest & en quantité, & si la charge est bien arrangée & pesante par elle-même, on pourra faire descendre le centre de gravité du navire jusqu'à celui de la carene, & peut-être plus bas; on pourra alors se relâcher sur le premier membre de la formule qui exprime la stabilité, & diminuer la largeur du vaisseau, lorsque d'autres raisons, étrangeres à notre sujet, le demandent.

CHAPITRE V.

Explication de la cause principale des roulis & des tangages, & de la meilleure maniere de les diminuer, relative à cette cause, avec quelques réflexions sur la nature des lames.

§. XLV. Nous avons considéré jusqu'ici la surface des eaux comme horizontale; ou plutôt, nous n'avons pas encore fait attention à la figure ondoyante des lames. C'est cependant dans cette figure ondoyante & variable

55 See the Introduction p. 105 h. v.

que consiste la principale cause des agitations du navire. L'équilibre des corps qui nagent au milieu de ces ondes, est tout différent de celui qui convient aux eaux dont la surface est unie & horizontale; quand le navire ne feroit que suivre constamment sa position d'équilibre, il seroit obligé de faire des balancemens très-considérables, que j'appellerai *balancemens d'équilibre*; & les balancemens absolus seront toujours plus grands que les simples balancemens d'équilibre. C'est ici un nouveau sujet de réflexion fort essentiel & très-important; mais ce sujet nous meneroit extrêmement loin, si nous voulions le traiter avec toute l'exactitude dont il est susceptible. Les loix hydrodynamiques connues l'éclairciroient suffisamment, moyennant des hypothèses physiques bien choisies & telles qui ne peuvent nous écarter sensiblement de la vérité. Mais il nous faudroit examiner la figure des lames, le mouvement des eaux, tant près la surface qu'à des profondeurs données, quelle est à chaque instant & à chaque profondeur la compression des eaux, qui n'est plus exactement proportionnelle à la colonne verticale terminée par la surface des eaux. C'est cette compression qui marque la poussée de l'eau contre chaque élément de la carene, & elle ne sauroit être déterminée sans connoître les accélérations & retardations du mouvement des eaux, ni par conséquent la vraie position d'équilibre du navire. Toutes ces recherches demandent un traité à part, & je dois me contenir dans les bornes d'un Mémoire, que j'ai peut-être déja passée. Je n'exposerai donc que le plus précis de mes idées sur cette matiere, & je simplifierai les hypothèses autant qu'on peut le faire, sans tomber dans des erreurs sensibles.

§. XLVI. Si nous supposons que le vent souffle constamment avec la même force, & sous la même direction; que la mer est à une très-grande étendue fort profonde & libre, & que le fond de la mer est uni & horizontal, il n'y a aucun doute que les lames excitées par un tel vent, & sous de telles circonstances, ne soient extrêmement uniformes & régulieres. C'est sous cette forme que nous allons le considérer.

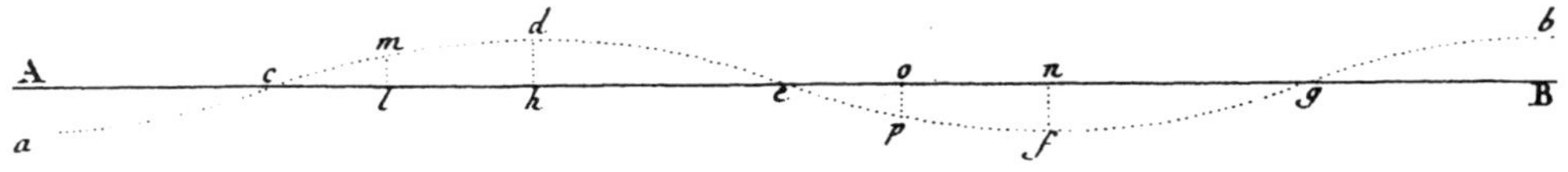

Fig. 5

Soit donc AB (Fig. 5) la surface horizontale de la mer, ou plutôt une ligne de cette surface prise parallelement à la direction du vent; & supposons que pendant les agitations des eaux, cette ligne se change en *acdefgb*. Toute cette ligne continuée de part & d'autre, formera une courbe continue qui pourra être exprimée par une seule & même équation; & pour peu qu'on fasse attention à toutes les propriétés que cette courbe doit avoir manifestement, on voit assez qu'elle ne sauroit s'éloigner beaucoup de la nature de celle que prend une longue corde tendue, lorsqu'elle fait des vibrations, en formant plusieurs nœuds, tels que sont ici les points c, e, g, &c. Ainsi, pour exprimer la nature de notre courbe ondoyante, je me servirai de l'équation $y = \alpha \sin \operatorname{Arc} \frac{x}{a}\pi$, en entendant par x une abscisse quelconque cl, par y l'appliquée lm, par α la plus grande appliquée hd prise au milieu de la lame, par a toute la largeur de la lame ce, & enfin par π le demi-cercle, dont le rayon est égal à l'unité. Suivant cette idée, toute la ligne $acdefgb$ prend au même instant la position du niveau AB; les nœuds, tels que c, e, g, restant immobiles; chaque onde, telle que cde, fait des balancemens alternativement au-dessus & au-dessous la droite che; à mesure qu'elle s'éleve ou se baisse, celle qui lui est voisine se baisse ou s'éleve. En différentiant notre équation, on trouve $dy = \frac{\pi\alpha}{a} dx \cos \operatorname{Arc} \frac{x}{a}\pi$. Cette équation différentielle marque qu'aux points de la plus grande élévation & de la plus grande dépression, tels que d, f, la surface est constamment horizontale, & que c'est dans les nœuds que la surface des eaux prend la plus grande pente. La tangente de chaque angle deh sera $\frac{\pi\alpha}{a}$. Si donc on suppose, par exemple, la plus grande hauteur d'une lame par-dessus la surface horizontale de la mer faire la sixieme partie de la largeur de la lame; c'est-à-dire, si $dh = \frac{1}{6}ce$, on aura l'angle $deh = 27° 40'$.

Je n'ai rien trouvé dans aucun auteur qui détermine ni la largeur ni la plus grande hauteur d'une lame; il y a cependant apparence qu'en général les lames sont d'autant plus grandes, que le vent est plus fort, & que la mer est plus profonde. Je remarquerai seulement qu'on sait dans la Physique que la moindre circonstance ou la plus petite cause imaginable suffit pour fixer les nœuds dans ces sortes de systêmes, & que la nature recherche avec un soin sans bornes le synchronisme dans les variations périodiques combinées. Ces raisons me font soupçonner, que dans la mer profonde la lame fera précisément un balancement entier dans le tems que le vent emploie pour parcourir la largeur de la lame; & qu'à mesure que la profondeur de la mer diminue, le nombre des balancemens augmentera, pour le même vent en raison des nombres ou naturels impairs; & j'ai plus de penchant pour la derniere opinion. Il en est comme des tons qu'on tire en variant l'embouchure d'un seul & même tuyau simple, & qui dans les tuyaux ouverts vont comme les nombres 1, 2, 3, &c.; & dans les tuyaux bouchés[56], comme les nombres 1, 3, 5, &c. Si donc un vent produit dans l'Océan des lames de 300 piés de largeur, ce même vent pourra produire dans d'autres mers moins libres & moins profondes des lames de 100 piés de largeur. Voici une expérience qui éclaircit ce sujet: une corde tendue sera ébranlée en pinçant une autre corde mise à l'unisson; partagez la premiere en trois parties égales, dont chacune donne la douzième, & chaque partie sera encore ébranlée. Les grandes lames de l'Océan répondent aux sons fondamentaux; & les petites, excitées par le même vent, aux sons accessoires, qu'on remarque dans chaque corde pincée. Quant à la durée de chaque lame, on ne sauroit la déterminer précisément; mais il est assez avéré que ces durées sont pour les lames semblables en raison sous-doublée des largeurs des lames.

§. XLVII. M. Newton, pour expliquer la nature des lames, suppose, en prenant par-tout $eo = cl$, un tuyau uniformément large, qui descende verticalement de chaque point l jusqu'à une certaine profondeur, qui se replie ensuite horizontalement, & qui enfin remonte verticalement jusqu'au point o; après quoi il considere le mouvement des lames comme formé par des balancemens de l'eau dans toute cette suite infinie de tuyaux communicans, & il démontre qu'en supposant la longueur de chacun de ces tuyaux $= L$, les balancemens deviennent isochrones avec ceux d'un pendule simple de la longueur[57] $\frac{1}{2}L$. De cette maniere, les balancemens des lames dépendent, quant à leur durée, de la profondeur des eaux agitées; & cette profondeur peut être supposée proportionnelle à la largeur de chaque lame. Supposons donc la largeur d'une lame $= x$, & la profondeur à laquelle les eaux sont agitées $= mx$; supposons aussi que le vent parcoure 12 piés dans une seconde, & donnons simplement 3 piés à la longueur du pendule simple à secondes. Dans ces suppositions, le tems que le vent emploie pour parcourir la largeur de la lame, sera $= \frac{x}{n}$ secondes, & le tems d'un balancement entier de la lame sera $= \sqrt{\frac{2mx + x}{6}}$ secondes, & si on fait $\frac{x}{n} = \sqrt{\frac{2mx + x}{6}}$, on obtient $x = \frac{2mnn + nn}{6}$ piés. Soit, par exemple, $n = 30$, & $m = 6$, la largeur de la lame sera $= 1950$ piés; mais dans les mers peu profondes, cette largeur peut, à mon avis, se réduire au tiers, ou à la cinquieme, ou à la septieme partie, &c.

56 This is a misleading analogy; see the Introduction p. 114 h. v.
57 See the Introduction p. 108 h. v.

Il est vrai cependant que cette maniere d'envisager les balancemens des lames, souffre beaucoup de difficultés; car tous ces tuyaux devroient se croiser; ce qui blesse absolument l'imagination. Outre cela, on voit que rien ne sauroit empêcher les eaux dans la partie *def* de rouler vers l'endroit le plus bas; de sorte que le mouvement alternatif des eaux doit être manifestement en partie horizontal & en partie vertical; les balancemens horizontaux seront les plus rapides vers les nœuds, parce que la pente y est la plus grande; & ils seront nuls au milieu, entre deux nœuds, parce que les eaux n'y ont point de pente. Les balancemens horizontaux doivent retarder les balancemens verticaux, parce que l'action de la pesanteur est partagée; & comme cependant tous les balancemens doivent être absolument isochrones & synchrones, il s'ensuit que les tuyaux communiquans ne sauroient être partout d'une longueur égale. Je considere donc plutôt chaque deux demi-lames *de* & *fe* comme composées de canaux ou tuyaux emboîtés les uns dans les autres, & que toutes ces paires de demi-lames alternatives forment chacune un systême à part, sans se troubler les unes les autres & sans jamais mêler leurs eaux. Les extrêmités de chacun de ces tuyaux sont également éloignées du point *e*; près des points *d* & *f*, le mouvement des eaux sera presque purement vertical, pendant que près du point *e*, ce mouvement est pour la plus grande partie horizontal. On voit donc que les extrêmités de chaque tuyau sont d'autant plus inclinées, qu'elles sont plus proches du point *e*, & que ces tuyaux sont plus courts; & c'est-là précisément la source de l'isochronisme commun à tous les tuyaux. Car si d'un côté les balancemens sont accélérés par la diminution de la longueur du tuyau, ils sont d'un autre côté retardés par l'inclinaison des extrêmités du même tuyau. Soit la longueur du tuyau uniformément large $= \lambda$ & le sinus de l'angle que le tuyau fait avec l'horizon vers ses extrêmités $= s$; on aura la longueur du pendule simple isochrone avec les balancemens des eaux dans ce tuyau[58] $= \frac{\lambda}{2s}$. Il arrivera donc, suivant ces idées, que s soit partout proportionnel à λ, pourvu qu'on suppose les tuyaux d'une même largeur dans toute leur longueur[59].

§. XLVIII. Il nous reste à examiner les loix hydrostatiques, suivant lesquelles se fait la poussée de l'eau, dont la surface n'est pas horizontale, pour soutenir les corps flottans. C'est ici la question la plus essentielle. Voici un problême dont la solution pourra éclaircir en quelque maniere notre question.

58 See the Introduction p. 110 h. v.
59 This hypothesis is approximately true; see the Introduction p. 113 h. v.

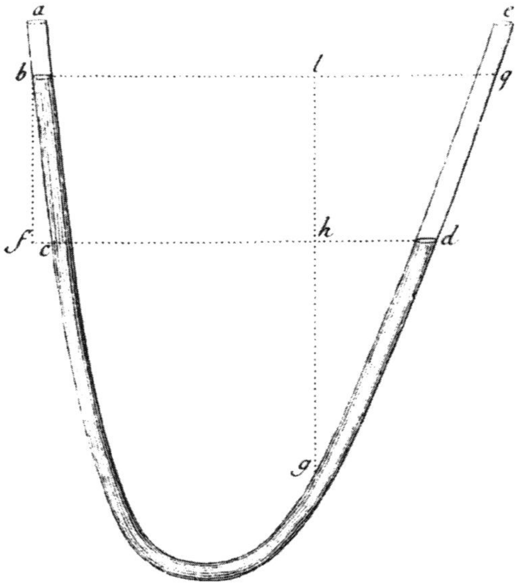

Fig. 6

Soit *acde* (Fig. 6) un canal uniformement large & recourbé, dont la partie *bgd* soit remplie d'eau qui se meut dans le sens *bcd*; il est question de déterminer la compression dans chaque point g.

Qu'on tire les horizontales *fd* & *bq*, & les verticales *bf* & *ghl*. Soit $bf = x$, $gh = y$; la longueur $bcd = l$; la longueur $gd = z$, & qu'on nomme π la hauteur verticale de la colonne qui indique la compression cherchée pour le point g: on aura la force accélératrice qui anime la partie $bcg = \frac{x + y - \pi}{l - z}$, & celle qui anime la partie $gd = \frac{\pi - y}{z}$; & il est clair que ces deux forces accélératrices doivent être égales entre elles, puisque les vitesses de l'une & de l'autre partie sont, durant tout le mouvement, égales entre elles. Nous avons donc

$$\frac{x + y - \pi}{l - z} = \frac{\pi - y}{z},$$

ce qui donne $\pi = y + \frac{z}{l}x$. Cette valeur nous fournit beaucoup de corollaires.

(a) Pour un tuyau simple & droit, la compression est partout nulle, de quelque façon qu'on l'incline; parce que $x = nl$, & $y = -nz$.

(b) La formule indique, comme il doit arriver, que la compression est nulle tant au point b qu'au point d, puisque pour le point b, il faut faire $y = -x$ & $z = l$; & que pour le point d, il faut supposer $y = o$ & $z = o$.

(c) Au moment que les points b & d sont de niveau, la compression est la même que dans les eaux calmes.

(d) La formule satisfait aussi à la conversibilité des extrêmités b & d; car si on veut rapporter à l'extrêmité d, ce que j'ai dit par rapport à l'extrêmité b, on voit bien qu'il faut substituer $-x$ à la place de x, $l - z$ à la place de z, & $y + x$ à la place de y. Si l'on substitue donc ces valeurs, on trouve, à la place de $y + \frac{z}{l}x$, la quantité $y + x - \frac{l-z}{l}x$; ce qui fait, comme auparavant, $y + \frac{z}{l}x$.

(e) Lorsque le point g est fort près du point b, & qu'on nomme cette petite distance α, on aura $l - z = \alpha$ & $y = -x + \beta$, en entendant par β la petite hauteur verticale du point b par-dessus le point g, pris fort près de l'autre; substituant donc ces valeurs, on trouve la compression $= \beta - \frac{x}{l}\alpha$. Cette compression est donc plus petite que suivant les loix de l'Hydrostatique ordinaires.

(f) Si au contraire, le point g est fort près de l'autre extrêmité, savoir du point d; & si on suppose la petite distance $gd = \alpha$: on aura $z = \alpha$ & $y = \beta$, en entendant par β la petite hauteur verticale du point d par-dessus le point g; sur cela la compression de l'eau devient $= \beta + \frac{x}{l}\alpha$, qui est par conséquent plus grande que suivant les loix hydrostatiques ordinaires.

§. XLIX. Pour appliquer, ce que nous venons de trouver, aux lames, il faut examiner quel peut être le tuyau qui passe par le point, pour lequel on cherche la compression; & quoiqu'on ne puisse rien déterminer de précis là-dessus, on voit cependant que toutes les eaux élevées par-dessus le niveau de la mer calme, sont moins comprimées que suivant les regles communes. La différence peut être sensible près la surface des eaux, comme on voit par les corollaires (e) & (f); car en supposant les deux branches verticales près la surface de l'eau, ce qui fait $\alpha = \beta$, la compression pourra différer en raison de 1 à $1 \pm \frac{x}{l}$; & comme $\frac{x}{l}$ pourroit fort bien faire $\frac{1}{6}$, il s'ensuit que ladite différence pourroit bien s'étendre jusqu'à un sixieme de la compression ordinaire. Cependant il n'en faut pas conclurre que le navire plonge plus ou moins qu'à l'ordinaire, lorsqu'il se trouve au-dessus ou au-dessous du niveau moyen de la mer, parce que le navire aura lui-même à-peu-près le même mouvement qu'auroit l'eau qu'il déplace. Les résultats de notre problême sont à la vérité fort utiles pour connoître à-peu-près ce qui doit arriver dans différens cas; mais ils n'admettent pas un calcul exact. Il faut surtout faire attention à la largeur des lames, & la comparer avec les dimensions du navire, puisqu'il est tout simple que les lames fort courtes doivent faire sur le navire un tout autre effet, que les grandes lames de l'Océan, sur lesquelles le navire ne représentera qu'un assez petit corps. C'est surtout ce dernier cas que nous devons examiner avec plus d'attention.

§. L. On voit bien que chaque point de la carene peut être considéré comme étant fort près de l'extrêmité du tuyau, qui passe par ce point; on voit aussi que lorsque le navire est censé occuper une petite étendue, chaque point de la carene est soutenu par une colonne d'eau, dont la hauteur verticale est, en vertu des corollaires (*e*) & (*f*) du précédent §., $= \beta \pm \frac{x}{l}\alpha$, en entendant par β la hauteur verticale depuis le point en question jusqu'à la surface de la lame, par α la position du tuyau intercepté entre la surface de la lame & le même point en question, par l la longueur entiere du tuyau, & par x la hauteur verticale entre les deux extrêmités du tuyau. Remarquons ici que α doit être à-peu-près proportionnelle à β, parce que la petite portion de tuyau peut être considérée comme une ligne droite inclinée, & tous les points de la carene répondront à-peu-près à des tuyaux dont les extrêmités seront également inclinées. La quantité ou le rapport de $\frac{x}{l}$ peut aussi être censé le même pour tous les tuyaux qui répondent aux différens points de la carene, tant à cause de la petite étendue de la carene par rapport à la grande largeur de la lame, que parce que la quantité x ne sauroit manquer d'être à-peu-près proportionnelle à l pour des tuyaux peu éloignés entre eux. Toutes ces considérations nous menent à la conclusion, qu'on peut supposer pour chaque point de la carene la quantité $\frac{x}{l}\alpha$ proportionnelle à β; de sorte que la quantité $\beta \pm \frac{x}{l}\alpha$ peut être censée $= n\beta$, en entendant par n un nombre un peu plus grand ou un peu plus petit que l'unité, suivant que le vaisseau se trouve au-dessous ou au-dessus du niveau de la mer calme. Nous voyons de-là que chaque point de la carene est soutenu à-peu-près par une force proportionnelle à la hauteur verticale prise depuis le point en question & terminée par la surface des eaux ondoyantes[60]; & c'est ici la propriété principale que je prétendois établir.

60 This turns out to be an oversimplification; see the Introduction p. 119 h. v.

§. LI. Il suit de la dite propriété, que quelque courbée & inclinée que soit la surface de l'eau, la poussée de l'eau qui soutient un corps flottant, peut toujours être considérée comme concentrée dans le centre de gravité de la partie submergée censée homogene, tout comme dans les eaux calmes, sans tomber dans aucune erreur sensible, pendant que la pesanteur du corps doit être considérée comme concentrée dans son centre de gravité. L'équilibre pour chaque moment demande encore que les deux dits centres de gravité se trouvent dans une seule & même verticale; mais nous allons voir qu'un tel équilibre doit nécessairement faire incliner le corps lorsque la surface de l'eau est inclinée, & nous ne pouvons nous dispenser de rechercher quelle sera la relation entre l'inclinaison de la surface d'eau, & celle du corps supposé en équilibre. Pour faciliter cette nouvelle recherche, nous supposerons l'inclinaison de la surface d'eau très-petite, & la surface d'eau, pour la petite étendue que le corps occupe, être plane. Ces suppositions ne pourront nous écarter sensiblement de la vérité, & les théorêmes qui résulteront seront également utiles. Outre ces suppositions, nous considérerons simplement un plan verticalement plongé dans les eaux, & parallelement au plan de la cinquieme figure. Nous simplifions ainsi la question, afin de pouvoir exprimer, par des formules analytiques, la stabilité ordinaire, sans employer aucun signe sommatoire.

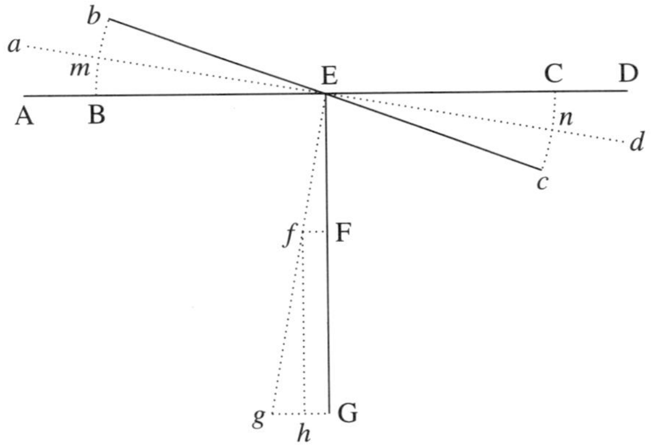

Fig. 7 (redrawn after the original)

§. LII. Soit donc à présent *AD* (Fig. 7) la surface horizontale des eaux calmes, & que *BC* marque l'intersection du plan vertical flottant avec ladite surface d'eau; soit le centre de gravité du plan flottant en *F*, & le centre de gravité de sa partie submergée censée homogene en G; qu'on tire par les deux points F & G la ligne EFG, qui, pour l'équilibre que nous supposons, sera verticale. Il est question de déterminer la position d'équilibre du même plan, lorsqu'on suppose la surface d'eau inclinée comme ad, ou de déterminer l'angle BEb, en supposant la ligne BC prendre la position bc.

Considérons dabord le plan flottant comme ayant pris la même inclinaison que la surface des eaux, desorte que BC soit parvenu dans la position mn, & que la ligne EFg ait pris la position Efg perpendiculaire à la ligne mn: l'angle GEg sera égal à l'angle AEa, le point g sera encore le centre de gravité de la partie submergée considérée comme homogene, puisque cette partie est la même qu'elle étoit avant l'inclinaison de la surface d'eau, & le point f sera le centre de gravité du plan. Mais cette position du plan ne sauroit être celle de l'équilibre, puisque la ligne fg n'est plus verticale; il faut donc que la ligne BC, après avoir pris la position mn, subisse un second changement, & qu'elle prenne la position bc, qui remette les points f & g dans la même verticale; & on remarquera à cet égard, que dans tous les corps qui ont naturellement un certain degré de stabilité, la ligne fg se rapproche de la verticale en augmentant l'angle BEm. Supposons maintenant l'angle cherché $mEb = \sigma$; l'angle donné $BEm = S$; $BC = q$; FG ou $fg = s$; la partie submergée du plan $= M$; qu'on tire du centre E les petits arcs de cercle fF & Gg, avec la verticale fh, il y aura entre l'angle mEb & l'angle gfh la même relation que nous avons trouvée dans le premier chapitre, puisque, par la petite inclinaison de la ligne ad, cette relation ne sauroit être changée; or en appliquant nos formules générales aux angles extrêmement petits, on trouve

$$gh = \Big(\frac{q^3}{12M} - s\Big)\sigma \ ,$$

ce qui donne le petit angle

$$gfh = \Big(\frac{q^3}{12Ms} - 1\Big)\sigma \ ,$$

& ce même angle étant aussi égal à l'angle BEm, ou $= S$, nous aurons

$$\Big(\frac{q^3}{12Ms} - 1\Big)\sigma = S \ ,$$

& par conséquent l'angle mEb, ou

$$\sigma = \frac{12Ms}{q^3 - 12Ms} S .$$

Si on ajoûte à l'angle mEb, l'angle BEm, on aura l'angle

$$BEb = \frac{q^3}{q^3 - 12Ms} S .$$

§. LIII. Les résultats que nous venons de trouver, méritent toute notre attention. Nous voyons que l'angle BEb est toujours plus grand que l'angle BEm, tant que le centre de gravité du corps plongé est plus haut que le centre de gravité de la partie submergée homogene. Il est vrai que nous n'avons considéré qu'un plan plongé verticalement; mais on n'a qu'à supposer le navire être un prisme construit sur un tel plan, pour pouvoir appliquer nos formules aux navires; remarquons d'ailleurs, que le dénominateur sera pour tous les vaisseaux, quelque figure qu'ils aient, proportionnel à la stabilité naturelle du vaisseau, & que cette stabilité est toujours composée de deux parties, dont l'une dépend uniquement de la section du vaisseau à fleur d'eau, & l'autre de la hauteur verticale du centre de gravité du vaisseau pardessus le centre de gravité de la carene. Si nous dénotons donc généralement la premiere partie par P, & l'autre par Q, nous aurons généralement l'angle $mEb = \frac{Q}{P - Q} S$, & l'angle $Beb = \frac{P}{P - Q} S$.

Suivant la construction des navires, leur charge & le lest qu'on emploie, le rapport de P à Q est tantôt plus grand, tantôt plus petit; mais il est certain que souvent la quantité Q va jusqu'à la moitié, & peut-être même jusqu'aux deux tiers de la quantité P. Supposons, par exemple, $Q = \frac{1}{2}P$: dans cet exemple l'angle BEb sera le double de l'angle BEm; c'est-à-dire, que l'inclinaison du navire sera deux fois aussi grande que l'inclinaison de la surface des eaux. Si donc les eaux étoient inclinées de 15 ou 20 degrés, le navire pourroit s'incliner de 30 ou 40 degrés, en faisant abstraction de l'erreur que la grandeur de tels angles peut jetter sur nos formules, & il est à remarquer que les vaisseaux seront quelquefois obligés de prendre de telles inclinaisons uniquement pour se mettre à l'équilibre; & si l'on fait attention qu'ils pourront s'élancer par l'impression reçue au-delà de l'équilibre, on doit être effrayé de ces énormes agitations qui peuvent arriver, surtout aux roulis; car par rapport au tangage, la quantité P est toujours beaucoup plus grande que Q. Il pourroit même y avoir des corps qui auroient un certain degré de stabilité dans les eaux calmes, & qui seroient renversés par une petite inclinaison des eaux.

§. LIV. Ce sont les angles BEb qui forment les *balancemens d'équilibres* dont j'ai fait mention au commencement de ce chapitre, & qui pourroient devenir excessifs, si on n'y apportoit aucun remede. Ces balancemens seront isochrones & synchrones avec les balancemens des eaux, qui par bonheur se font beaucoup plus lentement dans les grandes lames que les *balancemens naturels* du navire, sans quoi l'inclinaison BEb seroit encore extrêmement augmentée par les élancemens qui surviendroient, à cause de la vitesse des *balancemens d'équilibre*, & cela en raison de $\theta\theta$ à $\theta\theta - tt$, en entendant par θ le tems d'un balancement entier des lames, & par t le tems d'un *balancement naturel* du navire (§§. XXV & XXVI). Comme cependant ces *balancemens d'équilibre* dépendent de la proportion qu'on met entre les quantités P & Q, & qu'on est le maître de donner de grands changemens à cette proportion, il ne sera pas difficile d'y remédier après en avoir découvert la nature. On voit qu'une planche qui nageroit sur les eaux, formeroit ses balancemens d'équilibre parfaitement égaux aux balancemens des eaux, & que l'angle mEb deviendroit nul; on voit aussi qu'une très-longue poutre, chargée de plomb par une de ses extrêmités, conserveroit sa position verticale malgré les balancemens des eaux (car je fais abstraction des impulsions de l'eau, ne considérant ici que l'effet du principe de la pesanteur); de sorte que les angles BEb seroient entierement anéantis. Aussi l'un & l'autre cas sont-ils parfaitement bien indiqués par notre théorie; car dans le premier cas on peut supposer $Q = 0$, comme étant proportionnel à la hauteur verticale entre le centre de gravité de la planche & celui de sa partie submergée, laquelle hauteur peut être censée nulle; desorte que l'angle $mEb \left(= \frac{Q}{P-Q} S\right)$ devient nul; & dans le second cas, c'est P qui peut être censé nul, à cause de la petitesse de la section de la poutre par la surface des eaux, ce qui donne l'angle $BEb \left(= \frac{P}{P-Q} S\right) = 0$.

§. LV. Mais quel est le but qu'on doit se proposer à l'égard de ces balancemens? Doit-on tâcher d'imiter l'exemple de la planche, ou celui de la poutre, ou quelqu'autre état moyen? Il me semble que c'est absolument celui de la planche. Car si on vouloit anéantir les balancemens d'équilibre, ou les diminuer trop, il arriveroit toujours qu'une moitié du navire seroit comme noyée,

& que l'autre sortiroit trop des eaux; & comme le navire est toujours sujet à d'autres balancemens, ceux-ci pourroient devenir dangereux aux navires déja fortement noyés par un côté ou par une moitié. Je suis donc d'avis qu'on doit se proposer d'anéantir simplement l'angle mEB, & de rendre les *balancemens d'équilibre* égaux aux balancemens des eaux; c'est ce que les Marins apppellent *obéir à la lame*. M. Chauchot dit, à la fin de la page 38, *qu'il ne faut pas que le navire obéisse trop à la lame*; mais je ne doute pas qu'il ne soit de mon avis, quant aux lames que j'appelle régulieres; car quant aux lames irrégulieres, qui agissent brusquement & sans ordre ni régularité, l'inertie de la matiere du navire l'empêchera bien d'obéir entierement à cette espece de lames. Il est de conséquence de distinguer les lames & d'en examiner les diffférens effets, aussi bien que de distinguer les balancemens du navire, tels que les *balancemens naturels*, les *balancemens forcés* & les *balancemens d'équilibre*.

§. LVI. Après avoir bien établi la maxime qu'on doit se proposer à l'égard des *balancemens d'équilibre*, il est question d'examiner ce qu'on peut faire pour y satisfaire. J'ai déja dit qu'il faut faire $Q = 0$, & par conséquent $s = 0$, c'est-à-dire, qu'il faut, pour la position droite du navire, faire descendre le centre de gravité du navire jusqu'au centre de gravité de la carene censée homogene, & faire tomber précisément l'un sur l'autre. Je crois cet article si essentiel à tous égards, que je voudrois qu'on y mît toute l'attention réquise. Ordinairement le centre de gravité du navire est plus haut que celui de la carene, & il sera difficile de faire tomber ces deux points l'un sur l'autre, si on ne veut donner beaucoup de creux au navire, employer du bon lest & en quantité suffisante, arranger avec soin tout ce qui doit être mis sur le vaisseau, élargir le navire vers la flottaison afin de hausser le second centre, ménager la matiere en tout ce qui est hors de l'eau, & prodiguer en quelque façon le fer vers le fond du navire. Il est sûr qu'avec toutes ces attentions on pourra déprimer le centre de gravité du navire chargé jusqu'au centre de gravité de la carene, & encore plus bas si on le vouloit.

Après qu'on sera parvenu à mettre les deux centres de gravité dans un seul & même point, la stabilité du navire pour les petites inclinaisons ne dépendra plus que de la section du navire par la surface horizontale de l'eau; & si on appelle q la largeur de cette section, pour une abscisse x, prise dans la longueur, la stabilité sera simplement $= \sigma \int \frac{1}{12} q^3 dx$, en entendant par σ le petit angle d'inclinaison (§. XIII). Cette simple formule servira à reconnoître par une expérience facile à faire, si l'on aura bien réussi à mettre les deux centres de gravité à la même hauteur. Si ladite section étoit irréguliere, on pourroit trouver la quantité $\int \frac{1}{12} q^3 dx$ par parties, comme M. Bouguer le fait en plusieurs occasions. Si cette section étoit un rectangle, qui eût pour largeur a, & pour longueur b, la stabilité en largeur deviendroit $= \frac{1}{12}a^3b\sigma$; & j'ai conseillé de la rendre à-peu-près telle, & de ne s'en écarter qu'autant que l'arrondissement des angles le demande; mais pour la construction ordinaire des navires, on peut supposer la section formée par deux demi-ellipses, sans tomber dans aucune erreur sensible; & j'ai démontré, au milieu du §. XIII, qu'alors la stabilité en largeur devient $= \frac{11}{224}a^3bc$, en entendant par a & par b la plus grande largeur & la plus grande longueur de la section. Si l'on exprime en piés les quantités a & b, la formule donnera un certain nombre de piés cubes d'eau agissant sur un levier pris à volonté; & si on divise la formule par 28, on aura, au lieu de piés cubes d'eau, des tonneaux à raison de 2000 liv. chacun; le *momentum* de la stabilité sera donc, en tonneaux pour le poids, & en piés pour le levier, $\frac{a^3b\sigma}{570}$. Après avoir calculé cette quantité, on pourra appliquer horizontalement & perpendiculairement à la longueur du navire un certain nombre de leviers qui débordent le navire, & suspendre par les extrêmités de ces leviers des poids assez forts pour faire pencher le navire de côté d'un angle d'environ deux degrés; on mesurera cet angle avec beaucoup de précision, par le moyen d'un bon instrument, & on prendra la somme de tous les *momentum* des forces appliquées pour faire pencher le navire. Si on trouve cette somme égale à $\frac{a^3b\sigma}{570}$, ce sera une marque que les deux centres de gravité sont réunis dans un même point; si elle est plus petite ou plus grande, le centre de gravité du navire sera plus haut ou plus bas que le centre de gravité de la carene; & si on connoît la solidité de la carene, on pourra exactement déterminer de combien le premier centre est plus haut ou plus bas que l'autre; car si la somme des *momentum* en question est à la quantité $\frac{a^3b\sigma}{570}$ comme n à m, on aura $P : P - Q = m : n$, & par conséquent $Q = \frac{m-n}{m}P$, & comme $Q = ms\sigma$, en entendant par M la solidité de la carene, on aura $Ms\sigma = \frac{m-n}{m}P = \frac{m-n}{m} \times \frac{a^3b\sigma}{570}$, & par conséquent $s = \frac{m-n}{m} \times \frac{a^3b\sigma}{570 \times M}$; & dans cette formule, il faudra exprimer la solidité de la carene pareillement en tonneaux de 2000 livres. Une telle expérience demande que les eaux soient entierement calmes. Voici un exemple qui éclaircira ces regles.

Soit $a = 30$ piés; $b = 100$ piés; la solidité de la carene $= 750$ tonneaux; $\sigma = 0.03$ qui répond à une inclinaison de $1^{\mathrm{d}}\,43^{\mathrm{m}}$, on aura $\frac{a^3 b\sigma}{570} = 142\frac{2}{19}$; qu'on applique cinq leviers, chacun de 30 piés de longueur, depuis le milieu du tillac; si l'on avoit chargé l'extrêmité de chaque levier de $\frac{18}{19}$ de tonneau, on en conclurra, que le centre de gravité du navire est précisément au centre de gravité de la carene; mais si dans l'expérience on n'y avoit suspendu que $\frac{15}{19}$ de tonneau, on auroit $m = 6$ & $n = 5$, ce qui donneroit $s = \frac{20}{19}$ p. & d'où il faudroit inférer que le centre de gravité du navire est élevé par-dessus le centre de gravité de la carene d'un pié $7\frac{11}{19}$ lignes. En ce cas, je conseillerois de lester davantage le navire, quand même ce seroit au préjudice de la charge, à moins qu'on ne puisse faire descendre un peu le centre de gravité par un arrangement plus convenable de la charge; mais quand même on ne voudroit rien changer, il seroit toujours bon de faire cette sorte d'expérience pour chaque navire, car un Pilote entendu doit être exactement informé de l'état de son vaisseau. Si on ne trouvoit pas nécessaire de déprimer le centre de gravité du navire tout-à-fait jusqu'au centre de gravité de la carene, il faudroit du moins fixer à quelle hauteur on doit mettre le premier par-dessus l'autre, & cette hauteur doit toujours faire une partie aliquote de la plus grande largeur de la section du navire par la surface de la mer, comme sa vingtieme, sa vingt-cinquieme, ou sa trentieme partie, & par-là on se mettra en même tems en état de fixer le lest, qu'on ne détermine que d'une maniere fort vague; il ne faut donner au hazard que le moins qu'il est possible.

§. LVII. Nous avons considéré jusqu'ici le vaisseau comme beaucoup plus petit, tant en longueur qu'en largeur, que n'est toute la largeur d'une lame. Je ne m'arrêterai pas à détailler tous les autres cas qui peuvent arriver; ils sont en trop grand nombre. Il faudroit avoir égard à la largeur des lames, à la longueur & à la largeur du navire, à l'angle que le navire forme avec le plan de la cinquieme figure, à la vitesse du navire, à la durée de chaque balancement des lames, &c. & si on vouloit combiner toutes ces variations, il en résulteroit des questions infinies, que j'aime mieux abandonner à l'intelligence de mon lecteur; je me contenterai de remarquer qu'aux points d & f l'inclinaison des eaux est nulle; que cette inclinaison est la plus grande dans les nœuds; que les eaux sont d'autant moins comprimées qu'elles sont plus élevées, & d'autant plus comprimées qu'elles sont plus basses; que lorsque le vaisseau se trouve sur la convexité des eaux, il doit être soutenu plus par le milieu que par les extrémités; & que sur la concavité il arrive le contraire; que les inclinaisons des eaux sont tantôt affirmatives, tantôt négatives, à la même distance des nœuds; & qu'au milieu de chaque balancement la surface des eaux devient tout-à-fait plane. Je ne fais pas ces réflexions comme devant nous servir à trouver quelques nouvelles corrections, quant à la construction du navire; mais plutôt comme tendantes à expliquer la nature du roulis & du tangage, & leur variations suivant les différentes circonstances où le navire se trouve à chaque moment, & même à diminuer ces balancemens par quelques autres attentions; comme, par exemple, de tâcher de tenir le milieu entre deux nœuds, lorsque les lames prennent par le travers; de franchir les nœuds, lorsque la lame est prête à se mettre de niveau; de couper la lame sous un angle plus ou moins grand, suivant qu'on craint davantage le tangage ou le roulis; de faire force de voiles, lorsqu'on n'appréhende pas le vent, d'autant qu'un vaisseau cinglant sera toujours moins agité qu'un vaisseau flottant; de faire à plusieurs reprises de fausses routes, mais bien entendues, &c. L'art de bien gouverner ne sera pas sans succès dans ces occasions; & quoique les lames ne soient jamais aussi régulieres que je les ai supposées dans ce chapitre, nos réflexions n'en seront pas moins utiles. La maxime principale que j'ai prétendu établir, est celle de réunir, s'il est possible, les deux centres de gravité dans un seul & même point.

CHAPITRE VI.

Quelques réflexions particulieres sur le Tangage.

§. LVIII. Dans tout le cours de ce Mémoire, j'ai traité également du roulis & du tangage; & je n'ai pas manqué, toutes les fois que l'occasion le demandoit, de considérer en particulier l'une ou l'autre espece de balancemens.

J'ajoûterai cependant encore quelques réflexions particulieres sur le tangage. La stabilité est beaucoup plus grande pour le tangage que pour le roulis, si on prend des inclinaisons égales de part & d'autre. En supposant les centres de gravité du navire & de la carene à la même hauteur, en considérant la section horizontale du navire à fleur d'eau comme elliptique ou formée par deux demi-ellipses, & en ne considérant que de petites inclinaisons, les deux stabilités seront entre elles comme le quarré de la longueur au quarré de la largeur. Cette raison, jointe à la grande résistance de l'eau, fait que le mouvement angulaire du tangage est beaucoup plus petit que celui du roulis; mais la grande distance de la proue & de la pouppe, depuis l'axe de rotation, fait aussi que leur mouvement absolu ne laisse pas de surpasser souvent celui des flancs dans les roulis; aussi exhausse-t-on les bords vers la proue & la pouppe pour prévenir les suites funestes de ce mouvement absolu. Le mouvement du tangage est ordinairement arrêté brusquement par la grande résistance des eaux, qui ne sauroit être deplacée assez vite, & qui est frappée dans une très-grande surface. Il est sûr que la matiere & le corps du navire sont fort fatigués par ces rudes coups; mais comme ils arrivent dans un tems où l'une des deux extrêmités peut-être déja extrêmement baissée, pendant qu'il lui reste encore beaucoup de mouvement, on peut dire que ce sont précisément ces violens & subits efforts qui sauvent le navire. Le tangage ne sauroit répondre à la nature des *balancemens libres*, parce que la grande résistance des eaux arrêteroit aussitôt ces balancemens. Le *tangage d'équilibre* sera aussi tel qu'il doit être sur les grandes lames régulieres, parce que la partie de la stabilité qui dépend de la section du navire à fleur d'eau, est ici comme infiniment plus grande que celle qui dépend de la hauteur verticale entre les deux centres de gravité; je veux dire que le navire obéira assez bien à la lame par sa longueur, lorsque la lame est grande & réguliere. Il n'y a donc que le tangage *forcé* & formé par des causes irrégulieres, qui puisse mériter notre attention. L'inégalité du vent ne sauroit faire ici aucun effort sensible à cause de la grande stabilité; le mouvement horizontal des eaux contre le navire ne sauroit faire non-plus un grand effet pour le faire tanguer. En un mot, le tangage excessif ne sauroit être produit que par un mouvement subit & violent des eaux du bas en-haut, ou parce que l'eau cessera subitement de soutenir le navire dans toutes les parties de la carene. Les brisants d'eau peuvent occasionner ces deux causes; car lorsque deux lames viennent à se choquer, les eaux du milieu sont extrêmement comprimées, & cette compression lance les eaux en-haut avec beaucoup de force, après quoi les eaux des deux lames brisées se retirent avec un mouvement contraire & laissent un vuide au milieu. La partie du navire qui se trouve au milieu de ces eaux comprimées, sera lancée pareillement en-haut; & puis le poids de cette partie, faute d'être soutenu, causera un mouvement contraire également violent, jusqu'à ce que le navire retrouve les pleines eaux & en soit soutenu dans toute la carene; après quoi ce mouvement sera arrêté assez promptement.

§. LIX. Ces agitations irrégulieres du navire seront toujours composées de deux mouvemens; savoir, d'un mouvement de rotation autour de l'axe qui passe par le centre de gravité, & d'un autre par-lequel tout le corps du navire montera ou descendra un peu avec précipitation; le mouvement absolu se fait sur l'axe de rotation spontanée; mais comme le second mouvement allégué n'entre pas ici en compte[61], il faut, dans le tangage, toujours considérer la ligne qui passe par le centre de gravité comme l'axe de rotation.

Si donc cette ligne est plus proche de la proue que de la pouppe, le mouvement sera plus grand vers la pouppe que vers la proue, & cela arrive en plaçant le fort du navire plus près de la proue que de la pouppe; car le centre de gravité de la carene étant par-là approché de la proue, le centre de gravité du navire doit l'être pareillement. Je conseillerois donc de placer le fort du navire au milieu de la carene, à moins qu'on ne prétende exprès de jetter dans le tangage plus de mouvement sur la pouppe que sur la proue; il est sûr que le navire n'en silleroit que mieux, quoique quelques-uns prétendent le contraire. L'autorité de M. Bouguer devroit suffire à ceux-ci pour se désabuser de ce faux préjugé. Il est vrai, & ce même illustre Auteur l'a fort bien démontré, qu'en bien des cas le navire gouverne mieux, tant par les voiles que par le gouvernail, lorsqu'il est renflé vers la proue; mais je doute que cela soit généralement vrai. Je crois aussi que lorsqu'on ne réussit pas à bien gouverner le navire, c'est le plus souvent plutôt par la faute d'une manœuvre bien entendue, que par aucune mauvaise construction du navire; une manœuvre réussira quelquefois, & manquera une autrefois, simplement parce que la vitesse du navire est un peu plus grande. Je suis d'avis qu'il faudroit plutôt s'appliquer à perfectionner la manœuvre pour toutes les circonstances où l'on peut se trouver, qu'à gagner un peu plus de facilité à gouverner par une construction nuisible aux qualités les plus essentielles du navire; peut-être aussi qu'on ne feroit pas mal d'ajoûter à l'usage du gouvernail celui de longues rames, simplement mobiles dans le plan vertical, qui doit être aussi celui des pales, & perpendiculaires à la longueur du navire. Il est sûr qu'en plongeant ces pales dans l'eau, elles feroient plus d'effet que le gouvernail, pour peu qu'on leur donnât de surface, & qu'on en multipliât le nombre; un gouvernail de moyenne grandeur n'a qu'environ 50 piés quarrés de surface, & il faut, à cause de l'obliquité, lors même qu'elle est la plus avantageuse, en retrancher presque les deux tiers, & enfin le corps de la carene empêche encore l'eau d'agir de toute sa force sur le gouvernail; d'où l'on voit assez que l'effet des pales ne sauroit manquer d'être considérable, surtout si on fait duement incliner leurs plans verticaux à l'égard de celles qui ne seroient pas appliquées au milieu du navire. De pareilles rames pourront encore être employées utilement dans plusieurs autres occasions &

61 We refer again to D. Bernoulli St. 32, *De motibus oscillatoriis corporum humido insidentium* where it is proved that the vertical and the angular oscillations are independent, for small amplitudes.

avec un maniement convenable & bien exécuté; elles pourront même servir à modérer un peu les agitations du navire.

§. LX. C'est à-peu-près sur le même principe qu'on doit déterminer la *quête de l'estambot*, de laquelle les constructeurs ne conviennent pas. Je voudrois que l'inclinaison de l'estambot, qui est une piece droite, fût égale à l'inclinaison de la tangente de l'estrave à l'endroit qui est à fleur d'eau; on sait que cette estrave est une piece courbe faisant un arc de cercle de 70 degrés, & par conséquent inégalement inclinée dans toute sa longueur. L'élancement de l'estrave est sans doute très-nécessaire; & si on pouvoit l'augmenter sans nuire à la force de l'assemblage des pieces qui forment l'éperon, ou sans tomber dans d'autres inconvéniens, on devroit le faire à mon avis; on diminueroit la résistance de l'eau dans le sillage, & on augmenteroit la stabilité du navire contre le tangage, parce qu'on augmenteroit la section du navire à fleur d'eau; on éleveroit un peu le centre de gravité de la carene, & il seroit d'autant plus facile de le réunir avec celui du navire. Ces mêmes raisons, excepté la premiere, concourent pour faire augmenter la quête de l'estambot; d'ailleurs, il sera utile de rendre égales & semblables les deux moitiés de l'avant & de l'arriere de la section horizontale du navire à fleur d'eau, afin de distribuer également sur la proue & sur la pouppe le mouvement du tangage. En suivant ma regle sur la quête de l'estambot, cette égalité subsistera même dans le navire lorsqu'il sera un peu incliné, au lieu que si on vouloit élever l'estambot tout droit, & que la proue s'élevât par-dessus la surface de la mer, il arriveroit que ladite section seroit considérablement raccourcie, & que les deux moitiés en deviendroient d'autant plus dissemblables. M. Chauchot n'a pas manqué de conseiller pareillement de rendre la proue semblable à la pouppe un peu au-dessus & au-dessous de la ligne de flottaison.

§. LXI. Il est naturel de dire, qu'on diminuera encore le tangage en augmentant la longueur du navire. Ce qu'il y a de sûr, est que la stabilité en seroit considérablement augmentée. Cependant comme la quille n'est déja que trop sujette à s'arquer, il est à craindre qu'on n'augmente ce mal en augmentant trop la longueur de la quille; car quand même on pourroit distribuer la charge d'une maniere que le poids de chaque tranche verticale du navire perpendiculaire à sa longueur fût égale au poids de l'eau déplacée par cette tranche, ce qui seroit apparemment le meilleur moyen pour empêcher la quille de s'arquer, on voit assez que cet équilibre ne subsisteroit plus pendant les agitations de la mer & du navire, qui sont presque continuelles. Il se peut d'ailleurs qu'il y ait une certaine proportion entre les dimensions du navire & celles de la masse d'eau qui tend à agiter le navire, qu'on doive craindre davantage; & est-il bien sûr qu'on s'éloigne le plus communément de cette proportion en augmentant la longueur du navire? Je crois bien qu'on diminuera presque toujours par-là le mouvement angulaire du tangage, & que par conséquent la mâture en sera moins fatiguée; mais il n'est pas sûr qu'on diminue généralement parlant le mouvement absolu de la proue & de la pouppe. Il ne faut pas non plus que le navire se refuse trop à l'impression des lames, ce qu'il pourroit faire à cause de sa grande masse, surtout à l'égard des lames irrégulieres, qui agissent avec promptitude, & qui ne répliquent pas leurs efforts. Un navire qui conserveroit sa position droite à l'égard de l'horizon, & qui par conséquent n'auroit ni roulis ni tangage, seroit continuellement noyé soit par la proue soit par la pouppe, & courroit risque de se remplir d'eau & de couler à fond. Il y a tels articles sur lesquels il faut consulter l'expérience plutôt que la simple théorie.

§. LXII. On aidera le navire à obéir à la lame en le chargeant & le lestant beaucoup plus vers le milieu que vers les extrêmités; car de cette maniere la matiere du navire oppose moins d'inertie aux efforts de la lame. Si au contraire on remarquoit au navire trop de promptitude à obéir aux efforts des eaux agitées, il faudroit jetter plus de charge vers les extrêmités, & soulager le milieu. Les *balancement libres*, s'ils trouvoient lieu dans le tangage, se feroient plus rapidement dans le premier cas que dans le second. Cette réflexion ne doit pas nous prévenir contre la maxime de surcharger le milieu du navire, parce que les balancemens du tangage appartiennent presque uniquement à la classe des *balancemens forcés*, dont les durées dépendent simplement des agitations de l'eau. Je dirai à cette occasion que les *balancemens libres* du tangage ne sauroient manquer de s'achever beaucoup plus rapidement que ceux du roulis. Je trouve même qu'ils ne dépendent presque point de la longueur du navire, & que leur durée est pour la plus grande partie déterminée par la simple hauteur du tirant d'eau; le pendule isochrone sera tout au plus égal à cette hauteur; mais je ne l'estime pas si long, parce qu'il y a plus de matiere vers le milieu du navire, que vers les extrêmités, qui ont le plus de mouvement. Un parallelipipede, dont la longueur seroit beaucoup plus grande que sa largeur & sa hauteur, s'il étoit uniformement pesant dans toute sa longueur, feroit ses balancemens, dans le sens de [la] longueur, isochrones avec un pendule simple, dont la longueur seroit égale à la hauteur de son tirant d'eau[62]; & cette remarque suffit pour voir la vérité de ce que je viens de dire. Ainsi un navire qui a 18 piés de tirant d'eau,

62 See the Introduction, pp. 119–120 h. v.

fera ses *balancemens libres* du tangage en moins de $2\frac{1}{3}$ secondes. Ceux du roulis sont plus lents par plusieurs raisons, dont la principale est à mon avis la hauteur qu'on donne ordinairement au centre de gravité du navire par-dessus le centre de gravité de la carene, laquelle hauteur diminue trop sensiblement la stabilité du navire dans le sens de sa largeur, & par conséquent la force qui tend à redresser le navire. En réunissant les deux centres de gravité, comme j'ai conseillé de faire, autant qu'il est possible, les *roulis libres* se feront plus rapidement. Ce seroit là un inconvénient, si les roulis actuels n'étoient presque toujours des *roulis forcés* & non des *roulis libres*. Les roulis forcés, je le répete, sont isochrones avec le balancemens des lames, qui le plus souvent sont beaucoup plus lents que les *roulis libres* du navire. On gagnera donc par l'accélération des *roulis libres*, parce qu'on évitera presque toujours ce cas fâcheux duquel j'ai parlé aux §§. XXIII, XXVI & XLII, qui arrive lorsque les accès de forces qui agitent le navire sont isochrones avec les *balancemens libres* du navire.

§. LXIII. Voilà les principes que j'ai cru pouvoir servir à la solution de notre question; j'espere qu'ils seront de quelque utilité, tant aux Pilotes qu'aux Constructeurs de vaisseaux. C'est à eux que je laisse le soin d'en faire usage avec les précautions nécessaires, suivant leurs connoissances acquises, les priant seulement de les écouter sans ce préjugé qui retarde si souvent la perfection des arts & des sciences. Ils suppléeront facilement à ce que j'ai pu omettre, & sauront employer ces principes avec beaucoup plus de succès que je n'ai pu faire dans les circonstances où je me trouve.

St. 65

Expositio theoretica singularis Machinae hydraulicae Tiguri Helvetiorum exstructae

NCP XVII, 1772 (1773), pp. 251–271

§. 1. Machinam, de qua nobis sermo erit, non dubito inter ingeniosissimas referre; inventorem habuit virum egregium, *Andream Wirz* Tigurinum, Artificem stannarium, variis ornatum dignitatibus in civitate patria, suoque ingenio mechanico inter concives conspicuum: quin ipse inventor parvulam optatissimo cum successu exstruxit machinam in fluvio Limagi, cuius ope aquae ad notabilem elevantur altitudinem in usus officinae tinctoriae ad ripam sitae. Accuratiorem novae huius machinae descriptionem dedit Cl. *Ioh. Henricus Zigler* Med. D. in tertio Volumine Transactionum Tigurinarum, ad quam licet lectorem ablegare, cui perfunctoria nostra delineatio non satisfecerit.

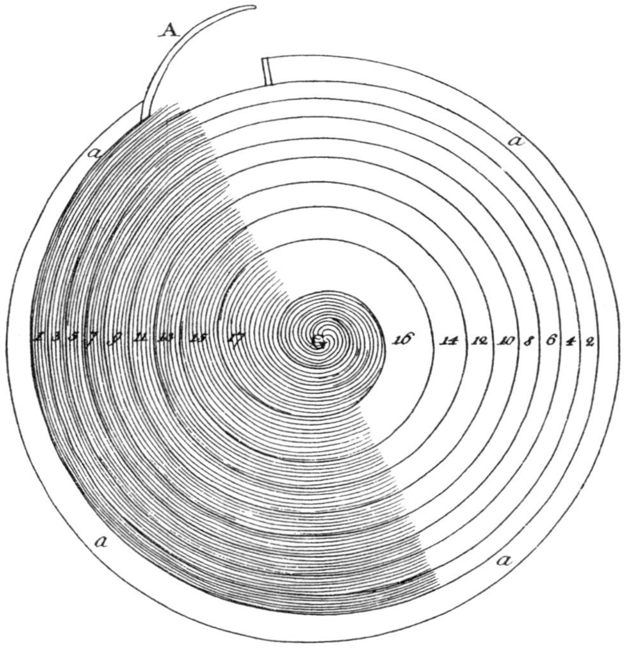

Fig. 1

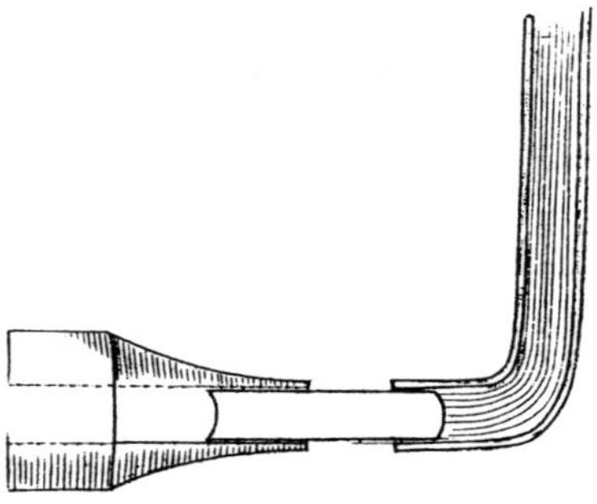

Fig. 2

§. 2. Interna machinae Wirzianae structura eadem perfecte est, quae tympani horologici, cui elastrum inclusum est spirale; figura nostra prima sectionem sistit tympani perpendicularem ad axem: involucrum habet designatum litteris *a*, *a*, *a*, *a*, quod definit in sitellam aquis hauriendis dicatam *A*; cavitati tympani inclusa est lamina stannea, dimidiam lineam crassa, atque ductu spirali ab interiori involucri superficie versus centrum intorta, relicto debito intervallo inter quamvis laminae duplicaturam; sic helices formantur 1, 2; 3, 4; 5, 6 etc. Utraque ora laminae intortae sit quaevis exactissime in eodem plano, ut helices a fundis superimpositis in utroque latere perfecte claudantur; ferruminatione hoc opus, in quo cardo rei vertitur, perficere docuit auctor. Ex medio alterutrius fundi exoriatur bulbus perforatus instar modioli rotae vel choenicii formatus, per quem aquae et aër ex intima helice lateraliter exprimantur et transfundantur in diabeten vel siphonem immobilem, primo horizontalem, mox incurvatum et verticaliter assurgentem; diabetes aquarum elevationi et effluxui per orificium superius dicatus est; modiolus autem sua cavitate extremitatem diabetis strictissime circumplectatur, ita ut in tota iunctura ne guttula quidem fluidi, utut compressi, perspirare possit; attamen opera detur, ut praefatum choenicium circa diabetis exortum ipsi insertum absque notabili impedimento rotari possit: figura secunda hanc machinationis partem exprimit.

§. 3. Usus machinae Wirzianae in eo consistit, ut aquis ad dimidiam propemodum altitudinem submersa sit, axe horizontaliter posito altera vero dimidietas extra aquas promineat; tum si rotetur machina, secundum ductum spirarum a centro versus peripheriam, fiet, ut alternis vicibus aquam et aërem hauriat, utrumque fluidum versus centrum profluat, per choenicium in diabeten transfundatur, in diabete assurgat, tandemque eiiciatur. Sic parvula machina ascensum aquarum obtinuit inventor octodecim pedum, quae machinae praestantia atque theoria omnibus erunt mirabiles: nec tamen omnis a machina obtinetur effectus, nisi partes eius singulae ad debitam conformationem atque proportionem fuerint reductae; ultima machinae perfectio in eo consistere videtur, ut quavis revolutione extima helix, ad dimidiam sui partem novis impleatur aquis, posteaque aëre naturali, ut ambo fluida secunda revolutione in helicem secundam, tertia in tertiam, et sic porro, progrediantur, donec ad intimam pervenerint helicem, ut ex hac per choenicium in diabeten transfundantur, eleventur ac denique effluant. Nisi omnis aqua in primam helicem hausta, suo ordine hanc viam perficiat, necesse est ut pars eius in supremo helicis loco retropellatur atque regurgitet in helicem praecedentem, unde venerat: ista vero regurgitatio effectum non solum diminuit, sed et aliquando integrum evertit, quod ipsi contigit. Inventori in primo suo tentamine, cum helices versus centrum nimis coarctabat, commissum errorem protinus animadvertit, factaque eius correctione plenissimum a machina obtinuit effectum. Constructio machinae non sine praevio circumstantiarum examine aggredienda est: cum vero interna aquarum in singulis helicibus dispositio atque ordinatio, a qua totus machinae effectus pendet, oculis discerni non possit, haec disquisitio *a priori* secundum veras leges hydrostaticas instituenda erit.

§. 4. Ante omnia existimo attendendum esse ad altitudinem, ad quam aquae sint elevandae, et quanta inde in infima diabetis parte oritura sit compressio, dispiciendum; ponamus primo nullam esse hanc altitudinem, ita ut aquae effluant per axem excavatum horizontalem; tunc utique in singulis helicibus aquae infimum occupabunt locum, aër supremum, nec aër iste erit compressus; servabit suam densitatem naturalem, placidusque fiet utriusque fluidi transfluxus; si modo orificium effluxus amplum fuerit, lentiusque machina circumagatur; nec tunc, si animus a resistentiis alienis abstrahatur, ulla vis ad machinam rotandam requiretur. Si vero diabetes aliquam habuerit altitudinem, tunc aquae in illo contentae repriment aquas effluxui proximas, simulque aërem subsequentem comprimènt, et in minus volumen condensabunt: inde facile videtur fore ut aquae in singulis helicibus eleventur in latere sinistro figurae nostrae, deprimanturque in latere dextro: sic singuli arcus aquei pressionem exercebunt versus centrum, atque summa omnium harum pressionum dabit proxime integram potentiam in elevationem aquarum, quae in diabete continentur, impendendam: vocabo deinceps pro quavis helice, *altitudinem arcus aquei* illam altitudinem verticalem, qua in illa helice extremitas posterior aquae elevata est supra extremitatem anteriorem; singuli autem arcus aërei comprimuntur vi, quae respondet summae altitudinum arcuum aqueorum a tergo subsequentium; quanto magis elevandae sunt aquae in diabete assurgente, tanto maiores oriri debent altitudines arcuum aqueorum, donec tandem summitatem helicis suae attigerint, qui ultimus est machinae propositae effectus: tunc arcus aquei integram capacitatem lateralem sinistram occupant, totumque latus dextrum aëre sit repletum oportet: figura nostra prima ultimo huic statui respondet, in qua numeri impares arcus aqueos, numeri pares arcus aëreos indicant; hunc ultimum statum ut singuli arcus aquei simul attingant, opera danda est, ne ulla in ordinatione perturbatio oriri possit: quod si altitudo diabetae assurgentis excedat suos limites, nullae profecto aquae effluent, nullaeque a sitella haurientur, nisi quantum requiritur, ut parvuli excitentur motus reciproci, dum alternis vicibus parum aquae sitellam ingreditur rursusque eiicitur.

§. 5. Cum nullus *arcus aqueus* maiorem possit habere *altitudinem*, quam est diameter helicis suae, sequitur aquas in diabete contentas nunquam maiorem exercere posse repulsum, quam qui respondet altitudini omnium diametrorum ad singulas helices pertinentium; imo ob multas rationes requiritur, ut aquae in diabete contentae minorem renisum faciant, quam qui deducitur ex praefata diametrorum summa; inter has rationes pro momentosissima habeo, quod helices in progressu suo versus centrum aliquam capacitatis suae diminutionem capessere debeant, uti mox demonstrabo, si omnia rite perficere velimus; sic igitur, si extima helix ad dimidium fuerit aquis repleta, fiet ut haec eadem aquarum quantitas in interioribus helicibus arcum efformet plus quam 180 graduum; tunc autem necesse est, ut *altitudo arcus aquei* minor sit quam diameter helicis; apparet exinde, non tam diametros helicum, quam *altitudines arcuum aqueorum* accipiendas esse, ex quarum summa aliquod de altitudine diabetis ferri potest iudicium; attamen cum non ita pridem parvulam obtinuissem machinam, protinus providi, fore ut machinula diabeten admitteret multo altiorem quam esset summa altitudinum arcuum aqueorum, immo quam summa diametrorum; id valde paradoxum videbatur amicis praesentibus, in quorum conspectu primum parabam experimentum; mox tamen, audita asserti ratione, in sententiam meam transierunt, nec exspectationem nostram fefellit experimentum. Iam dicam, quod res est.

§. 6. Machina Wirziana alternatim aquam et aërem haurit, haustaque per orificium diabetis superius reddit; diabetes in parvula machinula, quam prae manibus habebamus, vix duas lineas in diametro habebat, nec aquae in hoc siphone strictiori diffluere poterant: igitur in illo formari debebant columellae aqueae et aëreae alternatim positae, atque ascendere: cum vero pondus columellarum aërearum facile negligi possit, solae columnae aqueae pondere suo reactionem facient: postulat igitur aequilibrium, ut maxima diabetis altitudo aestimetur ex sola summa columellarum aquearum, neglecta altitudine columellarum aërearum: hoc modo altitudo diabetis absoluta erit multo maior quam altitudo hydrostatica. Erit autem altitudo hydrostatica ad altitudinem absolutam proxime ut 1 ad $1 + \frac{a}{A}\,\text{log. hyperb.}\left(1 + \frac{a}{A}\right)$, si per a intelligatur ipsamet altitudo hydrostatica, et per A altitudo aquae, cuius pressio eadem sit cum pressione atmosphaerae. At si aquae ductus ascendens amplior fuerit, facile eveniet, ut aër intermixtus erumpat, sicque pars aquarum confluat, quod si contingat, minorem ascensum machina permittet.

§. 7. Quae modo dixi de augenda aquarum elevatione ab intermixto aëre, iam diu mihi fuerant observata, cum viderem idem hoc artificium Parisiis fuisse excogitatum atque felici cum successu, adhibitum in antlia aspirante (*pompe aspirante*), cuius ope aquae ad altitudinem notabiliter maiorem quam 33 aut 32 pedum ex castello adducebantur, prouti videre est in posterioribus Commentariis Academiae Regiae Scientiarum Parisinae, huncque mirandum effectum obtinuit inventor hoc solo artificio, quod in inferiori parte machinae aëri per parvulum foraminulum accessum concederet: sic nempe aquae aëre permixtae veluti fluidum efformebant specifice levius quam est aqua pura, quod tanto altius quanto levius est a pressione atmosphaerae elevari poterat. Videantur hac de re Historia et Commentarii praelaudatae Academiae[1] ad annum 1766. pag. 150 et 431. Summam memorabilis inventi paucis exponam verbis. Machina exstat Parisiis in aedibus D. *Bellangé* composita ex tubo vel diabete verticaliter assurgente, extremitate sua inferiore aquis immerso; diametrum internam habet 10 lin. atque ascendit ad altitudinem 55 ped. ubi inseritur in antliam, 25 lin. pro diametro interna habertem, in qua embolus agitationes reciprocas 8 poll. permittit; embolus ipse perforatus est atque valvula instructus, quae ascensum aquis concedit, descensum aut regressum negat: in infima tubi parte similis est valvula aquis ex castello vel dolio ingressum in tubum, cum embolus in antlia elevatur, concedens, regressum, cum deprimitur embolus, vetans. Denique in parva distantia, veluti decem pollicum, a superficie aquae in dolio contentae, diabetes ascendens a latere parvulo foramine perforatus est, cuius

1 Nollet [M. l'Abbé], *Réflections sur quelques nouveaux phénomènes d'hydrostatique*: Mém. Paris 1766 (1769), pp. 431–445.
Nollet [M. l'Abbé], *Sur quelques nouveaux phénomènes d'hydrostatique*: Hist. Paris 1766 (1769), pp. 150–156 (a condensed version of the previous item).

diameter semilineam parum superat, interius valvula debilissima ac mobilissima obtecta, aëri externo, dum embolus elevatur, ingressum permittente, at egressum omnem, cum deprimitur embolus, impediente. Ope huius machinae, dum agitatur embolus, aquae ad totam diabetis altitudinem 55 ped. una cum permixto aëre, per foraminulum laterale ingresso, elevantur, antliam ingrediuntur, embolum dum depremitur traiiciunt, ob eodemque mox elevato ex antlia eiiciuntur. Cl. *Nolletus* vera huius effectus tradit principia mechanico-physica generalia; mensuras atque proportiones reconditas non addit, quae calculum paullo operosiorem requirunt, quamvis plane necessariae sint, ut pro diversis circumstantiis effectus non solum explicari sed et provideri possint. Commentationes, quas reliquas fecit Cl. D. *Nollet*, nunc nimis alienas, in aliud differam otium. Nunc in viam redeo.

§. 8. Quae adhuc diximus, generaliora sunt; sunt autem nonnulla specialius pertractanda. Quo maior adhibetur spirarum numerus, eo facilius continget, ut aquae in diversis helicibus situm obtineant valde inordinatum; in aliquibus helicibus *altitudines arcuum aqueorum* (§. 4.) poterunt esse maximae, in aliis minores, in aliis negativae, si nempe praeponderent in latere dextro; pro re nata fiet, ut effectus oriatur maximus, vel minor vel etiam nullus, quod posterius iam dixi §. 3. ipsi Inventori in primo suo tentamine contigisse. Ante omnia certiores simus, fieri non posse, quin singuli arcus aquei magis sint elevati a latere sinistro, quam a dextro; sic enim omnes ad communem inter se conspirabunt effectum; at enim quamvis tunc aquae in una vel altera helice imo in singulis helicibus sint multo minus elevatae, quam alias fieri potuisset, decrescet quidem altitudo, ad quam aquae per diabeten elevari possint, simul autem in eadem ratione decrescet potentia ad machinam circumagendam requisita, ita ut nihil de effectu utili perdatur.

§. 9. Cardo negotii in hoc verti videtur, ut cuivis helici debita concilietur capacitas; cum artifex noster, in secundo suo tentamine, singulis helicibus aequalem dedisset capacitatem, egregium habuit successum, nec tamen optimum, si aquae ad maiorem elevandae fuissent altitudinem. Optima erit methodus, quae ad ductum paragraphi quinti instituitur: arcus aquei suum utique volumen fervant, dum omnes helices percurrunt, si modo aquae nusquam regurgitare possint in helicem proximam unde venerant; at vero arcus aërei, dum cochleam percurrunt, continuo magis magisque comprimuntur, condensantur, atque in minus volumen coërcentur. Velim autem ego, ut uterque arcus, aqueus et aëreus, ubique integram expleant helicem; id enim ni fiat, perturbabitur situs, quem aquae et aër in helicibus singulis observare debent, atque in paragraphum praecedentem peccabitur. Necesse adeoque est, ut capacitas helicum tanto magis diminuatur, quanto propior facta est centro. In primo tentamine Inventor nimis constrinxit helices versus centrum, in secundo nimis aperuit. Quod si huic nostro satisfiat praecepto, eveniet ubique, ut arcus aqueus plusquam dimidiam helicis suae occupet partem, atque adeo ut *altitudo arcus* aquei ad integram helicis diametrum nunquam ascendere possit, etiamsi summitas arcus aquei in ipsa summitate helicis posita sit: semper enim deficiet ista altitudo ea parte, quae respondet sinui verso arculi superabundantis: at exiguus est iste defectus, nec ullum inde oritur *effectus utilis* decrementum, ratione habita ad potentiam motricem simul diminutam. Hos singulos arcus aqueos, quotcunque fuerint, certissime habebimus conspirantes apteque locatos, quae sola conditio essentialiter cum argumento nostro connexa videtur, atque optatum machinae successum certum reddit: nec tamen requiritur, ut omni accuratione expleatur: imo pro minoribus aquarum elevationibus parum derogabitur ex machinae perfectione, si eadem capacitas in singulis helicibus conservetur.

§. 10. Ut vera atque convenientissima determinetur amplitudo in singulis helicibus, ante omnia respiciendum est ad altitudinem diabetis per quem aquae elevandae sunt: vidimus autem §. 6 diabeten, praesertim si strictior fuerit, columellis impleri aqueis et aëreis alternatim positis atque solas priores pondere suo comprimere aërem infimum diabeti incarceratum, quia pondus columellarum aërearum facile negligi potest; eadem autem est compressio aëris in intima helice, quae est in infima parte diabetis; fuerit iam altitudo diabetis $= b$ et altitudo integra omnium columellarum aquearum $= a$, habebimus vi paragraphi sexti $a+A \log. \text{hyperb.}\left(1+\frac{a}{A}\right) = b$ quam proxime; atque haec aequatio, facillima appropinquatione, subministrabit valorem altitudinis a, sub qua aqua comprimit aërem in intima helice contentum; quia vero altitudo atmosphaerae concurrit ad comprimendum eundem aërem, haec in aquam conversa erit $= A$ et integra altitudo $= a + A$. In extima helice aër est naturalis, atque comprimitur sola altitudine A: erit adeoque densitas aëris naturalis ad densitatem aëris in intima helice contenti ut A ad $A + a$, et cum quantitas aëris eadem sit in utraque helice, erit spatium aëreum in extima helice ad idem spatium in intima ut $A + a$ ad A; quod si itaque spatium aëreum in extima helice sit $= c$, erit idem spatium in intima helice $= \frac{A}{A+a}c$ et cum extima helix aequali volumine aëris atque aquae impleta ponatur, erit capacitas extimae helicis $= 2c$; capacitas vero intimae helicis $= c + \frac{A}{A+a}c$; sunt itaque ambae capacitates ut 1 ad $\frac{2A+a}{2A+2a}$; erit quoque capacitas extimae helicis ad differentiam inter extimam et intimam ut 1 ad $\frac{a}{2A+2a}$.

Quod deinde attinet ad helices intermedias, sufficiet praefatam differentiam aequaliter distribuere, ita ut differentia inter quasvis duas helices proximas eadem sit, nempe $= \frac{1}{n-1} \times \frac{a}{2A+2a}$, si per n intelligatur numerus helicum. Inde facillime etiam deducitur pro quavis helice ratio inter arcum aqueum et arcum aëreum, modo unica quaevis helix uniformis amplitudinis censeatur.

§. 11. Lubet haec omnia unico illustrare exemplo: ponatur $a = A = 32$ ped. paris sic fiet, vi paragraphi praecedentis, altitudo diabetis

$$b = a + A \,\text{log. hyperb.}\left(1 + \frac{a}{A}\right) = 54 \text{ ped.}$$

neglecta fractione superaddenda; habebimus porro densitatem aëris in intima helice duplo maiorem, quam est densitas aëris naturalis: capacitatem intimae helicis = tribus quartis partibus capacitatis extimae helicis. Denique duae partes intimae helicis aqua, una pars aëre condensato replebitur: arcus nimirum aqueus erit 240 gr. aëreus 120 gr. Numerus autem spirarum vel helicum tantus sit oportet, ut si in summam colligantur *altitudines arcuum aqueorum* paragrapho quarto definitorum, fiat haec summa paulo maior, quam 32 ped. Caeterum praefatae mensurae conveniunt machinae non solum perfectissimae, sed et ad statum maximae suae efficacitatis reductae, pro quo nimirum singuli arcus aquei ad summitatem helicis suae pertingunt; quia vero in hoc statu facile evenit, ut aquae hanc summitatem praeterlabantur et in helicem praegressam regurgitent, imo ut aquae in diabete subsistant nullaeque effluant, velim ut in calculo altitudo b octava vel saltem decima sua parte maior adhibeatur, quam quae pro altitudine diabetis proposita est: convenientissimas puto allatas proportiones, cum aquae ad altitudinem 48 vel 50 pedum sunt elevandae. Imo si diabetes ita amplus ponatur, ut conservatio columellarum aërearum integra locum habere non possit, tunc altitudo eius parum superabit altitudinem a id est, 32 ped. haec omnia probe dispiciantur antequam ipsa machinae constructio tentetur.

§. 12. Quae dixi de augmento altitudinis, ad quam aquae data machina elevari possint, ab intersertis columellis aëreis oriundo, essentialiter ad machinae effectum pertinent, nec sine dispendio adhibitarum virium motricium negligi possunt; et enim machina nihil addit altitudini a, quod non acceperit a viribus motricibus adhibitis, et cum vires eaedem impendantur, sive negligatur augmentum sive studiose conservetur, caveamus, ne plus de viribus motricibus pereat, quam necesse est: vires motrices non solum in motum aquarum machinae inclusarum earumque in singulis helicibus elevationem, sed et in compressionem atque condensationem aëris interni impenduntur. Ostendi autem in *Hydrodynamica* mea, omni aëri condensato *vim vivam potentialem* inesse, cuius ope datum pondus, ad certam altitudinem, quam in *Hydrodynamica* definivi, elevari possit, antequam exhausta fuerit. In nostro autem argumento, quod hic tractamus, exhauritur, dum omnes aquae ulterius elevantur ad altitudinem $b - a$ sive altitudinem A log. hyperb. $\left(1 + \frac{a}{A}\right)$. Huc quadrat *Hydrodynamicae* sectionis decimae paragraphus quadragesimus. Igitur si nobis timeamus a nimia diabetis amplitudine, quam ut columellae aëreae manere possint integrae, existimo huic incommodo occurri posse, si plures diabetae coniungantur: nec tamen puto, fieri unquam posse, ut omnes aquae in diabete confluant, liberamque aëri eruptionem concedant; videtur enim id minime timendum esse de aquis celeriter satis transfluentibus, nec de aëre tot vinculis retento. Quin etiam si concedatur, aërem maxima parte in aquas proximas penetrare, an propterea se statim totus extricabit? non puto; videtur enim, in progressu suo aquas permansuras esse spumosas, atque adeo leviores, quam sint aquae purae; huiusmodi permixtio aëris cum aqua eundem propemodum effectum habebit, quem faciunt columellae aqueae et aëreae sibi invicem succedentes.

§. 13. Pauca supersunt monenda de diabete assurgente; equidem existimo non male actum fore, si amplitudo eius fiat uniformis atque tanta sit, quanta est amplitudo helicis intimae, sique machina ea circumagatur velocitate, ut omnis aqua omnisque aër ex helice in diabeten quavis revolutione transfundantur. Hac posita distributione erunt singulae columellae aqueae in diabete eiusdem altitudinis; ipsaeque etiam columellae aëreae, si in densitatem naturalem restitutae putentur, eadem illa altitudine gauderent in hypothesi nostra, qua nimirum singulis machinae revolutionibus aquam et aërem naturalem aequali volumine hauriri ponimus, cui soli hypothesi, brevitatis gratia, insistimus, ne in re facili molesti simus nimia prolixitate. Sit nunc altitudo cuiusvis columellae aqueae $= \alpha$, earumque numerus $= n$, atque putemus columellam aqueam esse in diabetis summitate positam; erit altitudo subsequentis columellae aëreae $= \frac{A\alpha}{A+\alpha}$; deinde similis altitudo pro secunda, tertia etc. columella aërea erit $\frac{A\alpha}{A+2\alpha}$; $\frac{A\alpha}{A+3\alpha}$ etc. usque dum perveniatur ad infimam columellam aëream, cuius altitudo erit $= \dfrac{A\alpha}{A + n\alpha}$. Igitur omnes et singulae columellae aëreae simul sumtae altitudinem habebunt, aequalem summae terminorum

$$\frac{A\alpha}{A+\alpha} + \frac{A\alpha}{A+2\alpha} + \frac{A\alpha}{A+3\alpha} \ldots\ldots\ldots\ldots + \frac{A\alpha}{A+n\alpha}\,.$$

Equidem notum est, huiuscemodi progressiones indefinitas perfecte summabiles non esse, posse tamen quam proxime earum summam indicari, quicunque numeri assumantur pro literis A, α et n. Sic peculiari appropinquatione invenio hanc summam propemodum

$$= \frac{A\alpha}{A+\alpha} + A \,\text{log. hyperb.}\, \frac{2A + 2n\alpha + \alpha}{2A + 3\alpha}\,,$$

quae expressio plus satis convenit argumento nostro; fuerit, exempli gratia, $A = 32$ ped. $\alpha = 1$ ped. et $n = 32$, erit summa omnium altitudinum aërearum $\frac{32}{33} + 32\,\text{log. hyperb.}\,\frac{129}{67} = 21.916$. At vero summa omnium columellarum aquearum altitudinem habet $n\alpha = 32$. Hae determinationes optime conveniunt cum paragrapho undecimo, ubi idem pertractavimus sub alia facie exemplum, cum litera a idem significet, quod $n\alpha$.

§. 14. Fieri utique potest, ut machina, dum rotatur, aquam et aërem naturalem inaequali volumine hauriat; tunc etiam columellae aqueae et aëreae in diabete mutabuntur: fuerint ambo volumina praefata α et β; fiet quaevis columna aquea, ut antea, $= \alpha$; at vero quaevis columna aërea suam altitudinem mutabit in ratione α ad β; igitur altitudo pro omnibus columellis aqueis manebit $= n\alpha = a$; at vero altitudo pro omnibus columellis aëreis fiet $= \dfrac{A\beta}{A+\alpha} + \dfrac{A\beta}{\alpha}\,\text{log. hyperb.}\,\dfrac{2A + 2n\alpha + \alpha}{2A + 3\alpha}$. Haec omnia ita se habere, observare potui in machinula typica cuius diabetes componebatur ex tubis vitreis decem pollices longis, in utraque extremitate cochlea orichalcea instructis, ut unus cum altero firmiter cohaereret atque perfecte iungeretur; sic altitudo diabetis pro lubitu augeri vel diminui poterat. Convenientissimum autem videbatur, cum machinula ita aquis submergebatur, ut aequali volumine aquam et aërem naturalem alternatim in se ingurgitaret, quod factum facile intelligitur ex comparatione columellarum aquearum et aërearum in diabete.

§. 15. Non difficile est, rem ita immutare, ut aqua et aër simul in diabeten transfluant, nec tum porro ullae formabuntur columellae in diabete, sed aqua habebitur insigniter spumosa vel aëre turgida: fluidum mixtum densius erit in inferiori diabetis parte quam in superiori; id quoque contingit in nova machina Parisina §. 7. descripta. Haec densitatum variatio pluribus quaestionibus physicis solvendis inservire potest: nunc vero aliud non monebo, quam quod ad argumentum nostrum pertinet. Dico igitur, quod pro nova hac hypothesi in duobus paragraphis praecedentibus litera α consideranda sit tanquam infinite parva, numerus n tanquam infinite magnus, ita ut pro $n\alpha$ semper poni possit altitudo a, quae denotat altitudinem omnis aquae in diabete contentae, si ab omni aëre liberaretur: hoc modo potest quantitas

$$\frac{A\beta}{A+\alpha}+\frac{A\beta}{\alpha}\,\text{log. hyperb.}\,\frac{2A+2n\alpha+\alpha}{2A+3\alpha}\ \text{censeri}=\frac{A\beta}{a}\,\text{log. hyperb.}\Big(1+\frac{a}{A}\Big):$$

est itaque altitudo, quam omnis aër inaequaliter densus occupat, ut a ad $\frac{A\beta}{\alpha}$ log. hyperb.$\left(1+\frac{a}{A}\right)$ vel, posito $\beta=\alpha$, ut a ad A log. hyperb.$\left(1+\frac{a}{A}\right)$. Sic igitur si ponatur, exempli gratia, $a=A$, erunt praefatae altitudines in ratione 1 ad log. hyperb. 2 sive proxime ut 100 ad 69; unde integra diabetis altitudo nunc fiet $=\frac{169}{100}\times 32=54.08$ ped. conferantur §.§. 6, 10 et 11. Suprema autem columella aërea altitudinem habebit duplam eius; quam infima habet.

§. 16. Atque haec sunt, quae partim ad novae cochleae explicationem, partim ad veram eiusdem œconomiam, quae optatum sucessum et certiorem et uberiorem reddat, inservire posse existimavi: sua novitate perinde placuit atque ingenio haec machina omnibus, in quorum notitiam venit, de pluribus autem eius momentis varii varia statuerunt; duo sunt quae mihi quidem in machina Wirziana minus placent; *primum* est, quod difficillimum putem, praesertim in machinis maiusculis, totum laminae intortae tractum utroque fundo ita cooperire, ut ne minima quidem rimula superfit; hinc factum est, accusante ipso Cl. *Zieglero*, ut cessante rotatione aquae in diabete mox subsiderent, quae certe suspensae haerere debebant. *Secundum* est, quod diametri helicum versus interiora insigniter decrescant, nimio virium motricium decremento; alium igitur suasi cochleae perficiendae modum praestantiorem simul atque sua constructione longe faciliorem, quem nunc quoque ipsi artifices tigurini sequuntur. Novam cochleam paucis exponam verbis.

§. 17. Tubus plumbeus sufficientis longitudinis, cuius diameter 8, 10 vel 12 lineas habuerit, tympano circumflectatur gyris iuxta se invicem positis; diameter tympani, pro re nata, sit unius, duorum pluriumve pedum, atque tubus efformet duodecim, sedecim aut plures circumvolutiones; tum vero versus axem tympani dirigatur ibique cum assurgente, ut ante, coniungatur. Huiusmodi machinae constructio utique facillima est, nec ulli periculo vel minimae eruptionis lateris obnoxia, quae in latentibus rimulis timeri possit, simulque maiores format arcus aqueos: sed et haec mutata constructio eandem ubique requirit helicum diminutionem, quam paragrapho nono et sequentibus definivi, qua sola effici potest, ut singuli arcus aquei atque aërei in toto plumbei decursu sua sponte debitam conservent dispositionem atque collocationem; patet vero, hanc helicum diminutionem duobus diversis modis posse obtineri, vel coarctando tubum plumbeum tympano cylindrico circumflectendum, vel adhibendo tympanum conicum pro tubo uniformiter amplo, cuiusmodi propemodum configuratio in horologiis portatilibus conciliatur parvulo cono truncato, quem catenula involvit, et quem galli *fusee* vocant. Sic pro exemplo, quod in paragrapho decimo atque undecimo pertractavimus, diametri in utraque basi rationem obtinerent ut 4 ad 3, si tubus plumbeus ab una basi ad alteram decurrat.

§. 18. Si lubeat praefato tympano, sive cylindrico sive conico, capsulam vel receptaculum cylindricum longitudinaliter et sub axe communi firmiter adiungere, cui tubus plumbeus pleno orificio inseratur, fiet ut alternis vicibus aqua et aër in istud receptaculum iniiciantur, ex quo demum utrumque fluidum in diabeten attollentem impelletur, sed alio modo, ac antea: nempe altera medietas receptaculi erit aquis repleta, altera aëre condensato; aër condensatus aquam in diabeten impellet, simulque et ipse, una cum aqua, iugiter in eundem iniicietur: quod receptaculum perdit, continue reparatur ab effluxu per tubum in cavitatem receptaculi hiantem, vid. Fig. 3.
Sic igitur traiectio, in utroque fluido simul facta, non sinet venam superius effluentem esse interruptam; aër, ni fallor, cum aqua permixtus manebit in diabete, at dum ascendit dilatabitur, et cum iam effluxui proximus est, ad densitatem naturalem reductus erit. In machina Parisina §. 7. descripta utrumque fluidum eodem modo manet permixtum, etiamsi in suprema parte massae aëreo-aqueae nonnisi sexta pars densitatis naturalis supersit, quae aequae purae inest. Atque sic etiam paragrapho decimo quinto satisfecimus, unde fortasse augmentum altitudinis, de quo paragrapho sexto mentionem feci, vel in siphonibus amplioribus, integrum manebit et omne virium in rotationem machinae adhibitarum dispendium vitabitur. Caeterum machina Wirziana, quae alis instructa est, facile ab ipso flumine, cui insidet, rotatur.

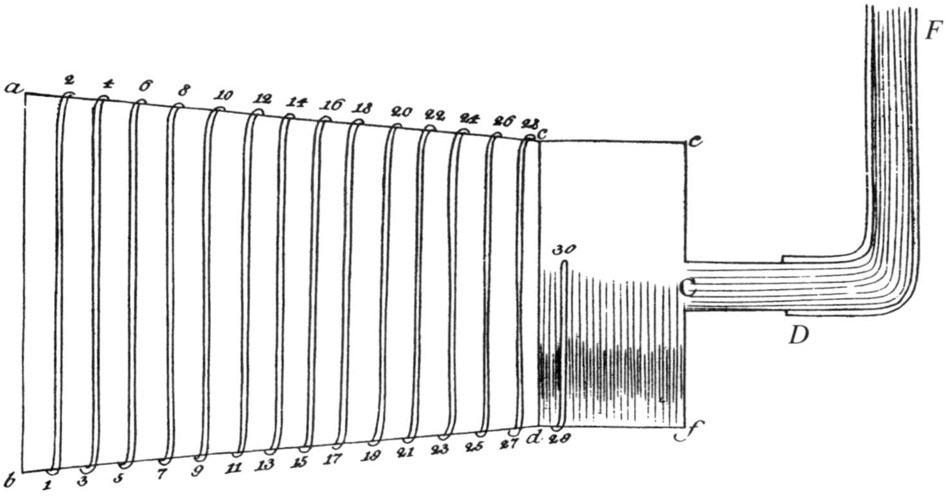

Fig. 3

EXPLICATIO FIGURARUM.

Figura prima et secunda pertinent ad paragraphum secundum atque indicant, quemadmodum, alternis vicibus, aqua et aër ex cavitate intimae helicis *C*, in figura prima, impellantur in canalem *C* figurae secundae, postquam choenicion fundo machinae, circa centrum perforato, superimpositum et exacte adaptatum fuit.

Figura tertia pertinet ad paragraphum 17. et 18. atque ideam vel configurationem machinae correctae sistit. Nempe *abcd* corticem sistit tympaniformem, cui canalis plumbeus

1. 2. 3. 4. 5. 6 .... 25. 26. 27. 28. 29 30

circumvolutus est. Praefato tympano *abcd* firmiter adiungitur receptaculum *cdef* ex ferro constructum, super quo extremitas tubi plumbei a 29 ad 30 decurrit, eidemque inseritur, ita ut, alternis vicibus, aqua et aër ex tubo in cavitatem receptaculi iniiciantur. Utrumque autem fluidum simul per ductum excretorium *CD*, continue cum integra machina rotatum, traiicietur in diabeten assurgentem et immotum *DF*.

Verzeichnisse

Indexes

Name Index

The following list contains, in alphabetic order, the names of all persons mentioned in this volume, in the source texts as well as in the commentary.

List of letters by and to Daniel Bernoulli referring to technology treated in volume 7 and 8

As in the other volumes of this series, we list the letters containing allusions to the works published in this volume or to related subjects.

The last column indicates either a printed version of the letter or the location of its manuscript. The following abbreviations are used:

AAN	Archive of the St. Petersburg Branch of the Russian Academy of Science
BPU Genève	Bibliothèque publique et universitaire de Genève
DSB Berlin	Deutsche Staatsbibliothek Berlin
Fuss 2	Paul-Heinrich Fuss (ed.), Correspondance mathématique et physique de quelques célèbres géomètres du XVIIIème siècle, T. II, St.-Pétersbourg 1843
ÖNB Wien	Österreichische Nationalbibliothek Wien
SH Genève	Societé d'Histoire de Genève
UB Basel	Öffentliche Bibliothek der Universität Basel
ZB Zürich	Zentralbibliothek Zürich

by	to	Date	Source	code
Bernoulli, D.	Mairan, J. J.	1725, April 26th	UB Basel	LIa 661, Nr. 14
Bernoulli, D.	Goldbach	1725, June 13th	Fuss 2	pp. 234–237
Bernoulli, D.	Mairan, J. J.	1725, July 8th	DSB Berlin	Slg. Darmstaedtes
Mairan, J. J.	Bernoulli, D.	1725, July 16th	UB Basel	LIa 709, fo. 120
Bernoulli, D.	Euler, L.	1733, Sept. 22nd	AAN	f. 136, op. 2, Nr. 7, 1. 8–8 R
Bernoulli, D.	Delisle, J.-N.	1733, July 17th	Paris Obs.	
Bernoulli, D.	Scheuchzer	1734, March 12th	ZB Zürich	Ms H 340, pp. 569–572
Bernoulli, D.	Mairan, J. J.	1737, June 4th	ÖNB Wien	Autogr. 5/7-1
Cramer, G.	Bernoulli, D.	1738, Sept. 23rd	UB Basel	LIa 686, fo. 789–794
Bernoulli, D.	Cramer, G.	1738, n. d.	UB Basel	LIa 676, fo. 79–82
Bernoulli, D.	Cramer, G.	1738, May 14th	BPU Genève	Ms Suppl. 384
Bernoulli, D.	Euler, L.	1738, May 24th	AAN	f. 136, op. 2, Nr. 7, 1. 33–34 R
Bernoulli, D.	Euler, L.	1738, Nov. 8th	AAN	f. 1, op. 3, Nr. 26, 1. 107–108 R
Bernoulli, D.	Euler, L.	1739, Aug. 29th	AAN	f. 1, op. 3, Nr. 28, 1. 143–145 R
Euler, L.	Bernoulli, D.	1739, Sept. 15th	AAN	f. 1, op. 3, Nr. 27, 1. 155–159 R
Bernoulli, D.	Jallabert, J.	1740, July 6th	SH Genève	243
Jallabert, J.	Bernoulli, D.	1741, n. d.	SH Genève	243
Bernoulli, D.	Jallabert, J.	1741, Nov. 3rd	SH Genève	243
Bernoulli, D.	Jallabert, J.	1742, Jan. 10th	SH Genève	243
Bernoulli, D.	Moula, F.	1743, Aug. 19th	AAN	f. V, op. 5–11, Nr. 3
Bernoulli, D.	Schumacher, J. D.	1744, Aug. 15th	AAN	f. 1, op. 3, Nr. 1
Bernoulli, D.	I. Acad. Sci. Pet.	1746, Nov. 12th	AAN	f. V, op. 15, Nr. 4, fo. 1–2
Bouger, P.	Bernoulli, D.	1751, April 28th	UB Basel	LIa 680, fo. 231–232
Bernoulli, D.	Bouger, P.	1754, Nov. 11th	UB Basel	LIa 53, Nr. 17
Bernoulli, D.	Acad. Sci. Paris	1755, Aug. 22nd	UB Basel	LIa 53, Nr. 16
Bouger, P.	Bernoulli, D.	1755, Sept. 27th	UB Basel	LIa 680, fo. 235–236
Bouger, P.	Bernoulli, D.	1756, Aug. 6th	UB Basel	LIa 680, fo. 237–238
Mallet, J. A.	Bernoulli, D.	1762, Oct. 27th	BPU Genève	Ms fr. 655, fo. 5
Bernoulli, D.	Mallet, J. A.	1763, Febr. 16th	BPU Genève	Ms fr. 655, fo. 7
Mallet, J. A.	Bernoulli, D.	1763, Dec. 30th	UB Basel	LIa 709, fo. 155

by	to	Date	Source	code
Bernoulli, D.	Mallet, J. A.	1764, Jan. 18th	BPU Genève	Ms fr. 655, fo. 10
Bernoulli, D.	Mallet, J. A.	1765, Nov. 23rd	BPU Genève	Coll. Autogr
Bernoulli, D.	Johann III	1766, June	UB Basel	L Ia 676, fo. 36–37
Bernoulli, D.	Johann III	1766, n. d.	UB Basel	L Ia 676, fo. 38–39
Bernoulli, D.	Mairan, J. J.	1768, July 1st	UB Basel	L Ia 709, fo. 122
Bernoulli, D.	Euler, J.-A.	1771, Febr. 16th	AAN	f. 1, op. 3, Nr. 58, fo. 11–12v
Euler, J.-A.	Bernoulli, D.	1771, May 7th	UB Basel	L Ia 679, fo. 193 u. 195
Bernoulli, D.	Mallet, J. A.	1771, May 18th	BPU Genève	Ms Inventaire 379
Bernoulli, D.	Euler, J.-A.	1771, June 8th	AAN	f. 1, op. 3, Nr. 58, fo. 53–54
Mallet, J. A.	Bernoulli, D.	1771, July 15th	UB Basel	L Ia 709, fo. 171
Bernoulli, D.	Mallet, J. A.	1771, Sept. 28th	BPU Genève	Ms fr. 655, fo. 58
Bernoulli, D.	Johann III	1771, n. d.	UB Basel	L Ia 676, fo. 41
Mallet, J. A.	Bernoulli, D.	1772, March 3rd	BPU Genève	Ms fr. 655, fo. 60
Bernoulli, D.	Mallet, J. A.	1772, April 15th	BPU Genève	Ms fr. 655, fo. 61
Bernoulli, D.	Johann III	1772, n. d.	UB Basel	L Ia 676, fo. 42
Bernoulli, D.	Euler, J.-A.	1775, Aug. 12th	AAN	f. 1, op. 3, Nr. 62, fo. 48–49v
Bernoulli, D.	Euler, J.-A.	1775, Nov. 11th	AAN	f. 1, op. 3, Nr. 62, fo. 180–181v

Verzeichnis der Werke Daniel Bernoullis

Dieses Verzeichnis wurde von H. Straub für den Artikel «Daniel Bernoulli» im *Dictionary of Scientific Biography* (vol. II, New York 1970, pp. 36–46) zusammengestellt und von P. Radelet-de Grave, V. Scheuber und M. Mattmüller ergänzt. Zu den Abkürzungen vgl. oben S. XVI–XVIII.

Gedruckte Werke

St. 1 *Dissertatio Inauguralis Physico-Medica De Respiratione, Quam Consensu & Auctoritate Gratiosissimi Medicorum Ordinis in Universitate Patria Pro Summis in Arte Medica Honoribus & Privilegiis Doctoralibus rite consequendis, Ad diem 2. Septembr. MDCCXXI L. H. Q. S. Publico Examini subjicit Daniel Bernoulli, Joh. Fil. Basil.*, o. O. [Basileae]: Typis Johannis Ludovici Brandmülleri, o. J. [1721] – *Werke* 1, pp. 61–83.

St. 2 *Positiones Miscellaneae Medico-Anatomico-Botanicae, Quas favente Deo Jussu & Auctoritate Amplissimi Senatus Academici Pro vacante Professione Anatomico-Botanica Solenni Eruditorum Examini subjiciet in Auditorio Hyberno Horis Consuetis Daniel Bernoulli, Med. Cand. Ad diem 24. Octobr. MDCCXXI*, Basileae: Typis Johannis Ludovici Brandmülleri, o. J. [1721] – *Werke* 1, pp. 85–91.

St. 3 *Theses Logicae sistentes Methodum examinandi Syllogismorum validitatem, Quas una cum Annexis Corollariis Miscellaneis Auspice Deo consentiente Magnifico & Sapientissimo Ordine Philosophico pro vacante Cathedra Logica Cl. Competitorum Examini subjicit Daniel Bernoulli, A. L. M. & Med. Cand. Respondente Juvene florentissimo Friderico Zwingero, Cel. Poliatri Fil. in Auditorio Juridico A. D. 4. Febr. MDCCXXII Horis consuetis pomer.*, Basileae: Typis Johannis Ludovici Brandmülleri, o. J. [1722] – *Werke* 1, pp. 257–264.

St. 4 *Exercitationes quaedam mathematicae*, Venetiis: Apud Dominicum Lovisam, 1724 – *Werke* 1, pp. 297–362.

St. 5 *Notata in praecedens Schediasma Ill. Co. Jacobi Riccati*: AE Suppl. Tom. VIII (1724), Sect. II, pp. 73–75 – *Werke* 1, pp. 272–274.

St. 6 *Danielis Bernoulli Explanatio Notationum suarum, quae exstant Supplem. Tomo VIII Sect. II*: AE Octobris 1725, pp. 470–473 – *Exercitationes quaedam mathematicae* [St. 4], pp. 73–77 – *Werke* 1, pp. 346–349.

St. 7 *Solutio Problematis Riccatiani propositi in Act. Lips. Suppl. Tom. VIII p. 73*: AE Octobris 1725, pp. 473–475 – *Exercitationes quaedam mathematicae* [St. 4], pp. 77–80 – *Werke* 1, pp. 349–351.

St. 8 *Discours sur la maniere la plus parfaite de conserver sur Mer l'égalité du mouvement des Clepsidres ou Sabliers*: Prix Paris 1725 (1725), pp. 3–21 – *Werke* 7, pp. 221–239.

St. 9 *Examen Principiorum Mechanicae, et Demonstrationes Geometricae de Compositione et Resolutione Virium*: CP I, 1726 (1728), pp. 126–142 – *Werke* 3, pp. 119–135 – [französische Übersetzung von J. Dhombres:] *Examen des Principes de la Mécanique et démonstrations géométriques de la composition et de la résolution des forces*: Sciences et techniques en perspective, vol. 11 (1986/7), Nantes 1987, pp. 39–58.

St. 10 *Tentamen novae de Motu Musculorum Theoriae*: CP I, 1726 (1728), pp. 297–313 – *Werke* 1, pp. 92–103.

St. 11 *Experimentum circa Nervum Opticum*: CP I, 1726 (1728), pp. 314–317 – *Werke* 1, pp. 104–106.

St. 12 *Theoria Nova de Motu Aquarum per Canales quoscunque fluentium*: CP II, 1727 (1729), pp. 111–125.

St. 13 *De mutua relatione Centri Virium, Centri Oscillationis et Centri Gravitatis Demonstrationes Geometricae*: CP II, 1727 (1729), pp. 208–216 – *Werke* 3, pp. 136–144.

St. 14a *Dissertatio de Actione Fluidorum in Corpora solida et Motu Solidorum in Fluidis*: CP II, 1727 (1729), pp. 304–342.

St. 14b *Dissertationis de Actione Fluidorum in Corpora solida et Motu Solidorum in Fluidis Continuatio*: CP III, 1728 (1732), pp. 214–229.

St. 15 *Methodus universalis Determinandae curvaturae fili a potentiis quamcunque legem inter se observantibus extensi, una cum solutione problematum quorundam novorum eo pertinentium*: CP III, 1728 (1732), pp. 62–69.

St. 16 *Observationes de Seriebus quae formantur ex additione vel subtractione quacunque terminorum se mutuo consequentium, ubi praesertim earundem insignis usus pro inveniendis radic[ibus] omnium Aequationum Algebraicarum ostenditur*: CP III, 1728 (1732), pp. 85–100 – *Werke* 2, pp. 49–64.

St. 17 *Problema astronomicum inveniendi altitudinem Poli una cum declinatione Stellae ejusdemque culminatione ex tribus altitudinibus Stellae et duobus temporum intervallis brevi calculo solutum*: CP IV, 1729 (1735), pp. 89–94 – *Werke* 1, pp. 443–447.

St. 18a *Theorema de Motu curvilineo corporum, quae resistentiam patiuntur velocitatis suae quadrato proprtionalem una cum Solutione Problematis in Act: Lips: M: Nov: 1728 propositi*: CP IV, 1729 (1735), pp. 136–143.

St. 18b *Additamentum ad Theoremata. De Motu Corporum curvilineo in mediis resistentibus, in quo resistentiae considerantur quae partim quadratis velocitatum partim momentis temporum proportionales sunt*: CP V, 1730/31 (1738), pp. 126–142.

St. 19 *Experimenta coram Societate instituta in confirmationem Theoriae pressionum quas latera Canalis ab aqua transfluente sustinent*: CP IV, 1729 (1735), pp. 194–201.

St. 20 *Notationes de Aequationibus, quae progrediuntur in infinitum, earumque resolutione per Methodum Serierum recurrentium: ut et de nova Serierum Specie*: CP V, 1730/31 (1738), pp. 63–82 – *Werke* 2, pp. 65–80.

St. 21 *Dissertatio brevis de Motibus Corporum reciprocis seu oscillatoriis, quae ubique resistentiam patiuntur quadrato velocitatis suae proportionalem. Ubi ostenduntur Theoremata, quorum ope expedite ad calculum revocantur motus pendulorum in mediis perfecte fluidis praesertim aëre ceu medio tenuissimo*: CP V, 1730/31 (1738), pp. 106–125.

St. 22 *Specimen Theoriae novae de Mensura Sortis*: CP V, 1730/31 (1738), pp. 175–192 – [deutsche Übersetzung von A. Pringsheim in:] *Die Grundlage der modernen Wertlehre: Daniel Bernoulli, Versuch einer neuen Theorie der Wertbestimmung von Glücksfällen*, Leipzig: Duncker & Humblot, 1896, pp. 21–60 – [englische Übersetzung von L. Sommer:] *Exposition of a New Theory on the Measurement of Risk*: Econometrica 22, n°. 1 (January, 1954), pp. 23–36 – *Werke* 2, pp. 223–234.

St. 23 *Theoremata de Oscillationibus corporum filo flexili connexorum et catenae verticaliter suspensae*: CP VI, 1732/33 (1739), pp. 108–122.

St. 24 *Recherches Physiques et Astronomiques sur le Probleme proposé pour la seconde fois par l'Academie Royale des Sciences de Paris. Quelle est la cause physique de l'inclinaison des Plans des Orbites des Planetes par rapport au plan de l'Équateur de la revolution du Soleil autour de son axe; Et d'où vient que les inclinaisons de ces Orbites sont differentes entre elles*: Prix Paris 1734 (1735), pp. 93–122 – [lateinische Erstfassung:] *Disquisitiones Physico-Astronomicae Problematis ab Inclyta Scientiarum Academia Regia, quae Parisiis floret, iterum propositi ...*: Prix Paris 1734 (1735), pp. 125–144 – *Werke* 3, pp. 303–326.

St. 25 *Danielis Bernoulli Demonstrationes Theorematum suorum de Oscillationibus corporum filo flexili connexorum et catenae verticaliter suspensae*: CP VII, 1734/35 (1740), pp. 162–173.

St. 26 *De Legibus quibusdam mechanicis, quas Natura constanter affectat, nondum descriptis, earumque Usu hydrodynamico, pro determinanda vi venae aqueae contra planum incurrentis*: CP VIII, 1736 (1741), pp. 99–127 – *Werke* 5, pp. 425–444.

St. 27 *De Variatione motuum a Percussione excentrica*: CP IX, 1737 (1744), pp. 189–206 – *Werke* 3, pp. 145–159.

St. 28 *Réfléxions sur la meilleure figure à donner aux Ancres, et la meilleure manière de les essayer*: Prix Paris 1737 (1737), pp. 49–84.

St. 29 *Commentationes de Immutatione et Extensione Principii Conservationis virium vivarum, quae pro Motu Corporum coelestium requiritur*: CP X, 1738 (1747), pp. 116–124 – *Werke* 3, pp. 160–169.

St. 30 *Commentationes de Statu Aequilibrii corporum humido insidentium*: CP X, 1738 (1747), pp. 147–163.

St. 31 *Hydrodynamica, sive De Viribus et Motibus Fluidorum Commentarii. Opus Academicum ab Auctore, dum Petropoli ageret, congestum*, Argentorati: Sumptibus Johannis Reinholdi Dulseckeri, Typis Joh. Henr. Deckeri, Typographi Basiliensis, 1738 – [russische Übersetzung von V. S. Gokhman:] *Gidrodinamika ili zapiski o silakh i dvizheni'akh zhidkostej*, o. O. [Leningrad]: Izdatel'stvo Akademii Nauk SSSR, 1959 – [deutsche Übersetzung von K. Flierl:] *Hydrodynamik oder Kommentare über die Kräfte und Bewegungen der Flüssigkeiten*, München: Forschungsinstitut des Deutschen Museums für die Geschichte der Naturwissenschaften und der Technik, 1965 – [englische Übersetzung von T. Carmody und H. Kobus:] *Hydrodynamics*, New York: Dover, 1968 – *Werke* 5, pp. 93–424.

St. 32 *De Motibus oscillatoriis corporum humido insidentium*: CP XI, 1739 (1750), pp. 100–115.

St. 33 *Traité sur le Flux et Reflux de la Mer*: Prix Paris 1740 (1741), pp. 55–191 – [nachgedruckt von Th. Le Seur und F. Jacquier in:] I. Newton, *Philosophiae Naturalis Principia Mathematica*, t. III, Pars I, Genevae 1742, pp. 133–246 [und verschiedene spätere Ausgaben] – *Werke* 3, pp. 327–438.

St. 34 *Commentationes de Oscillationibus compositis praesertim iis quae fiunt in corporibus ex filo flexili suspensis*: CP XII, 1740 (1750), pp. 97–108.

St. 35 *Excerpta ex Litteris a Daniele Bernoulli ad Leonhardum Euler*: CP XIII, 1741–43 (1751), pp. 3–15 – *Werke* 2, pp. 81–93.

St. 36 *De Motu mixto, quo corpora sphaeroidica super plano inclinato descendunt*: CP XIII, 1741–43 (1751), pp. 94–99 – *Werke* 3, pp. 170–177.

St. 37 *De Vibrationibus et Sono Laminarum Elasticarum Commentationes Physico-Geometricae*: CP XIII, 1741–43 (1751), pp. 105–120.

St. 38 *De Sonis multifariis quos Laminae Elasticae diversimode edunt Disquisitiones Mechanico-Geometricae Experimentis acusticis illustratae et confirmatae*: CP XIII, 1741–43 (1751), pp. 167–196.

St. 39 *Memoire sur la maniere de construire les Boussoles d'inclinaison, Pour faire, avec le plus de précision qu'il est possible, les Observations de l'Inclinaison de l'Aiguille Aimantée, tant sur Mer que sur Terre; ce qui suppose des Boussoles, qui, étant mises dans un même lieu, donneront sensiblement la même Inclinaison*: Prix Paris 1743 (1748), pp. 3–61 – *Werke* 7, pp. 67–111.

St. 40 *Nouveau Probleme de Mecanique resolu par Mr. Daniel Bernoulli*: Mém. Berlin 1745 (1746), pp. 54–70 – [deutsche Übersetzung von A. von Oettingen in:] *Abhandlungen über jene Grundsätze der Mechanik, die Integrale der Differentialgleichungen liefern* (Ostwalds Klassiker der exakten Wissenschaften, Nr. 191), Leipzig und Berlin: Engelmann, 1914, pp. 29–43 – *Werke* 3, pp. 179–196.

St. 41 {mit Johann II Bernoulli} *Nouveaux Principes de Méchanique et de Physique, tendans à expliquer la Nature & les Propriétés de l'Aiman*: Prix Paris 1744–46 (1748), pp. 117–144 – *Werke* 7, pp. 113–135.

St. 42a *Recherches Méchaniques et Astronomiques, Sur la Question proposée par l'Académie Royale des Sciences, pour l'année 1745. La meilleure maniere de trouver l'heure en Mer, par observation, soit dans la jour, soit dans les crépuscules, & sur-tout la nuit, quand on ne voit pas l'horison*: Prix Paris 1745–47 (1750), pp. 3–78 – *Werke* 7, pp. 241–293.

St. 42b *Suite des Recherches Méchaniques et Astronomiques, &c. … Qui tend principalement à fournir aux Navigateurs les moyens Méchaniques les plus sûrs pour faire en Mer, malgré l'agitation du vaisseau, les observations dont on peut conclurre l'heure*: Prix Paris 1745–47 (1750), pp. 79–110 – *Werke* 7, pp. 295–319.

St. 43 *Remarques sur le principe de la conservation des forces vives pris dans un sens général*: Mém. Berlin 1748 (1750), pp. 356–364 – [deutsche Übersetzung von A. von Oettingen in:] *Abhandlungen über jene Grundsätze der Mechanik, die Integrale der Differentialgleichungen liefern* (Ostwalds Klassiker der exakten Wissenschaften, Nr. 191), Leipzig und Berlin: Engelmann, 1914, pp. 65–75 – *Werke* 3, pp. 197–206.

St. 44 *Mémoire sur la nature et la cause des courans, et la meilleure maniere de les observer et de les déterminer*: Prix Paris 1749–51 (1769), pp. 3–108 – *Werke* 5, pp. 535–611.

St. 45 *Réflexions et eclaircissemens sur les nouvelles vibrations des cordes exposées dans les Mémoires de l'Académie de 1747 & 1748*: Mém. Berlin 1753 (1755), pp. 147–172.

St. 46 *Sur le mêlange de plusieurs especes de vibrations simples isochrones, qui peuvent coexister dans un même système de corps*: Mém. Berlin 1753 (1755), pp. 173–195.

St. 47 *Recherches sur la maniere la plus avantageuse de suppléer à l'action du Vent sur les grands Vaisseaux, soit en y appliquant les Rames, soit en y employant quelqu'autre moyen que ce puisse être*: Prix Paris 1753 (1769), pp. 3–99.

St. 48 *Mémoire sur la maniere de diminuer le Roulis & le Tangage d'un navire, sans qu'il perde sensiblement, par cette diminution, aucune des bonnes qualités que sa construction doit lui donner*: Prix Paris 1757 (1771), pp. 3–96.

St. 49 *Sur les nouvelles Aiguilles d'inclinaison, faites à Basle par M. Dietric*: JS 1757 (Janvier), pp. 14–24 = JSA t. XXV (Janvier 1757), pp. 27–50 [teilweise, unter dem Titel: *Memoire adressé à Messieurs les Auteurs du Journal des Sçavans*] – Acta Helvetica, vol. III (1758), pp. 233–249 – *Werke* 7, pp. 140–151.

St. 50 *Lettre de Monsieur Daniel Bernoulli, de l'Académie Royale des Sciences, à M. Clairaut de la même Académie, au sujet des nouvelles découvertes faites sur les vibrations des cordes tendues*: JS 1758 (Mars), pp. 157–166 = JSA XXXIV (Mars 1758), pp. 59–80.

St. 51 *Essai d'une nouvelle analyse de la mortalité causée par la petite Vérole, & des avantages de l'Inoculation pour la prévenir*: Mém. Paris 1760 (1766), pp. 1–45 – *Werke* 2, pp. 235–267.

St. 52 *Reflexions sur les avantages de l'Inoculation*: Mercure de France 1760 (Juin), pp. 173–190 – *Werke* 2, pp. 268–274.

St. 53 *Recherches physiques, mécaniques et analytiques, sur le son & sur les tons des tuyaux d'Orgues différemment construits*: Mém. Paris 1762 (1764), pp. 431–485 [Nachdruck: Bibliotheca Musica Bononiensis II, N. 74, Bologna: Forni, 1983].

St. 54 *Mémoire, sur les vibrations des cordes d'une épaisseur inégale*: Mém. Berlin 1765 (1767), pp. 281–306.

St. 55 *De Usu Algorithmi infinitesimalis in Arte Conjectandi Specimen*: NCP XII, 1766–67 (1768), pp. 87–98 – *Werke* 2, pp. 276–287.

St. 56 *De Duratione media Matrimoniorum, pro quacunque coniugum aetate, aliisque Quaestionibus affinibus*: NCP XII, 1766–67 (1768), pp. 99–126 – *Werke* 2, pp. 290–303.

St. 57 *Commentatio de utilissima ac commodissima Directione Potentiarum frictionibus mechanicis adhibendarum*: NCP XIII, 1768 (1769), pp. 242–256 – *Werke* 3, pp. 209–218.

St. 58 *Disquisitiones analyticae de novo Problemate coniecturali*: NCP XIV.1, 1769 (1770), pp. 3–25 – *Werke* 2, pp. 306–324.

St. 59a *Mensura Sortis ad fortuitam Successionem rerum naturaliter contingentium applicata*: NCP XIV.1, 1769 (1770), pp. 26–45 – *Werke* 2, pp. 326–338.

St. 59b *Continuatio argumenti De Mensura Sortis ad fortuitam Successionem rerum naturaliter contingentium applicata*: NCP XV, 1770 (1771), pp. 3–28 – *Werke* 2, pp. 341–360.

St. 60 *Commentationes Physico-Mechanicae de Frictionibus variis illustratae Exemplis*: NCP XIV.1, 1769 (1770), pp. 249–269 – *Werke* 3, pp. 221–238.

St. 61 *Examen Physico-Mechanicum De Motu mixto qui Laminis elasticis a Percussione simul imprimitur*: NCP XV, 1770 (1771), pp. 361–380.

St. 62 *De Summationibus Serierum quarundam incongrue veris earumque Interpretatione atque Usu*: NCP XVI, 1771 (1772), pp. 71–90 – *Werke* 2, pp. 101–116.

St. 63 *De Vibrationibus Chordarum, ex duabus partibus, tam longitudine quam crassitie, ab invicem diversis, compositarum*: NCP XVI, 1771 (1772), pp. 257–280.

St. 64 *De indole singulari Serierum infinitarum quas sinus vel cosinus angulorum arithmetice progredientium formant, earumque Summatione et Usu*: NCP XVII, 1772 (1773), pp. 3–23 – *Werke* 2, pp. 119–134.

St. 65 *Expositio theoretica singularis Machinae hydraulicae Tiguri Helvetiorum exstructae*: NCP XVII, 1772 (1773), pp. 251–271.

St. 66 *Theoria elementari[s] Serierum, ex sinibus atque cosinibus arcuum arithmetice progredientium diversimode compositarum, dilucidata*: NCP XVIII, 1773 (1774), pp. 3–23 – *Werke* 2, pp. 138–151.

St. 67 *Vera Determinatio Centri Oscillationis in corporibus qualibuscunque filo flexili suspensis eiusque ab Regula communi Discrepantia*: NCP XVIII, 1773 (1774), pp. 247–267.

St. 68 *Commentatio Physico-mechanica generalior Principii de Coexistentia vibrationum simplicium haud perturbatarum in Systemate composito*: NCP XIX, 1774 (1775), pp. 239–259.

St. 69 *Commentatio Physico-mechanica specialior de Motibus reciprocis compositis multifariis nondum exploratis qui in Pendulis bimembribus facilius observari possint in confirmationem Principii sui de coexistentia vibrationum simpliciorum*: NCP XIX, 1774 (1775), pp. 260–284.

St. 70 *Adversaria analytica miscellanea de Fractionibus continuis*: NCP XX, 1775 (1776), pp. 3–23 – *Werke* 2, pp. 156–172.

St. 71 *Disquisitiones ulteriores de indole Fractionum continuarum*: NCP XX, 1775 (1776), pp. 24–47 – *Werke* 2, pp. 175–194.

St. 72 *Diiudicatio maxime probabilis plurium Observationum discrepantium atque verisimillima Inductio inde formanda*: AP 1777.1 (1778), pp. 3–23 – *Werke* 2, pp. 361–375.

St. 73 *Specimen philosophicum de Compensationibus horologicis, et veriori mensura Temporis*: AP 1777.2 (1780), pp. 109–128 – *Werke* 2, pp. 376–390.

St. 74 {anonym} *Recherches physiques et matematiques sur la Theorie des vents reglés. Sujet proposé par l'Académie Royale des sciences pour 1746*: J. d'Alembert, *Reflexions sur la cause generale des Vents. Piéce qui a remporté le Prix proposé par l'Académie Royale des sciences et belles lettres pour l'Année MDCCXLVI. ... A laquelle on a joint les Pieçes qui ont concouru*, Berlin 1747, pp. 137–176 – *Werke* 5, pp. 509–535.

St. 75 {herausgegeben von O. Spiess und F. Verzár} *Oratio physiologica De Vita. Eine akademische Festrede von Daniel Bernoulli*: Verhandlungen der Naturforschenden Gesellschaft in Basel LII, 1940–41 (1941), pp. 189–266 [lateinischer Text pp. 219–234, deutsche Übersetzung pp. 203–218] – *Werke* 1, pp. 107–116.

R *Remarques sur les Aimans artificiels de Basle*: Nouvelle Bibliotheque Germanique, t. XVI.1 (Janvier, Fevrier & Mars 1755), pp. 225–229 – Journal Helvetique ou Recueil de Pieces fugitives de literature choisie (Fevrier 1755), pp. 201–207 – *Werke* 7, pp. 137–139.

S *Extrait d'une Lettre de Mr. Daniel Bernoulli à Mr. Garcin, Docteur en Médecine, sur les Elemens d'Algèbre de Mr. Clairaut, de l'Académie Roïale des Sciences de Paris*: Mercure Suisse ou Recueil de Nouvelles Historiques ... (Janvier 1747) pp. 57–65 – *Werke* 2, pp. 94–97.

Ta *Diverses Reflexions concernant la Physique generale*: Acta Helvetica, Physico-Mathematico-Botanico-Medica, vol. I (1751), pp. 33–42 – *Werke* 5, pp. 622–630.

Tb *Diverses Reflexions concernant la Physique generale. Second Memoire*: Acta Helvetica, Physico-Mathematico-Botanico-Medica, vol. II (1755), pp. 101–113 – *Werke* 5, pp. 631–640.

U *Lettre de Mr. Daniel Bernoulli, Professeur à Bâle & de l'Acad. Imp. des Sciences de Petersbourg, à Mr. Schoepflin, Professeur à Strasbourg & de l'Academie Royale des Inscriptions, sur un Livre que le prémier va publier*: Mercure Suisse ou Recueil de Nouvelles Historiques ... (Septembre 1734), pp. 42–50 – *Werke* 5, pp. 87–90.

V *Animadversiones in solutionem Problematis Lunular*[*um*] *quadrabilium*: G. Suzzi, *Disquisitiones mathematicae*, Venetiis: Apud Dominicum Lovisam, 1725, pp. 52–55 – *Werke* 1, pp. 292–294.

W [*Reponse de Mr. Bernoulli*]: J.-N. Delisle, *Discours Lû dans l'Assemblée publique de l'Academie des Sciences, Le 2. Mars 1728. Question: Si l'on peut demontrer par les seuls faits Astronomiques, quel est le vrai Systeme du monde? Et si la Terre tourne, ou non?*, A St. Petersbourg: De l'Imprimerie de l'Academie des Sciences, 1728, pp. 17–24.

X {anonym} *Remarques de Mr. ... sur les Observations Méteorologiques, inserées dans le Mercure Suisse de Janvier*: Mercure Suisse ou Recueil de Nouvelles Historiques ... (Mars 1734), pp. 80–87 – *Werke* 5, pp. 503–508.

Y *Abhandlung von der Höhe der in vorstehendem Stücke beschriebenen Orte*: D. Bruckner, *Versuch einer Beschreibung historischer und natürlicher Merkwürdigkeiten der Landschaft Basel*, XIII. Stück: *Von Waldenburg*, Basel 1755, pp. 1531–1539 – *Werke* 5, pp. 641–645.

Z *Observations Astronomiques sur la Comète, qui a paru l'année dernière*: Journal Helvetique ou Recueil de Pieces fugitives de literature choisie (Mars 1770), pp. 355–357 – *Werke* 4.

Ungedruckte Werke (Auswahl)

Nr. 76 *Methodus Isoperimetricorum ad novam problematum classem promota*: Ms UB Basel L Ia 751, C. 3 = fol. 27–31.

Nr. 77 *Solutio problematis inveniendi curvam, quae cum alia data sit tautochrona*: Ms UB Basel L Ia 751, C. 4 = fol. 32–35.

Nr. 78 *De Legibus motus mixti variati, quo Corpus sphaericum super plano aspero progreditur Commentatio Physico-Mechanica*: Ms UB Basel L Ia 751, C. 19 = fol. 154–164.

Nr. 79 *Discours sur la cause et la nature de la pesanteur adressé à Messrs de l'Acad. R. des sciences de Paris*: Ms UB Basel L Ia 752, D. 1 = fol. 1–8; D. 2 = fol. 9–29.

Nr. 80 *Discours sur la question de 1729: Quelle est la meilleure methode d'observer les hauteurs sur mer par le Soleil et par les Etoiles soit par des Instruments déja connus, soit par des Instruments de nouvelle Invention?*: Ms UB Basel L Ia 752, D. 3 = fol. 30–49.

Nr. 81 *Recherches Mechaniques et Astronomiques sur la theorie de Saturne et de Jupiter tendantes à expliquer et à déterminer parfaitement les inegalités que ces deux planetes se causent mutuellement tant vers le tems de leur conjonction que pendant tout leur mouvement*: Ms UB Basel L Ia 33, fol. 1–49.

Nr. 82 *Commentariorum De aequilibrio et motu fluidorum Sectio Praeliminaris* [St. Petersburger Erstfassung der *Hydrodynamica*]: Archiv der Akademie der Wissenschaften, St. Petersburg.

Nr. 83 *Notae ad experimentum in Anglia factum concernens principium conservationis virium vivarum*: Archiv der Akademie der Wissenschaften, St. Petersburg R. I op. 1, N. 1.

Chronologische Übersichtstafel der Werke Daniel Bernoullis

Die folgende Tafel ordnet die gedruckten Werke Daniel Bernoullis nach ihrem (mutmasslichen) Entstehungsjahr und nach der Bandaufteilung unserer Werkausgabe an.
Die arabischen Ziffern 1–75 («24» steht für die Arbeit St. 24) und die Siglen R–Y verweisen auf das vorstehende Werkverzeichnis.
Die für das Werkverzeichnis ausgewählten ungedruckten Schriften (Nr. 76–82) sind in eckigen Klammern in die Tabelle eingefügt; über ihre Datierung und ihre Aufnahme in die Editionsbände ist (mit Ausnahme von Nr. 82) nichts Definitives ausgesagt.
Die dreizehn Schriften, mit denen sich Bernoulli um den Preis der Pariser Académie des Sciences beworben hat (zehnmal davon erfolgreich), sowie seine einzige Berliner Preisschrift sind mit einem P gekennzeichnet.

Werke Band		1		2		3	4/5	6	7/8	
		Medizin, Physiologie	Mathematische Jugendschriften	Analysis	Wahrscheinlichkeit	Mechanik	Hydrodynamik	Elastizität	Magnetismus	Technologie I, II
Basel	1721	1 2								
	1722		3							
Venedig	1724		4 5 V							
	1725		6 7							8 P
St. Petersburg	1726	10 11				9				
	1727					13	12 14a			
	1728		[76]	16		[79 P] W*	14b	15		
	1729		17 [77] [80 P]				18a 19			
	1730			20			18b			
	1731				22		21			
	1732							23		
	1733						[82]			
Basel	1734					24 P	U X 31	25		

* Gedruckt im Band 8

Werke Band		1		2		3	4/5	6	7/8	
		Medizin, Physiologie	Mathematische Jugendschriften	Analysis	Wahrscheinlichkeit	Mechanik	Hydrodynamik	Elastizität	Magnetismus	Technologie I, II
[Basel]	1736						26			
	1737	75				27				28 P
	1738					29		30		
	1739							32		
	1740					33 P		34		
	1741			35		36		37		
	1742					[78]		38		
	1743								39 P	
	1744								41 P	
	1745					40				42a P
	1746						74 P			
	1747			S		[81 P]				42b P
	1748					43				
	1749						44 P			
	1751						Ta			
	1753							45		47 P
								46		
	1754						Tb			
	1755						Y		R	
	1757								49	48 P
	1758							50		
	1760				51					
					52					
	1762							53		
	1765							54		
	1766				55					
	1767				56					
	1768					57				
	1769				58	60				
					59a					
	1770				59b			61		
	1771			62				63		
	1772			64						65
	1773			66				67		
	1774							68		
								69		
	1775			70						
				71						
	1777				72					
					73					